An Introduction to

ANALYTIC GEOMETRY
AND CALCULUS

Revised Edition

A. C. Burdette

UNIVERSITY OF CALIFORNIA
DAVIS, CALIFORNIA

 ACADEMIC PRESS   New York   London

To my wife Emily

ACADEMIC PRESS, INC.
111 Fifth Avenue, New York, New York 10003

United Kingdom Edition published by
ACADEMIC PRESS, INC. (LONDON) LTD.
24/28 Oval Road, London NW1

LIBRARY OF CONGRESS CATALOG CARD NUMBER: 72-12444

PRINTED IN THE UNITED STATES OF AMERICA

An Introduction to ANALYTIC GEOMETRY AND CALCULUS

Revised Edition

Contents

13. FUNCTIONS OF SEVERAL VARIABLES

14. INFINITE SERIES

Preface

This edition follows the same plan and is intended to serve the same purposes as the original one. It was written to satisfy an unfilled need for a textbook suitable for an existing course at the University of California, Davis. The gratifying response from instructors at numerous colleges and universities indicates that others were in need of such a textbook for their courses. The changes in this edition are in response to my own experience in using the book and in response to recommendations from others who have used it.

New subject matter is to be found in the expansion of the chapter on infinite series. In response to a rather general demand, we have included a brief treatment of power series. The problem sets and illustrative examples have been augmented where experience indicated a need existed. For example, a complete new set of miscellaneous integration problems was inserted toward the end of Chapter 10 to fulfill a need expressed by both students and instructors. New problems, mostly in the easier category for drill purposes, have been included in many of the original problem sets.

Changes in the text have been made at numerous places. For example, the relationship between the indefinite integral and the antiderivative has been clarified. Or again, an improper integral, a matter not covered by this book, has been eliminated from one of the illustrative examples. In response to student demand, a short list of geometric formulas has been included in the appendices. Also the table of natural logarithms has been modified slightly to make it more convenient to use.

The book provides the student not only with a reasonable degree of understanding of the basic concepts, but also with a working knowledge of the elementary operations of calculus. The presentation is at the freshman level but it serves students of all ranks. Graduate students in various fields find it particularly useful for making up mathematical deficiencies. For many, it is their terminal course in mathematics.

Those looking for a completely novel treatment of calculus will be disappointed. It is true that the mean value theorem plays the central role it should, but seldom does; that the definite integral stands on its own, not as a mere adjunct to the antiderivative; that the choice of subject matter and the level at which it is presented are carefully coordinated with the objectives to be served; and that there is a wider than usual range of applications. On the whole, however, this is a traditional treatment of the subject. This is true, in particular, of the notation and the terminology, in the belief that this best serves the needs of the students involved. For example, the nonmathematics major, reading in his own field, will find the mathematical models of his problems stated in traditional notation, not the highly sophisticated notation of the modern mathematician.

In general, I have held to the conviction that the historical road of development of a discipline is the simplest path to its understanding. This philosophy explains why, for example, I have defined the logarithm function as the inverse of the exponential function, rather than the more direct, and perhaps more logical, definition as an integral. It is my opinion that this approach is more meaningful to the beginning student in spite of its disadvantages. These disadvantages worry him not at all at this point.

No attempt has been made to eliminate the need for an instructor, although there are numerous examples and, hopefully, the student can use the text to amplify and clarify classroom discussion. This book is intended to serve as a basic core of material upon which an instructor can develop a course with a considerable degree of latitude. For example, in Chapter 4 limits are defined, a few proofs of simple cases are given, and then the general theorems are stated without proof. The teacher who wishes to give a more rigorous course can elaborate the work here by presenting proofs of these theorems, while the one who wishes to give a more intuitive course can amplify the motivating examples and soft-pedal the ε-, δ-definitions.

The analytic geometry has been reduced to bare essentials in order to have more time to devote to calculus. Even so, there is probably somewhat more material than can be covered in the allotted time. Chapters 12, 13, 14 are independent of each other so parts or all of these chapters may be omitted without affecting the others. This permits a certain amount of flexibility to adjust to local needs. At Davis, many of our students expect to take physical chemistry so we cover Chapter 13 completely. On the other hand, we omit Chapter 14 entirely and include only parts of Chapter 12.

My thanks go to the many people, too numerous to try to list, who have made suggestions for improvements. Whenever feasible, within the limitations set on the book, these have been included.

Chapter 1

THE COORDINATE SYSTEM— FUNDAMENTAL RELATIONS

1-1. Introduction

This chapter, together with the next two, will be devoted to the task of developing a modest geometric background from the algebraic point of view. This is usually referred to as *analytic geometry* in contrast to *synthetic geometry* already studied in high school or elsewhere. Analytic geometry will serve us well as a basic tool as we proceed with our study of calculus.

1-2. Directed Lines

In elementary geometry a line segment is a portion of a line defined by the two points which mark its extremities. If the two end points are designated by A and B, we label a line segment AB or BA, making no distinction between the two notations. When the processes of algebra are applied to the problems of geometry, as they are in analytic geometry, we find it useful to define *directed lines*.

DEFINITION 1-1. A directed line is a line on which a positive direction has been assigned.

A line segment measured in the positive direction is considered positive and the same portion of the line measured in the opposite direction is considered negative. The positive direction may be assigned arbitrarily although, for consistency of notation and convenience in interpreting results, there are some situations in which certain positive directions are used ordinarily. For example, on horizontal and vertical lines the positive directions are usually taken to the right and upward, respectively.

DEFINITION 1-2. On a directed line, $BA = -AB$.

The following theorem regarding directed line segments will be useful in later work.

THEOREM 1-1. If A, B, and C are any three points on a directed line,

$$AB + BC = AC. \tag{1-1}$$

The truth of this theorem may be established by considering individually the six possible cases of arrangement. For example, let the points be arranged as indicated in Fig. 1-1, separated by the distances $a \geqslant 0$ and $b \geqslant 0$, and let

Fig. 1-1

the positive direction on the line be that of the arrow. Thus

$$AB = b, \qquad BC = -b - a, \qquad AC = -a$$

and (1-1) becomes

$$b + (-b - a) = -a,$$

which verifies the theorem for this case. The other possibilities can be treated in the same manner.

1-3. Cartesian Coordinates

The basis of the *cartesian*, or *rectangular*, coordinate system is a pair of mutually perpendicular directed lines, $X'X$ and $Y'Y$, on each of which a number scale has been chosen with the zero point at their intersection, O. This point is called the *origin* and the lines are called the *x-axis* and *y-axis*. The *x*-axis is usually taken in a horizontal position and, in that case, it is customary to take the positive directions on the two axes to the right and upward. This will be our choice unless specifically stated to the contrary.

Let P be any point in the plane and drop perpendiculars from it to each of the axes, thus determining the *projections* M and N of P on the *x*-axis and *y*-axis, respectively (Fig. 1-2). The measure of the directed segment OM in the units of the scale chosen on the *x*-axis is called the *x-coordinate*, or *abscissa*, of P. Similarly, the measure of the directed segment ON in the units of the scale chosen on the *y*-axis is called the *y-coordinate*, or *ordinate*, of P. Let the measures of OM and ON be x_1 and y_1, respectively.† Then the position of the point P with respect to the coordinate axes is described by the notation (x_1, y_1). As indicated, the abscissa is always written first in the parentheses. The pair of numbers (x_1, y_1) is referred to collectively as the coordinates of P.

When the coordinate axes have been selected and their corresponding number scales chosen, then to each point in the plane corresponds a unique pair of numbers and, conversely, to each pair of numbers corresponds a single point. Thus a coordinate system attaches a numerical property to points and

---

† Hereafter, for simplicity of statement, we will often speak of a segment AB when we mean the measure of AB. This should cause no confusion since the meaning will be clear from the context.

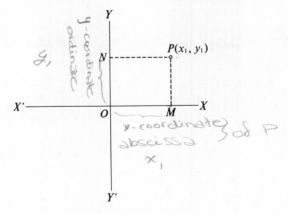

Fig. 1-2

this makes it possible to apply the processes of algebra to the study of geometric problems. This is the distinguishing characteristic of *analytic* geometry.

There are many physical situations in which the quantity measured by x is entirely different from that measured by y; for example, x and y may represent time and temperature, respectively. If such situations are recorded on a coordinate system, there is no need to use the same scale on the two axes unless inferences are to be drawn from geometric considerations. However, when geometric relationships such as angles between lines and distances between points are involved, it is important that the same scale be used on both axes. Since many physical applications of analytic geometry and calculus hinge on geometric properties, it will be convenient for us to make a blanket assumption on this subject:

Unless stated to the contrary, it is to be understood throughout this book that the same scale is used on both coordinate axes.

The coordinate axes divide the plane into four *quadrants.* These are numbered as shown in Fig. 1-3. Such a numbering is convenient for describing certain general situations.

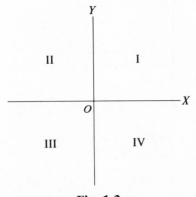

Fig. 1-3

1–4. Projections of a Line Segment on Horizontal and Vertical Lines

Let M_1 and M_2 be the projections of $P_1(x_1, y_1)$ and $P_2(x_2, y_2)$, respectively (Fig. 1-4), on the x-axis. The directed segment M_1M_2 is said to be the *projection of P_1P_2* on the x-axis. Clearly, the projection of P_1P_2 on any horizontal line is equal to M_1M_2. Moreover, from (1-1)

$$M_1M_2 = M_1O + OM_2 = OM_2 - OM_1.$$

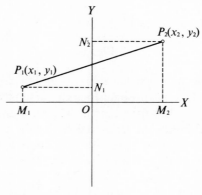

Fig. 1–4

But by definition

$$OM_1 = x_1, \qquad OM_2 = x_2,$$

and we have

$$M_1M_2 = x_2 - x_1. \tag{1-2}$$

Similarly, the projection of P_1P_2 on any vertical line is equal to its projection N_1N_2 on the y-axis and

$$N_1N_2 = N_1O + ON_2 = ON_2 - ON_1,$$

or

$$N_1N_2 = y_2 - y_1. \tag{1-3}$$

Example 1–1. Given the two points $A(3, -2)$, $B(-2, 5)$, find the projections of AB on the coordinate axes.

Choose A as P_1 and B as P_2 and formulas (1-2) and (1-3) give the required projections:

$$M_1M_2 = (-2) - (3) = -5,$$
$$N_1N_2 = (5) - (-2) = 7.$$

Note that the opposite choice of P_1 and P_2 would change the sign of these results. This is as it should be since this changes the direction.

1–5. Midpoint of a Line Segment

Let $P(x, y)$ be the midpoint of the line segment whose end points are $P_1(x_1, y_1)$, $P_2(x_2, y_2)$, and let M, M_1, M_2, respectively, be the projections of these points on the x-axis (Fig. 1-5). Then, making use of (1-2), we have

$$\frac{P_1 P}{P P_2} = \frac{M_1 M}{M M_2} = \frac{x - x_1}{x_2 - x},$$

and, since P is the midpoint of $P_1 P_2$,

$$\frac{P_1 P}{P P_2} = 1.$$

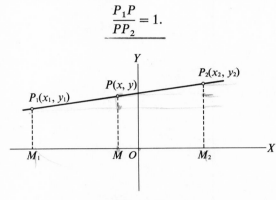

Fig. 1-5

Therefore

$$\frac{x - x_1}{x_2 - x} = 1,$$

from which, solving for x, we obtain the formula

$$x = \frac{x_1 + x_2}{2}. \tag{1-4}$$

Similarly, using the projections of P, P_1, P_2 on the y-axis, we have

$$y = \frac{y_1 + y_2}{2}. \tag{1-5}$$

Example I–2. Determine the point $P(x, y)$ which bisects the line segment joining the points $(2, -3)$, $(4, 5)$.

From (1-4) and (1-5),

$$x = \frac{2 + 4}{2} = 3, \qquad y = \frac{-3 + 5}{2} = 1.$$

Thus the required point is $(3, 1)$.

1-6. Distance between Two Points

Let $P_1(x_1, y_1)$ and $P_2(x_2, y_2)$ be any two points and let it be required to find the undirected distance between them. Construct the line through P_1 parallel to the x-axis and the line through P_2 parallel to the y-axis. Let their point of intersection be Q (Fig. 1-6). From the right triangle P_1P_2Q,

$$\overline{P_1P_2}^2 = \overline{P_1Q}^2 + \overline{QP_2}^2,$$

or by (1-2) and (1-3),

$$\overline{P_1P_2}^2 = (x_2 - x_1)^2 + (y_2 - y_1)^2,$$

or

$$|P_1P_2| = \sqrt{(x_2 - x_1)^2 + (y_2 - y_1)^2}. \tag{1-6}$$

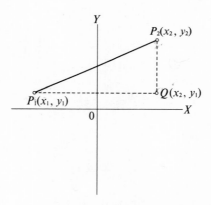

Fig. 1-6

By the symbol $|P_1P_2|$ we mean the magnitude of the distance from P_1 to P_2, that is, the undirected distance between the two points. We shall refer to (1-6) as the *distance formula*.

Example 1-3. Find the length of the line segment joining the points $(3, 2)$ and $(-4, 3)$.

If we call the first of these points $P_1(x_1, y_1)$ and the second $P_2(x_2, y_2)$ and apply (1-6), we have

$$|P_1P_2| = \sqrt{(-4 - 3)^2 + (3 - 2)^2} = \sqrt{50} = 5\sqrt{2}.$$

Note that if the opposite choice of P_1 and P_2 is made, it has no effect on the result.

Exercises 1-1

1. Plot the following pairs of points and find the distance between each pair:
   (a) $(-4, 5), (3, 2)$ (b) $(5, -3), (6, 6)$
   (c) $(-4, -4), (10, 2)$ (d) $(2, -7), (-6, 3)$

2. Find the midpoint of each line segment whose end points are the pairs of points in Ex. 1.

3. What property is common to all points of (a) the x-axis? (b) the y-axis?

4. State the properties common to all points in (a) quadrant III; (b) quadrant IV.

5. The ends of a diameter of a circle are $(-3, 4)$ and $(6, 2)$. What are the coordinates of the center?

6. One end of a diameter of a circle is at $(2, 3)$ and the center is at $(-2, 5)$. What are the coordinates of the other end of this diameter?

7. Show that the triangles whose vertices are given below are isosceles:
   (a) $(-3, -2), (4, -5), (5, 7)$ (b) $(0, 6), (-5, 3), (3, 1)$

8. Show that the triangles whose vertices are given below are right triangles:
   (a) $(-2, 2), (8, -2), (-4, -3)$ (b) $(7, 3), (10, -10), (2, -5)$

9. Given the points $P_1(2, 3)$, $P_2(-3, -1)$, $P_3(5, -4)$, show that the length of the line segment joining the midpoints of $P_1 P_2$ and $P_1 P_3$ is one-half the length of $P_2 P_3$.

10. Show by means of the distance formula that the points $(0, 3), (6, 0), (4, 1)$ are on a straight line.

11. Find the point of the y-axis which is equidistant from the two points $(-6, 2)$ and $(0, -2)$.

12. Determine x so that the points $(-8, 7), (-3, 3), (x, -1)$ lie on a straight line.

13. Find the center of the circle passing through the points $(2, 8), (5, -1), (6, 0)$.

14. Find the point whose coordinates are equal and which is equidistant from the two points $(-2, -3)$ and $(1, 4)$.

15. Find the area of the triangles whose vertices are the following:
    (a) $(-4, 2), (2, 2), (1, 8)$ (b) $(3, -1), (8, -1), (5, -7)$
    (c) $(2, 2), (-3, -3), (2, 10)$ (d) $(5, -1), (-2, -7), (-2, 8)$

16. The ends of the base of an isosceles triangle are $(-1, 2)$ and $(7, 2)$. Its altitude is 10 units. What are the coordinates of the third vertex? (Two solutions.)

1-7. Slope of a Line

We define the slope of a line as follows:

DEFINITION 1-3. Let $P_1(x_1, y_1)$, $P_2(x_2, y_2)$ be two points on a nonvertical line l. The slope of l, designated by m, is the ratio

$$m = \frac{y_2 - y_1}{x_2 - x_1}, \qquad x_1 \neq x_2. \tag{1-7}$$

If two different points $P_1'(x_1', y_1')$, $P_2'(x_2', y_2')$ are selected, we have, from Fig. 1-7,

$$m = \frac{y_2' - y_1'}{x_2' - x_1'} = \frac{y_2 - y_1}{x_2 - x_1},$$

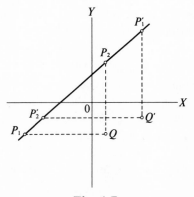

Fig. 1-7

since similar triangles are involved. Thus the slope of a line is independent of the points used to define it.

If $x_1 = x_2$, that is, if the line l is parallel to the y-axis, its slope is undefined. However, as the discussion of the straight line proceeds, we shall see that this does not present any real difficulty.

If the same scale is used on the two axes, as we have agreed to do (Sec. 1-3), we may define an *angle of inclination*, α, to be associated with the slope in a very useful way.

DEFINITION 1-4. The angle of inclination, α, of the line l is the least positive (counterclockwise) angle from the positive x-direction to l. If l is parallel to the x-axis, we choose $\alpha = 0$.†

---

† For simplicity of statement we often replace "angle of inclination" by "inclination."

Thus α satisfies the inequality

$$0° \leqslant \alpha < 180°, \tag{1-8}$$

and, from Fig. 1-8, we have

$$m = \tan \alpha. \tag{1-9}$$

Again we note that, although the inclination is defined, the slope of a line parallel to the y-axis is undefined. Except for such lines, taking (1-8) into account, the inclination of a line is uniquely determined by its slope, and conversely.

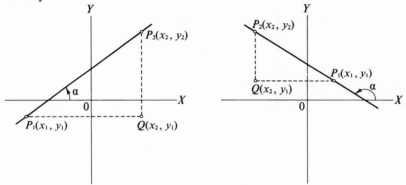

Fig. 1-8

Example I–4. Find the slope of the line determined by the two points $(3, -2)$ and $(-1, 5)$.

We label the points $(3, -2)$ and $(-1, 5)$ as P_2 and P_1, respectively. Then, by (1-7),

$$m = \frac{(-2) - (5)}{(3) - (-1)} = -\frac{7}{4}.$$

Note that a reverse labeling of the points would have no effect on the value of m.

Example I–5. Draw the line through $(-1, 2)$ with the slope $-\frac{1}{2}$.

One way to accomplish this is to determine a second point $(\bar{x}, \bar{y})$ such that

$$\frac{2 - \bar{y}}{-1 - \bar{x}} = -\frac{1}{2}.$$

This equation, since it contains two unknowns, has an infinite number of solutions, any one of which will serve us equally well. If we take $\bar{x} = 3$, we find $\bar{y} = 0$. Then the required line may be drawn by joining the points $(-1, 2)$ and $(3, 0)$.

A simple geometric construction will also determine a suitable point $(\bar{x}, \bar{y})$. Proceed two units from $(-1, 2)$ in the positive x-direction; then move one unit in the negative y-direction (Fig. 1-9). This clearly locates a point $(\bar{x}, \bar{y})$ satisfying the conditions. The basic requirements of this construction are: (1) the two directions must be opposite in sign, and (2) the distance in the x-direction must be twice that in the y-direction.

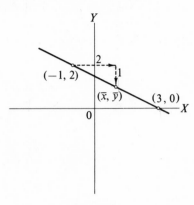

Fig. 1-9

1–8. Parallel and Perpendicular Lines

If two lines are parallel, their inclinations are equal, and thus their slopes are equal. Conversely, if the slopes of two lines are equal, their inclinations are equal and consequently the lines are parallel. Lines parallel to the y-axis are exceptional cases since their slopes are undefined.

THEOREM 1–2. Except for lines parallel to the y-axis, two lines will be parallel if and only if their slopes are equal.

If two lines are perpendicular, their inclinations α_1 and α_2 satisfy the relation (Fig. 1-10)

$$\alpha_2 = \alpha_1 + 90°. \tag{1-10}$$

Thus $\tan \alpha_2 = \tan(\alpha_1 + 90°) = -\cot \alpha_1 = -\dfrac{1}{\tan \alpha_1}$, or

$$m_2 = -\dfrac{1}{m_1}, \tag{1-11}$$

provided $\alpha_1 \neq 90°$, $\alpha_2 \neq 90°$. Moreover, if (1-11) is satisfied together with (1-8), relation (1-10) holds. Therefore we may state the theorem that follows.

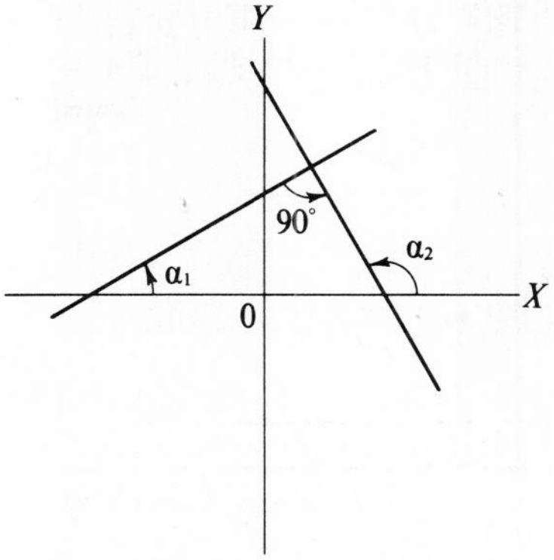

Fig. 1-10

THEOREM 1-3. Except for lines parallel to the coordinate axes, two lines are perpendicular if and only if the slope of one is the negative reciprocal of the slope of the other.

Example 1-6. Show that the line joining $(2, 1)$ and $(6, -1)$ is perpendicular to the line joining $(2, -2)$ and $(4, 2)$.

By (1-7) the slope of the first line is

$$m_1 = \frac{1 - (-1)}{2 - 6} = -\frac{1}{2},$$

and that of the second is

$$m_2 = \frac{-2 - 2}{2 - 4} = 2.$$

Therefore Theorem 1-2, or (1-11), is satisfied and the lines are perpendicular.

The next example gives an indication of how we deal with situations involving lines parallel to the coordinate axes.

Example 1-7. Given the points $(2, -3)$, $(2, 4)$ on the line l_1 and the point $(-4, 1)$ on the line l_2. Find a second point on l_2 if (a) l_2 is parallel to l_1; (b) l_2 is perpendicular to l_1.

On examination of the points on l_1 we discover $x_1 = x_2$; consequently l_1 is parallel to the y-axis. Therefore we cannot approach this problem by means of the slope of l_1. But parallelism to l_1 implies parallelism to the y-axis. Thus, in (a), we are seeking a line through $(-4, 1)$ parallel to the y-axis. Every point on this line will have its abscissa equal to -4. Hence the point $(-4, c)$, $c \neq 1$, is another point of l_2 satisfying (a).

In (b), perpendicularity to l_1 implies parallelism to the x-axis; that is, l_2 has the slope zero. The slope of l_2 will be zero if a second point on it has its ordinate equal to 1. Then (b) will be satisfied by the point $(d, 1)$, $d \neq -4$.

1-9. Angle Formed by Intersecting Lines

Let θ be the positive angle from l_1 to l_2, where neither l_1 nor l_2 is parallel to the y-axis, and let the inclinations of these lines be α_1 and α_2, respectively.

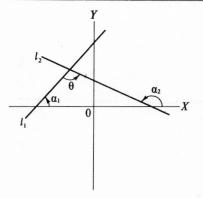

Then, from Fig. 1-11,

$$\theta = \alpha_2 - \alpha_1,$$

or

$$\tan \theta = \tan(\alpha_2 - \alpha_1) = \frac{\tan \alpha_2 - \tan \alpha_1}{1 + \tan \alpha_1 \tan \alpha_2}.$$

Thus

$$\tan \theta = \frac{m_2 - m_1}{1 + m_1 m_2}, \qquad (1\text{-}12)$$

where m_1 and m_2 are the slopes of l_1 and l_2, respectively.

Fig. 1-11

Example 1-8. A triangle has the vertices $A(-2, 1)$, $B(2, 3)$, $C(-2, -4)$. Find (a) $\angle ABC$; (b) $\angle ACB$.

(a) In accordance with the notation defined above we choose AB for l_1 and CB for l_2 in order that (1-12) may give the required angle. The opposite choice of l_1 and l_2 would give the angle β (Fig. 1-12). Then

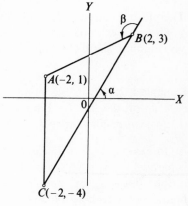

$$m_1 = \frac{1 - 3}{-2 - 2} = \frac{1}{2},$$

$$m_2 = \frac{-4 - 3}{-2 - 2} = \frac{7}{4},$$

and

$$\tan \angle ABC = \frac{\frac{7}{4} - \frac{1}{2}}{1 + (\frac{1}{2})(\frac{7}{4})} \cong 0.667.$$

Thus

$$\angle ABC \cong 33°42'.$$

Fig. 1-12

(b) Since AC is parallel to the y-axis its slope is undefined and consequently (1-12) cannot be used to calculate $\angle ACB$. However, an examination of Fig. 1-12 reveals that

$$\angle ACB = 90° - \alpha$$

where α is the inclination of CB. From above, the slope of CB is 1.75. Therefore $\alpha \cong 60°15'$ and

$$\angle ACB \cong 90° - 60°15' = 29°45'.$$

Exercises 1-2

1. What can be said regarding a line if its slope is (a) positive? (b) negative?

2. What is the slope of a line whose inclination is (a) $0°$? (b) $60°$? (c) $\pi/2$? (d) $3\pi/4$?

3. Find the slope of the line through the points
   (a) $(-1, 2), (3, 2)$ (b) $(0, -5), (-6, 1)$
   (c) $(5, -7), (-3, -7)$ (d) $(4, -1), (-3, 5)$

4. Find the inclination of the line through the points
   (a) $(1, 2), (-6, -5)$ (b) $(-3, 5), (6, -4)$

   (c) $(0, 2), \left(-\dfrac{2\sqrt{3}}{3}, 0\right)$ (d) $(5, 4), (3, 0)$

5. Show that the line joining $(2, -3)$ and $(-5, 1)$ is
   (a) parallel to the line joining $(7, -1)$ and $(0, 3)$
   (b) perpendicular to the line joining $(4, 5)$, and $(0, -2)$

6. Construct the line passing through the given point and having the given number as its slope:
   (a) $(-1, 5), 2$ (b) $(4, -3), 2/3$
   (c) $(2, 6), -(3/4)$ (d) $(-7, -2), -5$

7. The points $(2, -1)$ and $(3, 7)$ are the end points of the base of a triangle. What is the slope of the altitude of the triangle drawn to this base?

8. Show that $(-2, 1)$, $(3, -1)$ and $(7, 9)$ are the vertices of a right triangle.

9. Show that $(6, 1)$, $(3, -1)$, $(-4, 2)$ and $(-1, 4)$ are the vertices of a parallelogram.

10. Show that $(-5, 3)$, $(-4, -2)$, $(1, -1)$ and $(0, 4)$ are the vertices of a square.

11. Show by means of slopes that the points $(3, 0)$, $(6, 2)$ and $(-3, -4)$ lie on a straight line.

12. Determine y so that the line joining $(7, -5)$ with $(3, y)$ has the slope $-(3/2)$.

13. Determine x so that the points $(4, 0)$, $(-4, 6)$ and $(x, -3)$ lie on a straight line.

14. Show that the point $(-7, 4)$ lies on the perpendicular bisector of the line segment whose end points are $(-1, -4)$ and $(3, 4)$.

15. A quadrilateral has the points $(-4, 2)$, $(2, 6)$, $(8, 5)$ and $(9, -7)$ for its vertices. Show that the midpoints of the sides of this quadrilateral are the vertices of a parallelogram.

16. Three vertices of a parallelogram are $(2, 3)$, $(0, -1)$ and $(7, -4)$. Find the coordinates of the fourth vertex. (Three solutions.)

17. Find to the nearest degree the acute angle of intersection of a line with slope $-\frac{3}{2}$ and the line through the points $(1, -3)$ and $(7, 9)$.

18. Find the acute angle of intersection between
(a) the lines (a) and (b) of Exercise 3
(b) the lines (c) and (d) of Exercise 3

19. The slope of the hypotenuse of an isosceles right triangle is -3. What are the slopes of the other two sides?

20. The angle between two lines is $45°$ and the slope of one line is $\frac{1}{2}$. Find the slope of the other line. (Two solutions.)

21. A circle has a diameter with ends at $(-4, 1)$ and $(8, -3)$. Show that the point $(4, 5)$ lies on this circle.

Chapter 2

THE STRAIGHT LINE

2–1. Equation and Locus

The term *locus* is defined as follows:

DEFINITION 2–1. A locus is the totality of points which satisfy a given condition or set of conditions.

Thus a locus consists of some sort of a geometric configuration such as a collection of discrete points, a line, or a curve. If a coordinate system is defined, the conditions determining a particular locus may be expressible as an equation or set of equations involving the coordinates x and y of a point. Thus, when this is possible, a relationship is established between loci and equations. This relationship is the basis of two fundamental problems.

(1) Given a locus, required to find its equation.
(2) Given an equation, required to find the corresponding locus.

We shall consider these two problems first with respect to one of the simplest loci—the straight line. In this chapter we shall be concerned first with determining the line when given its equation; and second with finding an equation for a line when we are given the conditions that define it. This order of studying these two problems is chosen, contrary to what might appear to be the logical order, because, in our opinion, the derivation of equations of loci will be much more meaningful to the student if he has first gained some experience with the opposite problem.

2–2. The Graph of an Equation

We shall use the word "graph" in two senses. First, we shall use it interchangeably with locus; that is, the graph of an equation is the locus represented by the equation. Second, we shall use the word "graph" as the name for a representation of the locus of an equation. This representation may be a very good approximation of the locus, as is usually the case for a straight line, or it may be a very rough approximation in more complicated cases.

This dual use of the term will cause no difficulty because the context will make clear the meaning intended. The process of sketching a graph is called *graphing an equation*.

In order to graph an equation in x and y, we first note any information about the locus that may be available due to the special nature of the equation. Then we find a number of points whose coordinates are solutions of the equation, the number and general location of these points being influenced by the first step, and plot them. A smooth curve drawn through these points, again being influenced by the first step, will constitute the graph. The number of points needed in any given case will depend upon two things; first, the particular equation being graphed; and second, the purpose the graph is to serve. For many purposes a crude graph, that is, a rough approximation of the locus, is all that is needed.

2–3. The Graph of a Linear Equation

It will be shown later in this chapter that any first degree equation in x and y is the equation of a straight line. Let us assume this to be a fact for the moment. Thus, if we have a linear equation to graph, our first step is to observe that it is linear and therefore its locus is a straight line. Having this fact in mind, we note that two points are all that are needed to determine the line. We usually calculate three points so that the third point may provide a check on the computation.

Example 2–1. Graph the equation $2x - 3y = 9$.

To graph this equation we set $x = 0, 3, 6$ successively and obtain the values $y = -3, -1, 1$. Thus the three points $(0, -3)$, $(3, -1)$, $(6, 1)$ lie on the graph of this equation. Of course there are infinitely many points whose coordinates satisfy the given equation, any two of which may be used equally well to obtain the graph. Plotting these points and joining them with a straight line, we obtain the required graph (Fig. 2-1).

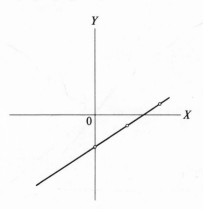

Fig. 2-1

Two points frequently used in graphing a linear equation are the points where the line crosses the coordinate axes. The point where it crosses the x-axis is obtained by setting $y = 0$ and solving for x; similarly for the y-axis, we set $x = 0$ and solve for y. If these points are designated by $(a, 0)$ and $(0, b)$, the numbers a and b are called the *x-intercept* and *y-intercept*, respectively. The x- and y-intercepts for the line graphed in Fig. 2-1 are 4.5 and -3, respectively. If the line goes through the origin, obviously both intercepts are zero and an additional point is necessary to obtain the graph. A vertical or horizontal line has only one intercept.

Exercises 2-1

1. Find at least six solutions of each of the following equations and show graphically that they lie on a straight line:

   (a) $2x - y = 3$ (b) $3x + 2y = 6$ (c) $4x + 3y = 0$

   (d) $3x - y + 6 = 0$ (e) $2x = 3$ (f) $y - 2x = 7$

2. Draw graphs of the following equations, making use of the intercepts on the coordinate axes:

   (a) $5x - y = 5$ (b) $4x + 3y = 12$ (c) $7x + 2y = 4$

   (d) $5x - 4y + 16 = 0$ (e) $2y = 5$ (f) $x - 2y = 0$

3. What lines do the following equations represent:

   (a) $x = 0$?   (b) $y = 0$?   (c) $x = 5$?   (d) $y = -7$?   (e) $y = x$?

4. Determine which of the following points lie on the line whose equation is $4x - 3y = 6$: (a) $(0, -2)$, (b) $(3, 2)$, (c) $(2, 0)$, (d) $(-3, 6)$, (e) $(6, 6)$, (f) $(-3, -6)$.

5. Write the equations of all lines parallel to the y-axis and four units distant from it.

6. Write the equations of all lines parallel to the x-axis and seven units distant from it.

7. Find the length of the segment cut off by the coordinate axes from the line whose equation is $7x - 24y + 168 = 0$.

2–4. The Point-Slope Equation

Let $P_1(x_1, y_1)$ be a fixed point and let m be a given slope. There is one and only one line passing through P_1 with the slope m. In order to obtain its equation let $P(x, y)$ be any other point on the line and consider what condition this imposes on its coordinates x and y. Clearly the slope of P_1P (Fig. 2-2) must be m. Thus

$$\frac{y - y_1}{x - x_1} = m,$$

or

$$y - y_1 = m(x - x_1) \tag{2-1}$$

is the required condition. Any point whose coordinates satisfy (2-1) will be on the given line and only points on the line will have this property. Therefore (2-1) is the equation of the line determined by the point $P_1(x_1, y_1)$ and the slope m. It will be referred to as the *point-slope equation*.

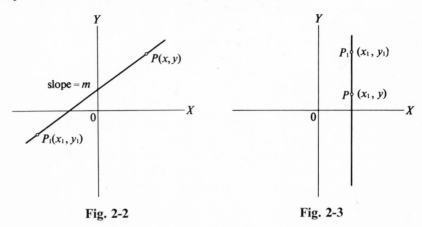

Fig. 2-2   Fig. 2-3

The slope m is undefined for lines parallel to the y-axis. Hence the point-slope equation will not serve to give the equation of a line through the point $P_1(x_1, y_1)$ parallel to the y-axis (Fig. 2-3). However, this presents no difficulty since the equation of such a line is obviously

$$x = x_1. \tag{2-2}$$

Example 2–2. Determine the equation of the line passing through the point $(2, -3)$ with the slope $-(4/5)$.

From (2-1) we have

$$y - (-3) = -\frac{4}{5}(x - 2).$$

Simplifying this we obtain

$$4x + 5y + 7 = 0$$

as the required equation.

Example 2–3. Determine the equation of the line through the point $(2, -3)$ parallel to the x-axis.

A line parallel to the x-axis has slope zero. Therefore (2-1) gives

$$y + 3 = 0.$$

Example 2–4. Determine the equation of the line through the point $(2, -3)$ parallel to the y-axis.

From (2-2) the required equation in this case is

$$x = 2.$$

2–5. The Two-Point Equation

Two distinct points $P_1(x_1, y_1)$, $P_2(x_2, y_2)$ determine a line. The equation of such a line is readily obtained from the point-slope equation discussed in the preceding section. The slope of the line is

$$m = \frac{y_2 - y_1}{x_2 - x_1};$$

thus, from (2-1), its equation may be written

$$y - y_1 = \frac{y_2 - y_1}{x_2 - x_1}(x - x_1). \tag{2-3}$$

This equation will be called the *two-point equation*.

If $x_1 = x_2$, that is, if P_1P_2 is parallel to the y-axis, (2-3) cannot be used because the right member is undefined. In this case, however, we know that the equation of the line is $x = x_1$ and therefore the use of (2-3) is unnecessary.

Example 2–5. Write the equation of the line determined by the points $(2, -3)$, $(1, 7)$.

Applying (2-3) we have

$$y - (-3) = \frac{7 - (-3)}{1 - 2}(x - 2)$$

which simplifies to

$$10x + y - 17 = 0.$$

The opposite choice of P_1 and P_2 would lead to precisely the same equation, as the student may verify.

2–6. The Slope-Intercept Equation

If a line is not parallel to the y-axis, it may be determined by its y-intercept b and slope m. If a line has the y-intercept b it passes through the point $(0, b)$. Hence we may use the point-slope equation to obtain the equation of a line in terms of these quantities. We have (Fig. 2-4)

$$y - b = m(x - 0),$$

or

$$y = mx + b. \tag{2-4}$$

Equation (2-4) will be called the *slope-intercept equation*.

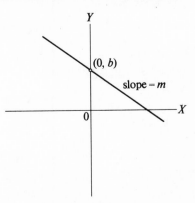

Fig. 2-4

2–7. The First-Degree Equation

We shall now justify the assumption made in Sec. 2-3 regarding first-degree equations. Consider the general first-degree equation

$$Ax + By + C = 0. \tag{2-5}$$

First let us assume that $B = 0$. In this case we may suppose $A \neq 0$; otherwise $C = 0$ and (2-5) is trivial. Hence we may write

$$x = -\frac{C}{A}. \tag{2-6}$$

This equation is true for all points which lie on a line parallel to the y-axis at a distance of $-(C/A)$ units from it, and for no other points. Hence (2-6) is the equation of a straight line.

Now let us assume that $B \neq 0$. In this case (2-5) may be written in the form

$$y = -\frac{A}{B}x - \frac{C}{B}. \qquad \text{— } y\text{-intercept} \tag{2-7}$$

This, according to (2-4), is true for points which lie on a line with slope $-(A/B)$ and y-intercept $-(C/B)$ and for no others. Hence (2-7) is the equation of a straight line.

Combining the preceding results we may state the following theorem.

__THEOREM 2–1.__ Every first-degree equation in x and y represents a straight line.

The reduction of (2-5) to (2-7) is important because it places in evidence the slope of the line represented by the equation.

Example 2–6. Obtain the equation of the line passing through $(-4, 1)$ which is perpendicular to the line $2x - 3y + 7 = 0$.†

Solving the given equation for y, thereby reducing it to the slope-intercept form, we have

$$y = \frac{2}{3}x + \frac{7}{3},$$

from which we observe the slope to be 2/3. Therefore the required perpendicular line will have the slope $-(3/2)$. Its equation, using the point-slope equation, is

$$y - 1 = -\frac{3}{2}(x + 4),$$

or

$$3x + 2y + 10 = 0.$$

---

† As a matter of convenience we shall henceforth speak of the line $2x - 3y + 7 = 0$ instead of the line represented by $2x - 3y + 7 = 0$, or the line which is the locus of $2x - 3y + 7 = 0$.

2–8. Intersection of Lines

A line is composed of the points whose coordinates satisfy the equation of the line. Hence the point of intersection of two lines may be obtained by finding the simultaneous solution of their equations.

Example 2–7. Find the point of intersection of the line $2x - y + 14 = 0$ with the line determined by the two points $(0, 1)$, $(5, -2)$.

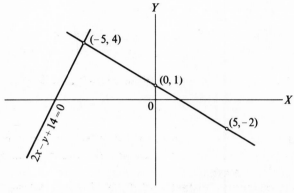

Fig. 2-5

The slope of the line through the two given points (Fig. 2-5) is $-(3/5)$. Hence its equation is given by

$$y - 1 = -\frac{3}{5}(x - 0),$$

or

$$3x + 5y - 5 = 0.$$

Thus the point of intersection of the two lines is given by the simultaneous solution of this equation with

$$2x - y + 14 = 0.$$

The student can readily verify that this point is $(-5, 4)$.

Example 2–8. Find the foot of the perpendicular drawn from the point $(5, -4)$ to the line $3x + y - 21 = 0$.

The slope of the given line is -3. Hence the slope of the perpendicular line (Fig. 2-6) is $1/3$. Thus the equation of the perpendicular line may be written

$$y + 4 = \tfrac{1}{3}(x - 5),$$

or

$$x - 3y - 17 = 0.$$

The solution of this equation with

$$3x + y - 21 = 0$$

gives the required point $(8, -3)$.

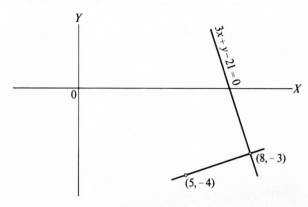

Fig. 2-6

This example also illustrates how one may find the distance from a point to a line. It is merely the distance from the given point to the foot of the perpendicular drawn from that point to the given line. In terms of this example it is the distance from $(5, -4)$ to $(8, -3)$. We have

$$d = \sqrt{(5 - 8)^2 + (-4 - (-3))^2} = \sqrt{10} \text{ unit.}$$

Exercises 2-2

1. Find the equation of the line through the given point having the given slope:
   (a) $(-2, 1)$, $m = 3$ (b) $(3, 5)$, $m = -(2/3)$
   (c) $(-7, -4)$, $m = -2$ (d) $(2, -7)$, $m = 5/2$

2. Find the equation of the line through the point $(3, -2)$ having an inclination of $60°$.

3. Find the equation of the line through the two given points:
   (a) $(3, 1)$, $(-2, 4)$ (b) $(6, -5)$, $(4, 3)$
   (c) $(-7, -4)$, $(2, -7)$ (d) $(5, 3)$, $(5, -5)$

4. Find the equation of the line through $(2, 4)$ and (a) parallel to the x-axis; (b) perpendicular to the x-axis.

5. Find the equations of the sides of the triangle whose vertices are $(-1, 8)$, $(4, -2)$, $(-5, -3)$.

6. Find the equations of the medians of the triangle given in Ex. 5.

7. Find the equations of the altitudes of the triangle given in Ex. 5.

8. Find the equation of the line which has -3 and 5 for its x- and y-intercepts, respectively.

9. Find the equation of the line which has the y-intercept 4 and is parallel to the line $2x - 3y = 7$.

10. Find the equation of the line which has the x-intercept -3 and is perpendicular to the line $3x + 5y - 4 = 0$.

11. The perpendicular from the origin to a line meets it at the point $(-2, 9)$. Find the equation of the line.

12. By using equations of lines, prove that the three points $(3, 0)$, $(-2, -2)$, $(8, 2)$ are collinear.

13. Find the equation of the perpendicular bisector of the segment of the line $3x - 2y = 12$ which is intercepted by the coordinate axes.

14. Find the point at which the line through the points $(6, 1)$ and $(1, 4)$ intersects the line $2x - 3y + 29 = 0$.

15. Find the point on the line $2x - y + 12 = 0$ which is equidistant from the two points $(-3, -1)$ and $(5, 9)$.

16. Find the foot of the perpendicular drawn from $(-4, -2)$ to the line joining the points $(-3, 7)$ and $(2, 11)$.

17. Find the equation of the line through the intersection of the two lines $7x + 9y + 3 = 0$ and $2x - 5y + 16 = 0$ and through the point $(7, -3)$.

18. The midpoints of the sides of a triangle are $(2, 1)$, $(-5, 7)$, $(-5, -5)$. Find the equations of the sides.

19. Find to the nearest minute the acute angle between the two lines $2x + 4y - 5 = 0$ and $7x - 3y + 2 = 0$.

20. Find to the nearest minute the angles of the triangle whose sides have the equations $x + 2y - 2 = 0$, $x - 2y + 2 = 0$, $x - 4y + 2 = 0$.

21. The hypotenuse of an isosceles right triangle has its ends at the points $(1, 3)$ and $(-4, 1)$. Find the equations of the legs of the triangle.

Chapter 3

NONLINEAR EQUATIONS AND GRAPHS

3–1. Introduction

This chapter will be devoted to a brief study of nonlinear equations and their graphs. The emphasis will be on obtaining the graph of an equation rather than a study of the loci themselves.

The first part of the chapter will be concerned with amplifying the general remarks on graphing in Sec. 2-2. We shall discuss some special properties of curves which may be discovered from examination of their equations and which will materially aid in drawing their graphs, both as to the number of points needed and as to the accuracy of the result.

In the later part of the chapter we shall consider briefly two particular loci: the circle and the parabola. We single out these two loci, together with the straight line of the preceding chapter, because of the frequency with which they occur in later work. It will be of material assistance in the remaining chapters if we can sketch these curves rapidly and accurately.

3–2. Intercepts

DEFINITION 3–1. The x-intercepts of a curve are the abscissas of the points where it crosses the x-axis; the y-intercepts of a curve are the ordinates of the points where it crosses the y-axis.

As noted in the discussion of the straight line, x-intercepts are obtained by setting $y = 0$ in the equation of the curve and solving for x; similarly, for y-intercepts, set $x = 0$ and solve for y. These operations, either one or both, are often easily performed and give us points of a critical nature with respect to the curve. For example, the x-intercepts provide us with the points at which the ordinates of points on the curve may change sign, that is, the points at which the curve may change from lying above the x-axis to lying below it, or vice versa. This is an important consideration in some problems.

Example 3–1. Find the intercepts of the curve represented by $2y^2 - x - 8 = 0$.

We set $y = 0$ obtaining $-x - 8 = 0$, or $x = -8$, the x-intercept. We set $x = 0$ and obtain $2y^2 - 8 = 0$. Thus there are two y-intercepts, 2, -2.

3–3. Symmetry

DEFINITION 3–2. A curve is said to be symmetric with respect to a line if for each point on the curve there is another point on the curve so situated that the line segment joining the two points has the given line as its perpendicular bisector.

Example 3–2. A circle is symmetric to any line through its center (Fig. 3-1).

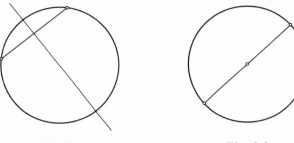

Fig. 3-1   Fig. 3-2

DEFINITION 3–3. A curve is said to be symmetric with respect to a point if for each point on the curve there is another point on the curve so situated that the given point is the midpoint of the line segment joining the two points on the curve.

Example 3–3. A circle is symmetric with respect to its center (Fig. 3-2).

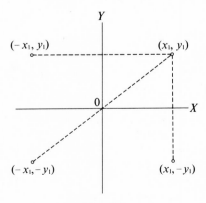

Fig. 3-3

It is a simple matter to test a curve for symmetry with respect to the coordinate axes and the origin. If a curve is symmetric with respect to the y-axis, and if (x_1, y_1) is a point of the curve, it follows from Fig. 3-3 that $(-x_1, y_1)$ is also a point on the curve. This means that both (x_1, y_1) and

$(-x_1, y_1)$ satisfy the equation of the curve; that is, changing the sign of x_1 does not have any bearing on its satisfaction of the equation of the curve. Now if we ask that all points of the curve have this property, we are led to the following theorem.

THEOREM 3–1. A curve is symmetric with respect to the y-axis if and only if the substitution of $-x$ for x in its equation yields an equivalent† equation.

Similarly, by referring to Fig. 3-3 we may state:

THEOREM 3–2. A curve is symmetric with respect to the x-axis if and only if the substitution of $-y$ for y in its equation yields an equivalent equation.

THEOREM 3–3. A curve is symmetric with respect to the origin if and only if the simultaneous substitution of $-x$ for x and $-y$ for y in its equation yields an equivalent equation.

Example 3–4. Test $x^2 y - y^2 = 5$ for symmetry.‡

If we replace x by $-x$ in this equation it is unaltered. Hence, by Theorem 3-1, the curve is symmetric to the y-axis.§ But if we substitute $-y$ for y, we obtain $-x^2 y - y^2 = 5$, which is not equivalent to the original equation. Therefore, by Theorem 3-2, this curve is not symmetric to the x-axis. Also, if we simultaneously replace x by $-x$ and y by $-y$, we obtain again $-x^2 y - y^2 = 5$. Thus, by Theorem 3-3, this curve is not symmetric to the origin.

Example 3–5. Test $x^2 y - 3x + y = 0$ for symmetry.

The student can easily verify that the simultaneous substitution of $-x$ for x and $-y$ for y leads to an equivalent equation, but either of these changes alone produces a nonequivalent equation. Hence this curve is symmetric to the origin but not to either axis.

3–4. Extent

A valuable aid in drawing the graph of an equation is knowledge of the *extent* of the curve, that is, what portion or portions of the plane are occupied

---

† The equivalence in this and later cases will be indicated by obtaining either the original equation or the original equation multiplied by -1.

‡ As indicated earlier in connection with the straight line, when there is no chance for misinterpretation we shall use "curve" and "equation" interchangeably. Thus we say here: "Test $x^2 y - y^2 = 5$," etc. instead of "Test the curve represented by $x^2 y - y^2 = 5$," etc.

§ For simplicity of statement we usually replace "with respect to" by "to." Thus we say "a curve is symmetric to the y-axis" rather than "a curve is symmetric with respect to the y-axis."

by the curve. For example, it would help in drawing the graph if we knew the curve was entirely above the x-axis, or had no points in a strip of two-unit width on either side of the y-axis, or was contained completely within a certain rectangle. In many cases information of this nature can be obtained rather easily.

When we are drawing the graph of an equation we are interested only in points with real numbers for coordinates. Information on the extent of a curve can be obtained by examining the restrictions that must be placed on a real x (or y) in order that the corresponding values of y (or x) be real also. The following examples will illustrate how this may be done.

Example 3–6. Discuss the extent of $y^2 + x = 4$.

If we write this equation in the form

$$x = 4 - y^2, \qquad\qquad (3\text{-}1)$$

we note that any real y substituted into it will give a real value for x. Hence there are no restrictions to be placed on y, or stated differently, the curve is infinite in extent in the y-direction.

Let us now solve the equation for y. We obtain

$$y = \pm\sqrt{4 - x}.$$

If $4 - x$ is negative, that is, if $x > 4$, y is not real. Hence, for y to be real, x must be real and satisfy the additional restriction $x \leqslant 4$. Thus no part of the curve lies to the right of the line $x = 4$. This same result could have been obtained by observing in (3-1) that the maximum value of x is 4.

Example 3–7. Discuss the extent of $x^2 + y^2 = 25$.

First we solve the equation for y and obtain

$$y = \pm\sqrt{25 - x^2}.$$

If $25 - x^2$ is negative, y is not real. This occurs when $x^2 > 25$, that is, when $x < -5$ or $x > 5$. Therefore, for y to be real, we must have $-5 \leqslant x \leqslant 5$. Expressed graphically, the curve is bounded by the lines $x = \pm 5$.

Solving for x, we obtain

$$x = \pm\sqrt{25 - y^2},$$

and by precisely the same reasoning we conclude that the curve is bounded by the lines $y = \pm 5$. Thus the curve is contained in a square ten units on a side, the sides being parallel to the coordinate axes and symmetric to them.

Example 3–8. Discuss the extent of $x^2 - y^2 = 4$.

We solve for x and obtain

$$x = \pm\sqrt{y^2 + 4}.$$

Since $y^2 + 4$ is positive for all real y, this imposes no restriction on y. Hence the curve is infinite in extent in the y-direction.

Solving for y, we obtain

$$y = \pm\sqrt{x^2 - 4}.$$

Since $x^2 - 4$ is negative when $-2 < x < 2$,† we see that no points on the curve lie between the lines $x = -2$ and $x = 2$.

The results in the preceding examples could be expressed in terms of *excluded values* of x and y. Thus, in Example 3-6 we could say that there are no excluded values of y but values of $x > 4$ are excluded. Example 3-7 has $|x| > 5$ and $|y| > 5$ excluded; Example 3-8 has $|x| < 2$ excluded. This is often the simplest method of describing the extent of a curve.

3–5. Graphing Equations

If we combine the general remarks of Sec. 2-2 with the special results of Secs. 3-2, 3-3, 3-4, we have a fairly sound working basis for drawing the graphs of many equations. In some cases, certain parts of the general discussion may be omitted because the labor of carrying them out is so tedious, or difficult, as to make them impractical. These results are not an end in themselves but rather an aid to the overall problem of drawing a graph. When they fail to serve this purpose in a reasonable fashion, they are not justified.

Example 3–9. Graph $y^2 + x = 4$.

(a) Intercepts: If $x = 0$, $y^2 = 4$ or $y = \pm 2$. If $y = 0$, $x = 4$.
(b) Symmetry: Theorem 3-2 gives symmetry to the x-axis.
(c) Extent: From Example 3-6, $x > 4$ is excluded.
(d) Additional points: Taking (a), (b), (c) into account, the brief table of additional points

x	2	-2	-5
y	$\pm\sqrt{2}$	$\pm\sqrt{6}$	± 3

is sufficient to give a graph of this equation adequate for ordinary purposes (Fig. 3-4).‡

---

† This condition may be expressed in the form $|x| < 2$. The symbol $|x|$ signifies the magnitude of x, ignoring the algebraic sign, or more precisely, $|x| = x$, $x \geq 0$; $|x| = -x$, $x < 0$.
‡ In order to emphasize the role of excluded values, that portion of the plane from which the curve is excluded is shaded.

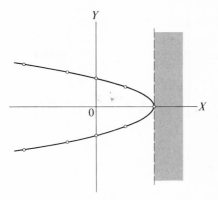

Fig. 3-4

Example 3–10. Graph $x^2 - y^2 = 4$.

(a) Intercepts: If $x = 0$, $y = \pm 2i$. Since these values of y are not real, we conclude that the curve does not intersect the y-axis. This is consistent with (c) below, where it is seen that $x = 0$ is in the excluded portion of the plane. If $y = 0$, $x = \pm 2$.

(b) Symmetry: Theorems 3-1, 3-2, 3-3 all indicate symmetry. Hence this curve is symmetric to both axes and the origin.†

(c) Extent: From Example 3-8, $|x| < 2$ is excluded.

(d) Additional points: The results of (a), (b), (c) make it possible to sketch the required graph from very few additional points. Observe that, owing to symmetry, the calculation of one point gives three others with no further calculation. The following points serve our purpose very well (Fig. 3-5):

x	3	5
y	$\pm\sqrt{5}$	$\pm\sqrt{21}$

Other properly chosen points would do equally well.

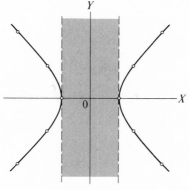

Fig. 3-5

---

† Note that if any two of Theorems 3-1, 3-2, 3-3 indicate symmetry, the third automatically does likewise.

Example 3–11. Graph $xy - 2y - 4 = 0$.

We have chosen this example to illustrate some points not covered by our discussion of intercepts, symmetry, and extent, but which occasionally are very useful.

(a) Intercepts: If $x = 0$, $y = -2$. If $y = 0$, the equation has no solution and consequently no x-intercept.

(b) Symmetry: No symmetry to the coordinate axes or origin is indicated.

(c) Extent: We solve for x and obtain

$$x = \frac{2y + 4}{y}.$$

Any real value of $y \neq 0$ gives a real value for x; if $y = 0$, x is undefined. However, as y gets closer and closer to 0, while always remaining positive, we observe that x gets larger and larger. Thus, as the curve approaches the x-axis from above, points on the curve move farther and farther out to the right. On the other hand, as y gets closer and closer to 0, while always remaining negative, we note that x is negative for $y > -2$, and $|x|$ gets greater and greater. Hence, as the curve approaches the x-axis from below, points on the curve move farther and farther out to the left.

Next we solve for y and obtain

$$y = \frac{4}{x - 2}.$$

Any real value of $x \neq 2$ gives a real value for y; if $x = 2$, y is undefined. If we examine, as above, what happens to y as x gets close to 2, we find that as the curve approaches the line $x = 2$ from the left, the curve falls away indefinitely; as the curve approaches the line $x = 2$ from the right, the curve rises indefinitely.

One more thing may be observed regarding the extent of this curve if we write its equation in the form

$$y(x - 2) = 4.$$

The right member of this equation is positive; hence the left member must be positive also. This will be true if and only if the two factors of the left member have the same sign. Both will be positive if $y > 0$, $x > 2$, and both negative if $y < 0$, $x < 2$. Thus the curve is confined to the upper-right and lower-left quadrants formed by the lines $x = 2$, $y = 0$.

(d) Additional points: The following points supply enough additional information to sketch a reasonably accurate graph (Fig. 3-6):

x	-3	-1	0	1	3	4	5	7
y	$-(4/5)$	$-(4/3)$	-2	-4	4	2	4/3	4/5

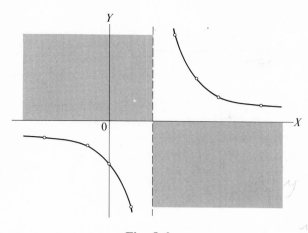

$$y = 2 - x^{\frac{2}{2}}$$

$$y = \frac{6 - v^3}{3}$$

Fig. 3-6

The observant student will note that this curve is symmetric to the point of intersection of the two lines $x = 2$, $y = 0$.

$$3y = 6 - x^2$$

Exercises 3-1

Discuss and sketch the graphs of the following equations:

1. $x^2 - 2y = 0$ 2. $y^2 + 2x = 0$ 3. $x^2 + 3y - 6 = 0$
4. $y^2 - 4x + 12 = 0$ 5. $x^2 + y^2 = 16$ 6. $9x^2 + 4y^2 = 36$
7. $16x^2 + 25y^2 = 144$ 8. $4y^2 - 9x^2 = 36$ 9. $16x^2 - 25y^2 = 144$
10. $x^2 + y^2 - 2y = 0$ 11. $x^2 + y^2 + 10x = 0$ 12. $x^2 + y^2 - 6x = 0$
13. $xy = 5$ 14. $xy + 2x - y - 1 = 0$ 15. $xy - 2x - 3y + 6 = 0$ *watch*
16. $xy^2 = 12$ 17. $xy^2 - 4x = 16$ 18. $x^2y + x^2 = 9y$

3-6. The Circle

DEFINITION 3-4. A circle consists of all points at a given distance from a fixed point. The given distance and fixed point are called the radius and center, respectively.

Let $P(x, y)$ be any point on the circle with center at $C(h, k)$ and radius r. Then, from (1-6) and Fig. 3-7, we have

$$\sqrt{(x - h)^2 + (y - k)^2} = r,$$

or

$$(x - h)^2 + (y - k)^2 = r^2, \qquad (3\text{-}2)$$

> equation of circle

the condition that P must satisfy if it is on the circle. Furthermore, if the

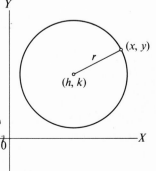

Fig. 3-7

coordinates x and y of a point satisfy (3-2), it is r units distant from (h, k) and therefore is a point on the circle. Thus (3-2) is an equation of the circle with center at (h, k) and radius r.

If we square the terms in the left member of (3-2) and simplify, we obtain

$$x^2 + y^2 - 2hx - 2ky + d = 0, \qquad (3-3)$$

another equation of this circle. Equations (3-2) and (3-3) enable us to recognize equations of circles and to obtain the center and radius of the circle represented by a given equation.

Example 3–12. Discuss and graph the equation

$$4x^2 + 4y^2 - 12x + 32y - 27 = 0.$$

We divide both sides of this equation by 4 and obtain

$$x^2 + y^2 - 3x + 8y - \frac{27}{4} = 0.$$

This equation is of the form (3-3). We may say, then, that this is the equation of a circle with center at $(3/2, -4)$. We find it convenient to reduce this equation to form (3-2) in order to obtain the radius. We have, completing the square,

$$\left(x^2 - 3x + \frac{9}{4}\right) + (y^2 + 8y + 16) = \frac{27}{4} + \frac{9}{4} + 16$$

or

$$(x-h)^2 + (y-k)^2 = r^2$$

$$\left(x - \frac{3}{2}\right)^2 + (y + 4)^2 = 25.$$

Thus we reconfirm the fact that the center is at $(3/2, -4)$ and discover that the radius is 5. Nothing more is needed to draw the graph in Fig. 3-8.

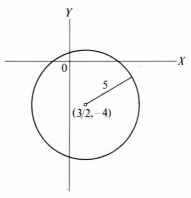

Fig. 3-8

Any equation of the form

$$\boxed{Ax^2 + Ay^2 + Bx + Cy + D = 0} \qquad (3-4)$$

Equation of circle

can be reduced to the form (3-2), so we say that (3-4) is the equation of a circle. However, certain cases are of no interest to us here because r^2 may turn out to be 0 or negative. In the first case the locus is real but consists of a single point; in the second case the radius is complex so the locus is not real.

Example 3–13. Discuss and graph the equation $x^2 + y^2 - 4x + 6y + 22 = 0$.

We reduce this equation to the form of (3-2) by completing squares and obtain

$$(x^2 - 4x + 4) + (y^2 + 6y + 9) = -22 + 4 + 9$$

or

$$(x - 2)^2 + (y + 3)^2 = -9.$$

Since $r^2 = -9$, r is not real and this equation has no real locus.

Example 3–14. Find an equation for the circle which is tangent to the line $y = -1$ and whose center is at $(-3, 2)$.

From (3-2) we note that the equation of a circle can be written down immediately if we know the center and radius. Since the center is given, we need only to determine the radius. In order to do this, we make use of the property of circles that a radius drawn to the point of tangency is perpendicular to the tangent. We observe that $y = -1$ is parallel to the x-axis; hence the radius drawn to the point of contact of this tangent is parallel to the y-axis and lies on the line $x = -3$ (Fig. 3-9). Thus the point of tangency is $(-3, -1)$

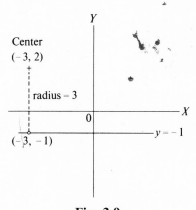

Fig. 3-9

and the radius is 3. This, together with the coordinates of the center, substituted in (3-2) gives us the required equation

$$(x + 3)^2 + (y - 2)^2 = 9,$$

or

$$x^2 + y^2 + 6x - 4y + 4 = 0.$$

Exercises 3-2

1. Find the radius and center of the circles whose equations follow, and sketch:

    (a) $x^2 + y^2 - 4x + 6y + 9 = 0$ (b) $x^2 + y^2 - 6x - 2y - 6 = 0$
    (c) $2x^2 + 2y^2 + 16x + 8y + 22 = 0$ (d) $3x^2 + 3y^2 - 6x + 4y - 4 = 0$
    (e) $9x^2 + 9y^2 + 12x - 18y - 32 = 0$ (f) $x^2 + y^2 + 5x = 0$
    (g) $4x^2 + 4y^2 - 4y - 39 = 0$

2. Determine equations for the circles having the following properties:

    (a) Center at $(2, -5)$, radius 7
    (b) Center at $(-3, -1)$, radius 5
    (c) A diameter has the end points $(5, -4)$ and $(-3, 2)$
    (d) Center at $(4, -2)$ and tangent to $x = 5$
    (e) Center at $(-7, 2)$ and tangent to the x-axis
    (f) Radius 5 and tangent to the y-axis at $(0, -1)$
    (g) Center at $(1, 3)$ and passing through the origin
    (h) Center at $(0, 3)$ and tangent to $2x + y - 13 = 0$
    (i) Center on x- axis, passing through $(1, 3)$ and $(5, 5)$
    (j) Center on y-axis, passing through $(5, 4)$ and $(13, -8)$
    (k) Center on $x + y = 3$, passing through $(2, 3)$ and $(-1, 5)$
    (l) Radius 1, passing through $(4, 0)$ and $(3, 1)$

3. Find the points of intersection of

    (a) $x - y = 8$ and $x^2 + y^2 + 6x - 12y - 244 = 0$
    (b) $y - 3x = 5$ and $x^2 + y^2 + 12x - 4y - 5 = 0$
    (c) $x^2 + y^2 - 2x + 18y = 87$ and $x^2 + y^2 + 2x - 2y = 11$

4. A point moves in such a manner that its distance from $(-4, -2)$ is twice its distance from $(2, 0)$. Show that this point moves along a circle. Find the center and radius of this circle.

5. A point P moves so that the distance from $(-1, 2)$ to the midpoint of the line segment joining P to $(3, -4)$ is always 5. Show that P traces a circle and find the radius and center.

3–7. The Parabola

DEFINITION 3–5. A parabola consists of all points equally distant from a fixed point and a fixed line. The fixed point and fixed line are called the focus and directrix, respectively.

Let us specialize our coordinate system so that the directrix is parallel to the y-axis and let its equation be $x = p$. Further, let the point (r, s), $r > p$, be the focus. Then, from Fig. 3-10, if P is a point on the parabola,

$$|PQ| = |PF|$$

by definition. But

$$|PQ| = |x - p|$$

and

$$|PF| = \sqrt{(x - r)^2 + (y - s)^2},$$

so we have

$$|x - p| = \sqrt{(x - r)^2 + (y - s)^2}$$

or

$$(x - p)^2 = (x - r)^2 + (y - s)^2,$$

or

$$(x - p)^2 - (x - r)^2 = (y - s)^2,$$

or

$$2(r - p)x - (r^2 - p^2) = (y - s)^2,$$

or

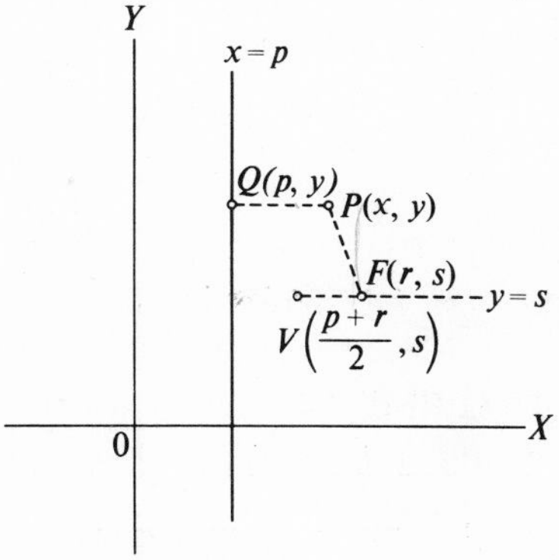

$$2(r - p)\left(x - \frac{r + p}{2}\right) = (y - s)^2. \tag{3-5}$$

Fig. 3-10

This is an equation which the coordinates of P must satisfy if it is a point on the parabola.

Now we shall show that if the coordinates of $P(x, y)$ satisfy (3-5), $|PQ| = |PF|$ and therefore P is a point on the parabola. We have

$$|PF| = \sqrt{(x - r)^2 + (y - s)^2}.$$

Since the coordinates of P satisfy (3-5) we may write

$$|PF| = \sqrt{(x - r)^2 + 2(r - p)\left(x - \frac{r + p}{2}\right)}$$

$$= \sqrt{x^2 - 2xp + p^2} = |x - p| = |PQ|,$$

as was to be shown. Hence we may state that (3-5) is an equation of the parabola described.

It is clear from the definition that the point V, midway between the focus and directrix, is a point on the parabola; moreover, under the assumption that $r > p$, it is the left-most point on the curve. Its coordinates, from (1-4) and (1-5), are $[(r + p)/2, s]$. This point, whose coordinates may be read directly from (3-5), is called the *vertex* of the parabola.

It is also clear from the definition that the parabola is symmetric to the line through the focus perpendicular to the directrix. This line is called the *axis* of the parabola and its equation is $y = s$ and may be obtained by inspection from (3-5).

What would be the effect on (3-5) if the focus were to the left of the directrix instead of as shown in Fig. 3-7, that is, if $r < p$? The only visible effect on the equation would be a change in sign of the factor $r - p$. However, the parabola would be reversed and the vertex would become the right-most point on the curve.

If we combine constants,† (3-5) takes the form

$$k(x - t) = (y - s)^2 \qquad (3\text{-}6)$$

and we may summarize the preceding observations as follows.

THEOREM 3–4. An equation reducible to the form (3-6) is the equation of a parabola having the following properties:

    (a) Its vertex is at (t, s).

    (b) If k is positive (negative) it extends to the right (left) from the vertex.

    (c) It is symmetric to its axis $y = s$.

If we choose our coordinate system so that the directrix is parallel to the x-axis and make a similar analysis, we arrive at the equation

$$k(y - s) = (x - t)^2, \qquad (3\text{-}7)$$

and the following conclusions.

THEOREM 3–5. An equation reducible to the form (3-7) is the equation of a parabola having the following properties:

    (a) Its vertex is at (t, s).

    (b) If k is positive (negative) it extends upward (downward) from the vertex.

    (c) It is symmetric to its axis $x = t$.

We note that any second-degree equation in x and y which has no xy term and in which either x^2 or y^2 is missing can be reduced to one of these two

---

† If p and r are needed for any reason, they may be obtained readily by solving the equations

$$2(r - p) = k, \qquad \frac{r + p}{2} = t.$$

forms and therefore represents a parabola. Thus

$$x^2 + ax + by + c = 0 \qquad (3\text{-}8)$$

is an equation of a parabola with its axis <u>parallel to the y-axis</u> and

$$y^2 + ax + by + c = 0 \qquad (3\text{-}9)$$

is an equation of a parabola with its axis <u>parallel to the x-axis.</u>

Example 3–15. Discuss and graph $x^2 + 6x - 4y + 13 = 0$.

We first note that this equation is of the form (3-8) and therefore represents a parabola. Next we reduce it to form (3-7) by writing

$$4y - 13 = x^2 + 6x,$$

and then, completing the square of the right member, obtain

$$4y - 4 = x^2 + 6x + 9,$$

or finally

$$4(y - 1) = (x + 3)^2.$$

From this form of the equation, making use of Theorem 3-5, we conclude that the parabola extends upward from its vertex $(-3, 1)$, symmetric to its axis $x = -3$. We combine this information with the points

x	-2	0	2
y	5/4	13/4	29/4

and obtain the graph in Fig. 3-11. Note that, due to symmetry to the line $x = -3$, these three points provide six points on the curve.

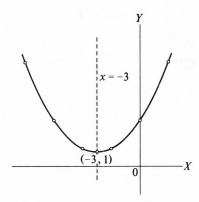

Fig. 3-11

Example 3–16. Discuss and graph $4y^2 + 12y + 3x + 3 = 0$.

After noting that this equation is essentially of form (3-9) we write it in the form

$$-3x - 3 = 4y^2 + 12y$$

or

$$-3x - 3 = 4(y^2 + 3y).$$

We complete the square on the right and obtain

$$-3x + 6 = 4\left(y^2 + 3y + \frac{9}{4}\right)$$

or

$$-\frac{3}{4}(x - 2) = \left(y + \frac{3}{2}\right)^2.$$

This, from Theorem 3-4, is the equation of a parabola extending to the left from its vertex $(2, -(3/2))$, symmetric to its axis $y = -(3/2)$. We calculate the additional points

x	-1	$5/3$	$-19/3$
y	0	-1	-4

and draw the graph in Fig. 3-12.

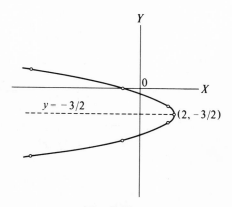

Fig. 3-12

Example 3–17. Determine an equation of the parabola with vertex at the origin, with axis on $y = 0$, and passing through the point $(1, 3)$.

Equation (3-6) takes the form

$$kx = y^2$$

in this case. If the parabola passes through $(1, 3)$, these coordinates satisfy this equation and we have

$$k(1) = (3)^2.$$

Thus the required equation is

$$9x = y^2.$$

Example 3–18. Determine an equation of the parabola with its axis parallel to the y-axis and passing through the three points $(0, 1)$, $(1, 0)$, $(-2, -1)$.

A parabola with its axis parallel to the y-axis has an equation of the form

$$x^2 + ax + by + c = 0.$$

We impose the condition that the coordinates of each of the three points must satisfy this equation, and obtain the system of linear equations

$$b + c = 0,$$

$$a + c = -1,$$

$$2a + b - c = 4.$$

The solution of this system of equations is readily found to be $a = 1/2$, $b = 3/2$, $c = -(3/2)$. We substitute these values in the general equation and obtain the required equation

$$x^2 + \frac{1}{2}x + \frac{3}{2}y - \frac{3}{2} = 0,$$

or

$$2x^2 + x + 3y - 3 = 0.$$

Example 3–19. The towers of a suspension bridge have their tops 110 ft above the roadway and are 500 ft apart. If the cable is 10 ft above the roadway at the center of the bridge, find the length of the vertical supporting cable 100 ft from the center.

The cables of a suspension bridge are known to hang in the form of a parabola, with its vertex at the lowest point and its axis vertical. Let us choose the coordinate axes as shown in Fig. 3-13. Then the equation of the parabola takes the form

$$x^2 = kv.$$

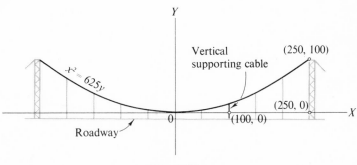

Fig. 3-13

Since this parabola passes through the point (250, 100), we have

$$62,500 = 100k,$$

whence $k = 625$. Hence, the length of the specified vertical supporting cable is given by

$$y + 10,$$

where y is the ordinate of the point on the parabola

$$x^2 = 625y$$

whose abscissa is 100. Thus

$$100^2 = 625y$$

or

$$y = 16,$$

and the required length is

$$y + 10 = 26 \text{ ft}.$$

Exercises 3-3

1. Graph the following equations by the methods of this section:
   (a) $x^2 - 2y = 0$ (b) $y^2 + 2x = 0$
   (c) $x^2 + 3y - 6 = 0$ (d) $y^2 - 4x + 12 = 0$
   (e) $x^2 + 4x + 3y - 8 = 0$ (f) $x^2 + x + y + 1 = 0$
   (g) $4y^2 + 12y - 8x + 17 = 0$ (h) $4x^2 - 4x - y = 0$
   (i) $y^2 + 3y + 2x - 1 = 0$ (j) $3y^2 - 2x + 2y - 7 = 0$

2. Determine equations of the parabolas with vertex at the origin and
   (a) Axis on the x-axis and passing through $(-1, 6)$
   (b) Axis on the x-axis and passing through $(2, -5)$
   (c) Axis on the y-axis and passing through $(4, -3)$
   (d) Axis on the y-axis and passing through $(-7, -2)$

3. Determine equations of the parabolas with axes parallel to the y-axis and passing through the following points:

(a) $(2, -2), (3, -9), (-1, 7)$

(b) $(5, 1), (-1, 2), (-3, 5)$

(c) $(-1, 2), (1, -1), (2, 3)$

4. Determine equations of the parabolas with axes parallel to the x-axis and passing through the following points:

(a) $(4, 2), (2, -1), (4, 1)$

(b) $(6, -2), (1, 3), (-8, 5)$

(c) $(-1, 2), (1, -1), (2, 3)$

5. An arch is in the form of an arc of a parabola with its axis vertical. The arch is 20 ft high and is 15 ft wide at the base. Choose a suitable set of coordinate axes and determine an equation for the parabola of which this arch is a segment.

6. The roof of a highway tunnel is in the form of a parabolic arch 25 ft high in the center. The tunnel is 50 ft wide. Can a truck 12 ft high use this tunnel if the driver must keep the right side of his truck no farther than 5 ft from the side of the tunnel?

7. A suspension bridge is 200 ft long with towers 45 ft high. If the roadway is 5 ft below the lowest point on the cable, how high is the cable above the roadway at a point 25 ft from the end of the bridge?

8. Water squirts from the end of a horizontal pipe 18 ft above the ground. At a point 6 ft below the pipe the stream of water is 8 ft beyond its end. What horizontal distance does the stream travel before it strikes the ground? (Assume that the water flows in the arc of a parabola with vertical axis and vertex at the end of the pipe.)

Chapter 4

FUNCTIONS AND LIMITS

4–1. Introduction

The subject matter of this chapter lies at the very foundation of calculus. A completely rigorous treatment of it would require both an extensive study of the real number system and a degree of mathematical maturity not reasonably to be expected in those using this book.

We shall assume that the student has an intuitive understanding of real numbers and their properties and content ourselves with only a few remarks and definitions related to this subject. Moreover, some of the statements labeled as definitions will not be precise mathematical definitions but rather descriptive statements of fundamental concepts. However, these "informal" definitions are not incompatible with their precise counterparts and, hopefully, will carry the message to the beginning calculus student in a more understandable language.

The first of these informal definitions follows.

DEFINITION 4–1. A set of real numbers is a collection of these numbers.

There are many kinds of sets, but we shall be concerned only with those composed of real numbers, and the use of the unmodified term _set_ will imply automatically _set of real numbers._

Some examples of sets are: the collection of all real numbers; the positive integers; the numbers $-1, 0, 1$; all numbers x satisfying the inequality $2 \leqslant x \leqslant 5$. We shall be particularly interested in sets like the first and last examples. Sets such as these are also called _intervals_.

If a and b are two real numbers, $a < b$, they may be used to define four different intervals (Fig. 4-1). The first of these, which includes both end values, is called a _closed interval_; the last one, which omits both end values, is called an _open interval_. The others, which omit one end value, are called _half-open_, or _half-closed_. The set of all real numbers is often indicated by $-\infty < x < \infty$, and is called an _infinite interval_ in contrast to _finite intervals_ such as those in Fig. 4-1.

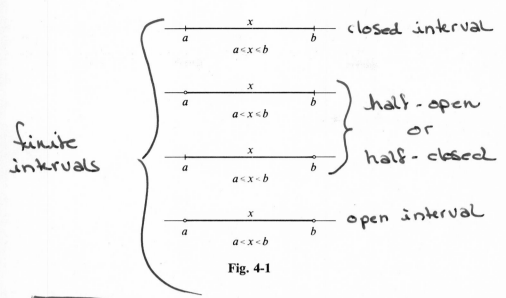

Fig. 4-1

DEFINITION 4–2. A variable is a symbol which may assume any value in a set S. The set S is called the range of the variable.

Thus, in the inequality $2 \leqslant x \leqslant 5$, x is a variable which may take on any value between 2 and 5 and hence its range is this set.

The symbol $|x|$, read *absolute value of x*, is defined in a footnote to Example 3-8. However, since we wish to remark on this notation here, we repeat the definition.

DEFINITION 4–3

$$|x| = x, \qquad x \geqslant 0;$$

$$|x| = -x, \qquad x < 0.$$

For example,

$$|2| = 2, \qquad |-2| = 2, \qquad |0| = 0, \qquad |x| = |-x|.$$

This notation is used frequently when a representation of the magnitude of the difference between two numbers, neglecting algebraic sign, is desired. We may wish, for example, to restrain a variable x so that the numerical difference between it and 10 may not be as great as 2. This may be expressed in the form

$$|x - 10| < 2$$

or, in the equivalent form,

$$|10 - x| < 2.$$

If we avoid the "absolute value" notation, this restriction may be expressed in the form

$$-2 < x - 10 < 2$$

or, equivalently,

$$8 < x < 12.$$

4–2. Function

There are countless examples in our everyday life of the dependence of one quantity on another. The area of a circle depends on the length of the radius; the length of time required to drive a given distance depends on the speed at which the driving is done; the boiling point of water depends on the atmospheric pressure; the number of years required for a given sum of money to accumulate to a specified amount depends on the interest rate, etc. All these examples have one point in common. If we specify one of the quantities involved, the other is uniquely determined. Thus, if the radius of a circle is required to be 2 in., the area is uniquely 4π square inches. In all such situations we say that *one quantity is a function of the other*, that is, *if one is given, the other is determined.*

There are many ways in which the relationship between two quantities may be defined. It may be that a simple table of values establishes a correspondence. Consider the following table:

x	-3	-2	-1	0	1	2	3
y	0	2	1	-1	-2	-5	-9

Let us agree that when we choose an x in the first row, we associate with it the number y directly below it. In this way a value of y is associated with each x in the first row. Thus *y is a function of x*. Note that once x is chosen, there is no ambiguity in the value of y to be used.

The functional relationship exists only for x belonging to the set S (-3, -2, -1, 0, 1, 2, 3) since the association between x and y is defined only for these values. For this reason, S is said to be the *domain of definition* or, more simply, the *domain* of the function. The set of associated values of y is called the *range* of the function.

If we use the concept of ordered pairs of numbers (x, y),† this relationship is completely expressed by the set of pairs

$$(-3, 0), (-2, 2), (-1, 1), (0, -1), (1, -2), (2, -5), (3, -9).$$

---

† To say that (x, y) is an ordered pair is to say that (x, y) is different from (y, x). The rectangular coordinates of a point is an example of an ordered pair. The point $(2, 3)$ is not the same as $(3, 2)$.

The set of first members of the pairs is the domain, the set of second members constitutes the range, and the associated numbers are defined by being members of the same pair.

A more common way of expressing a relationship between two quantities is through an equation. Consider

$$y = x^3 + 1.$$

If any x is selected, $-\infty < x < \infty$, this equation associates with it a unique value of y; for example, if we take $x = 2$, then our choice of y is given by

$$y = 2^3 + 1 = 9.$$

Thus this equation defines y as a function of x, the domain being the set of all real numbers. Moreover, the range of the function, that is, the set of values which y may assume, is also the set of all real numbers. This last statement is true because the given equation has a unique real solution x for any real y.

This relationship between x and y can also be represented by a set of ordered number pairs (x, y), the first member of which is chosen at will within the domain and the second calculated from the equation. However, it is not quite so simple in this case because there is an infinite number of pairs, and consequently they cannot be written down individually. Nevertheless they can be expressed in the form

$$(x, x^3 + 1), \qquad -\infty < x < \infty.$$

In both preceding examples, the choice of x was made at will (within the domain of definition), and for this reason it is called the *independent* variable. Similarly, since the value of y is dependent on the choice of x, we call it the *dependent* variable.

Now we would like to formalize the preceding discussion of the function concept into some sort of definition. It appears that we have two choices. One is that of a set of ordered number pairs. The other consists of a domain and a rule of association. In view of the type of functions with which we are going to deal, and in the hope that it will be more meaningful to the student, we elect the latter.

DEFINITION 4-4. A real valued function consists of two things:

(a) A set of real numbers, the domain of definition.

(b) A rule for associating one and only one real number with each number in the domain of definition.

In order to continue our discussion we need to define a notation. We shall use single letters such as f, g, F, etc., to represent a function. If a is a number from the domain of f, we shall use the symbol $f(a)$ to represent the number which the rule of f associates with a and we call $f(a)$ the function value of f at a. Thus, if f represents the first example discussed above, we write

$$f(2) = -5.$$

The symbol $f(5)$ has no meaning because 5 is not a number from the domain of f.

In the second example above, if we denote the function by F, the rule is

$$F(x) = x^3 + 1;$$

for example,

$$F(2) = 2^3 + 1 = 9.$$

Another way of expressing this same idea is to say that the symbol $\underline{F(x)}$ represents the value of F at x.

A function may be visualized as a "machine," the input for which are the numbers in the domain and its output the corresponding function values. The function F, defined above, may be represented schematically by Fig. 4-2.

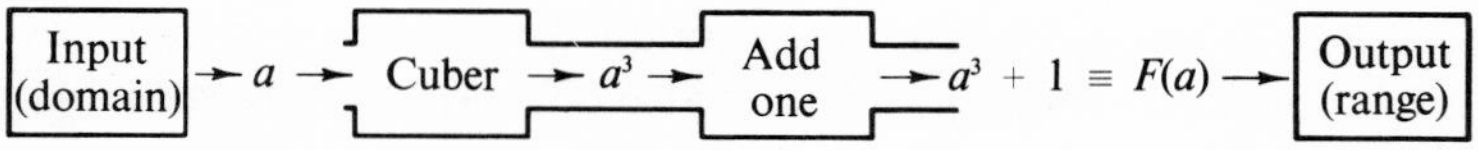

contains all real numbers

Input (domain) → a → Cuber → a^3 → Add one → $a^3 + 1 \equiv F(a)$ → Output (range)

Fig. 4-2

The input "bin" for this function contains all the real numbers. If we were to represent the function f, discussed above, in a similar manner, the input bin would contain only the numbers $-3, -2, -1, 0, 1, 2, 3$.

We shall deal exclusively with functions for which the "rule" is expressible as one or more equations. In this case we may not always state the domain. However, the student should not lose sight of the fact that the domain is an essential part of the definition and is, when not restricted otherwise, implied by the rule. Thus, if we speak of the function g defined by

$$g(x) = \sqrt{1 - x},$$

the domain $x \leq 1$ is implied automatically because these are the only values of x for which the rule gives real values for $g(x)$. Or, if we consider the function h defined by

$$h(x) = \frac{x + 1}{x - 1},$$

the implied domain is the set of all real numbers except $x = 1$.

DEFINITION 4–5. The set of number pairs defined by a function f, used as coordinates of points in the plane, constitute the graph of f.

Stated in the language of the previous chapters, *the graph of f is the graph of the equation*

$$y = f(x).$$

This concept will prove to be useful on many occasions.

Example 4–1. Discuss the function g where $g(x) = \sqrt{4 + x^2}$.

It is clear that any real x gives a real value $g(x) \geqslant 2$. Hence the domain of g is $-\infty < x < \infty$ and the range is $g(x) \geqslant 2$. The graph of g is shown in Fig. 4-3(a).

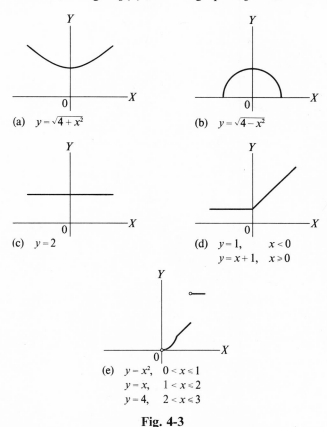

(a) $y = \sqrt{4 + x^2}$

(b) $y = \sqrt{4 - x^2}$

(c) $y = 2$

(d) $y = 1, \quad x < 0$
$y = x + 1, \quad x \geqslant 0$

(e) $y = x^2, \quad 0 < x \leqslant 1$
$y = x, \quad 1 < x \leqslant 2$
$y = 4, \quad 2 < x \leqslant 3$

Fig. 4-3

Example 4–2. Discuss the function h where $h(x) = \sqrt{4 - x^2}$.

In this case $h(x)$ is real only when $-2 \leqslant x \leqslant 2$, that is, when $|x| \leqslant 2$. Hence the domain of h is this interval. $h(x)$ is nonnegative and clearly cannot exceed 2. Therefore the range of h is $0 \leqslant h(x) \leqslant 2$. See Fig. 4-3(b) for the graph of h.

Example 4–3. Discuss the function ϕ where $\phi(x) = 2$.

The domain of ϕ is obviously $-\infty < x < \infty$ and the range is the single number 2. See Fig. 4-3(c) for the graph of ϕ.

Example 4–4. Discuss the function c where $c(x) = \cos x$.

$\cos x$ is defined for all real x and its values range from 1 to -1. Therefore the domain is $-\infty < x < \infty$ and the range is $-1 \leqslant c(x) \leqslant 1$.

In all preceding examples the correspondence between x and the function values was defined by a single formula. This need not be true. The functions in the following two examples illustrate this fact.

Example 4–5. Discuss the function f where

$$f(x) = 1, \qquad x < 0;$$
$$f(x) = 1 + x, \qquad x \geqslant 0.$$

This function has the set of all real numbers as its domain since the two statements provide a rule for calculating a real $f(x)$ when any real x is specified, the first for negative values of x, the second for nonnegative values. The range of f is $f(x) \geqslant 1$. The graph of f is shown in Fig. 4-3(d).

Example 4–6. Discuss the function g where

$$g(x) = x^2, \qquad 0 < x \leqslant 1;$$
$$g(x) = x, \qquad 1 < x \leqslant 2;$$
$$g(x) = 4, \qquad 2 < x \leqslant 3.$$

The domain of g is clearly the interval $0 < x \leqslant 3$ and its range consists of the interval $0 < g(x) \leqslant 2$ and $g(x) = 4$. The graph of g is shown in Fig. 4-3(e).

The next two examples are presented to familiarize the student with the functional notation.

Example 4–7. Given $F(x) = 2x^2 - 3x$. Find $F(2)$, $F(-2)$, $F(0)$, $F(a)$, $F(x + h)$, $F(x + h) - F(x)$.

Since the domain of F is $-\infty < x < \infty$, all required values are defined. We have

$$F(2) = 2(2)^2 - 3(2) = 2;$$
$$F(-2) = 2(-2)^2 - 3(-2) = 14;$$
$$F(0) = 2(0)^2 - 3(0) = 0;$$
$$F(a) = 2(a)^2 - 3(a) = 2a^2 - 3a;$$
$$F(x + h) = 2(x + h)^2 - 3(x + h) = 2x^2 + 4xh + 2h^2 - 3x - 3h;$$
$$F(x + h) - F(x) = (2x^2 + 4xh + 2h^2 - 3x - 3h) - (2x^2 - 3x)$$
$$= 4xh + 2h^2 - 3h.$$

Example 4–8. Given $f(x) = 2x + 3$, $g(x) = x^2$. Find $[f(2)][g(3)]$, $f(a)/g(a)$, $g[f(-2)]$.

The domains of f and g are both $-\infty < x < \infty$, so these values are all defined. We have $f(2) = 2(2) + 3 = 7$, $g(3) = 3^2 = 9$; thus

$$[f(2)][g(3)] = 63;$$
$$\frac{f(a)}{g(a)} = \frac{2a + 3}{a^2};$$
$$g[f(-2)] = [2(-2) + 3]^2 = (-1)^2 = 1.$$

Not all equations in x and y define a correspondence which fits our definition of function. Our next example illustrates this problem.

Example 4–9. Consider the equation $x^2 - y^2 = 1$.

We solve this equation for y and obtain $y = \pm\sqrt{x^2 - 1}$. Thus, to each value of x, $|x| \geqslant 1$, there corresponds two values of y. This is contrary to our definition of function, which requires that to each value of x there shall correspond only one value of y. In order to eliminate this difficulty we say that the given equation defines two functions f and g such that

$$f(x) = \sqrt{x^2 - 1}, \qquad g(x) = -\sqrt{x^2 - 1}.$$

Both functions have the domain $|x| \geqslant 1$. The range of f is $f(x) \geqslant 0$ and that of g is $g(x) \leqslant 0$.

Exercises 4-1

1. Use absolute value signs to write statements equivalent to the following:
   (a) $-4 \leqslant x \leqslant 4$ (b) $-5 < x - 3 < 5$
   (c) $-\varepsilon < x - a < \varepsilon, \quad \varepsilon > 0$ (d) $-3 \leqslant x \leqslant 5$

2. Write statements equivalent to the following without using absolute value signs:
   (a) $|x| < 3$ (b) $|x + 1| \leqslant 6$ (c) $|x - 4| \leqslant 10$ (d) $|x - h| < \varepsilon$

3. A governor regulates the speed S of an engine to 600 rpm with a maximum error of 5 rpm. Express this statement by means of:
   (a) absolute values
   (b) inequalities not involving absolute values

4. Let $h(x) = 2x - 7$. Find $h(1)$, $h(-\frac{1}{2})$, $h(4)$, $h(-5)$, $h(a + b)$, $h(h(7))$.

5. Let $g(x) = x - x^2$. Find $g(0)$, $g(2)$, $g(-3)$, $g(a)$, $g(a^2)$, $g(x + h)$.

6. Let $f(x) = x^2 - 1$ and $g(x) = -3x + 1$. Find $f(3)$, $g(-2)$, $f(g(-2))$, $g(f(3))$.

7. Give the domain and range of the functions defined by the following:
   (a) $f(x) = \sqrt{x^2 + 4}$ (b) $f(x) = \sqrt{x^2 - 4}$ (c) $f(x) = \sqrt{4 - x^2}$

   (d) $f(x) = x^2 - 4$ (e) $f(x) = \dfrac{x^2 - 4}{x + 2}$

8. Let $f(x) = |x|$. Give the domain and range of f and draw its graph.

9. Let $F(x) = -1, \quad x < -1$;
   $F(x) = x, \quad -1 \leqslant x \leqslant 1$;
   $F(x) = 1, \quad x > 1$.

   (a) Give the range and domain of F.
   (b) Find $F(-3)$, $F(-1)$, $F(\frac{1}{2})$, $F(1)$, $F(3)$.
   (c) Draw the graph of F.

10. Let $\phi(x) = \dfrac{1}{x}, \quad -2 \leqslant x < -\dfrac{1}{2};$

$\qquad \phi(x) = \dfrac{4(x-1)}{3}, \quad -\dfrac{1}{2} \leqslant x < 1;$

$\qquad \phi(x) = 1, \quad 1 \leqslant x \leqslant 2.$

    (a) Give the domain and range of ϕ.

    (b) Find $\phi(-1), \phi(-\tfrac{1}{2}), \phi(0), \phi(1)$.

    (c) Draw the graph of ϕ.

11. Let $f(x) = 1 - 3x$. Find a function F such that $f[F(x)] = x$. Does $F[f(x)] = x$?

4–3. Tangent to a Curve

The student has already encountered the idea of a tangent to a curve in the study of circles in elementary geometry. We propose to define tangents to curves in general in a manner consistent with our experience with tangents to circles. We shall use as a guide in arriving at this definition the idea that a line L tangent to a curve C at the point P should have the same "direction" as C at P. That is, if one point is moving along C and another is moving along L, and if they pass through P simultaneously, at that instant they are both going in the same direction or in opposite directions.

How can we determine a line which appears to have this property? Consider the secant line PQ where Q is a point of C near P (Fig. 4-4(a)). Let Q move along C in the direction of P. It may be that the secant line PQ *approaches*† a definite position as Q *approaches* P from either side (Fig. 4-4(b)). If this happens, we say that the line occupying this definite position has the same direction as C at the point P and we define it‡ to be the tangent to C at P.

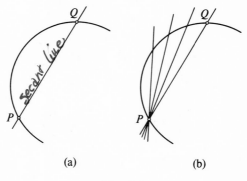

(a)   (b)

Fig. 4-4

---

† The term *approaches* will be dealt with later. For the moment its descriptive properties will serve us.

‡ This, of course, amounts to defining "direction" on C.

We note that if C is a straight line, all secant lines PQ coincide with C, so we conclude that a straight line is its own tangent at every point. If C is not a straight line, in general none of the lines PQ will be the tangent. If Q is very close to P, the line PQ will be a close approximation to the tangent, but however close, as long as Q is different from P, it will be only an approximation. If P and Q coincide, PQ is undefined. Thus we have to find other means of determining the tangent line described above.

In order to pursue this problem let us define a new notation. We shall use the symbol ΔS to represent a change in the variable S, that is, the difference between two values of S. The symbol Δ standing alone is meaningless. If we wish to center our discussion about values of S in the vicinity of a particular value of S, say $S = S_1$, we may represent these values by the notation $S = S_1 + \Delta S$ where ΔS varies and may be either positive or negative. If (x_1, y_1) are the coordinates of a point P, any neighboring point may be represented by $(x_1 + \Delta x, y_1 + \Delta y)$.

Now let us continue our discussion of tangents by considering the parabola

$$y = x^2 \tag{4-1}$$

or, stated differently, the graph of f where

$$f(x) = x^2. \tag{4-2}$$

We use this overworked example because it is one of the simplest nonlinear curves. Let us attempt to determine the tangent to this curve at the point $P(\frac{1}{2}, \frac{1}{4})$. A nearby point on the curve may be expressed in the form $Q(\frac{1}{2} + \Delta x, \frac{1}{4} + \Delta y)$ where (Fig. 4-5)

$$\frac{1}{4} + \Delta y = (\frac{1}{2} + \Delta x)^2$$

or

$$\Delta y = \Delta x + (\Delta x)^2. \tag{4-3}$$

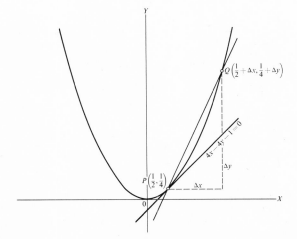

Fig. 4-5

Since Δx may be either positive or negative, Q may be on either side of P. The slope of PQ, from (1-7), is $\Delta y / \Delta x$, and from (4-3),

$$\frac{\Delta y}{\Delta x} = \frac{\Delta x + (\Delta x)^2}{\Delta x} = 1 + \Delta x. \tag{4-4}$$

What happens to the slope of PQ as Q moves along the curve toward P from either side? When Q behaves in this fashion, $|\Delta x|$ decreases. In fact $|\Delta x|$ decreases in such a manner that it can be made as small as we please by taking Q sufficiently close to P. Thus we see that the slope of PQ can be made to differ from 1 by as little as we please by taking Q sufficiently close to P or, what is the equivalent, by taking $|\Delta x|$ sufficiently small. Hence, as Q approaches P from either side, the line PQ approaches a line through P with the slope 1. We define this line to be the tangent to the parabola at P. The equation of this tangent is

$$4x - 4y - 1 = 0.$$

Thus this approach to the tangent problem, in this case at least, provides us with a definite line to call the tangent, although the line PQ never assumes this position.

Suppose we look at a more general problem. Let $P(x_1, y_1)$ and $Q(x_1 + \Delta x, y_1 + \Delta y)$ be two points on the curve $y = f(x)$ (Fig. 4-6). Then

$$y_1 = f(x_1)$$

and

$$y_1 + \Delta y = f(x_1 + \Delta x),$$

from which we obtain

$$\Delta y = f(x_1 + \Delta x) - f(x_1).$$

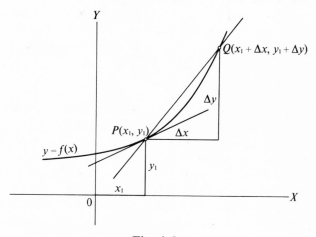

Fig. 4-6

Thus the slope of PQ is given by

$$\frac{\Delta y}{\Delta x} = \frac{f(x_1 + \Delta x) - f(x_1)}{\Delta x}. \qquad (4\text{-}5)$$

If this slope approaches a definite value m as Q approaches P from either side, that is, as Δx approaches zero, we define the tangent to $y = f(x)$ at $P(x_1, y_1)$ to be the line

$$y - y_1 = m(x - x_1)$$

and we say the slope of the curve at this point is m. Therefore, if we are to apply this method to the tangent problem, we must look more carefully at what meaning we shall attach to expressions like "Δx approaches zero," and "$(f(x_1 + \Delta x) - f(x_1))/\Delta x$ approaches a definite value m as Δx approaches zero." The discussion of these matters leads us to, the central idea of the calculus—*the concept of limit*.

4–4. Limit of a Sequence

The notion of sequence is vital to the development of calculus. Intuitively, *a sequence is an ordered set of numbers*. If *finite*, there is a first term, a second term, etc., and a last term. For example, the numbers

$$2, 4, 3, 6, 5, 10$$

constitute a finite sequence with first term 2, second term 4, etc., and last term 10. It must be emphasized that the order in which the numbers occur is an essential part of the concept. The numbers

$$2, 10, 4, 5, 3, 6$$

form the same set but not the same sequence.

Of more interest to us are *infinite sequences.* They have the same basic characteristic of order as do finite sequences, that is, a first term, a second term, etc., but *no last term.* An example of an infinite sequence is

$$1, \frac{1}{2}, \frac{1}{3}, \ldots, \frac{1}{n}, \ldots,$$

where the dots between $1/3$ and $1/n$ indicate an indefinite number of omitted terms; that is, n is not specified, and the dots after $1/n$ indicate that the sequence continues indefinitely according to the indicated law of formation. This law of formation is, in words: *the nth term is the reciprocal of the integer n.* Thus the 100th term is $1/100$. There are no limitations placed on n except that it is a positive integer, so there is no end to the sequence.

Any infinite sequence has a first term and associates this number (the first term) with the integer 1. Likewise there is a second term and this provides a unique association between a number and the integer 2. Continuing in this manner, we see that an infinite sequence associates a unique number with each of the positive integers, and therefore is actually a function. Hence we may formalize our definition of an infinite sequence.

DEFINITION 4–6. An infinite sequence is a function whose domain is the set of all positive integers.†

Although not explicitly stated in the definition, the essential characteristic of ordering is implied since there is an ordering associated with the positive integers.

As a general notation we shall use a letter with subscripts attached to indicate position in the sequence. Thus we write a general sequence

$$s_1, s_2, s_3, \ldots, s_n, \ldots,$$

where s_n is the nth term and represents the law of formation. We shall also use the abbreviated notation $\{s_n\}$ to indicate this sequence. The example considered above has

$$s_1 = 1, \quad s_2 = \frac{1}{2}, \quad s_3 = \frac{1}{3}, \ldots, \quad s_n = \frac{1}{n}, \ldots,$$

and may be represented by

$$\{s_n\} = \left\{\frac{1}{n}\right\}.$$

Or, we may write

$$s_n = \frac{1}{n}, \qquad n = 1, 2, 3, \ldots.$$

Since the domain is the same for all infinite sequences, we often omit any specific mention of it.

Let us continue with this example and explore another aspect of infinite sequences. It is obvious that as n increases, $s_n = 1/n$ becomes smaller. As a matter of fact, we can find an n such that s_n is as small as we please, this smallness, however, having been determined in advance. If we wish s_n to be less than 0.01, we readily see that this will be so when n is greater than 100. This is usually expressed by writing

$$s_n = \frac{1}{n} < 0.01, \qquad n > 100.$$

Similarly, it is clear that

$$s_n = \frac{1}{n} < 0.001, \qquad n > 1000.$$

---

† We are going to be interested solely in infinite sequences, so we state our definition accordingly.

In general, if we require s_n to be less than any previously assigned positive number, say ε, we have

$$s_n = \frac{1}{n} < \varepsilon, \qquad n > \frac{1}{\varepsilon}.$$

A very important point to note here is that it is possible to find not only a term in the sequence that is less than ε, *but also that every succeeding term in the sequence is less than ε.* Thus, if we have chosen $\varepsilon = 0.01$, not only is

$$s_{101} < 0.01,$$

but so is every s_n with $n > 100$.

When a sequence has these properties we say that *it converges to the limit zero* or merely that *it has the limit zero,* and we write

$$\lim_{n \to \infty} s_n = 0.$$

These symbols are read "the limit of s_n as n approaches infinity is zero." Alternate notations are s_n *approaches zero as n approaches* ∞, or $s_n \to 0$ as $n \to \infty$.

One way of describing this property of this particular sequence is to say that its terms are approximations of zero and that an approximation of any desired degree of accuracy may be found by going sufficiently far into the sequence. If we generalize this idea to numbers other than zero, we are led to the following definition of the limit of a sequence:

DEFINITION 4–7. The sequence $\{s_n\}$ is said to converge to the limit l if, for every positive ε, however small, it is possible to find a positive number N such that

$$|s_n - l| < \varepsilon$$

for all $n > N$.

This simply says that the terms in the sequence approximate the number l in the sense that a term in the sequence can be found which approximates l with any required degree of accuracy and such that every term beyond it is an approximation of l of at least equal accuracy.

Example 4–10. Show that the sequence

$$2, \frac{3}{2}, \frac{4}{3}, \ldots, \frac{n+1}{n}, \ldots$$

converges to the limit 1.

What we have to show is that for any $\varepsilon > 0$, we can find an N such that

$$|s_n - 1| < \varepsilon$$

for all $n > N$. We have

$$|s_n - 1| = \left| \frac{n+1}{n} - 1 \right| = \frac{1}{n},$$

so

$$|s_n - 1| < \varepsilon$$

requires

$$\frac{1}{n} < \varepsilon.$$

This is true if $n > 1/\varepsilon$. Therefore we take

$$N = \frac{1}{\varepsilon} \qquad \text{and} \qquad |s_n - 1| < \varepsilon$$

for all $n > N$. This proves the proposition.
    To illustrate, if we choose $\varepsilon = 0.001$, then we take

$$N = \frac{1}{0.001} = 1000,$$

and

$$|s_n - 1| = \left| \frac{n+1}{n} - 1 \right| = \frac{1}{n} < 0.001$$

for all $n > 1000$.

Example 4–11. Show that the sequence

$$\frac{1}{2}, \frac{1}{4}, \frac{1}{8}, \dots, \frac{1}{2^n}, \dots$$

converges to the limit zero.

    What we have to show in this example is that, for any $\varepsilon > 0$, there exists an N such that

$$|s_n - 0| = |s_n| = \frac{1}{2^n} < \varepsilon,$$

for $n > N$. First we note that $1/2^n < \varepsilon$ for all n† if $\varepsilon \geqslant 1$, so we may choose

---

† We remind the student here that, as elsewhere in connection with sequences, n is restricted to the positive integers.

N to be anything we please, say, $N = 0$. Therefore the following analysis is needed only for $\varepsilon < 1$.

We need to satisfy $1/2^n < \varepsilon$, or, equivalently, $2^n > 1/\varepsilon$. To resolve this we take the logarithm to the base 10 of both members and obtain

$$n \log_{10} 2 > \log_{10}\left(\frac{1}{\varepsilon}\right) = -\log_{10} \varepsilon,$$

whence

$$n > \frac{-\log_{10} \varepsilon}{\log_{10} 2}.$$

Thus, if

$$N = \frac{-\log_{10} \varepsilon}{\log_{10} 2},$$

our original inequality will be satisfied for $n > N$ and our proof is complete.

To illustrate numerically, if we choose $\varepsilon = 0.01$, then

$$N = \frac{-\log_{10} 0.01}{\log_{10} 2} \cong 6.6.$$

Then

$$|s_n| = \frac{1}{2^n} < 0.01$$

for all $n > 6.6$, that is, for $n \geqslant 7$. Thus every term in the sequence beyond the sixth will be less than 0.01.

Not all sequences converge to a limit. Consider the following example.

Example 4–12. Show that the sequence

$$-1, 1, -1, \ldots, (-1)^n, \ldots$$

does not converge to a limit.

This can be seen by noting that the numerical difference between any two successive terms in this sequence is 2. Thus, if any term is a close approximation to a number l, the next term must differ from l by almost two units. This violates the definition of convergence.

When a sequence fails to converge, it is said to *diverge*.

Limits of specific sequences are usually difficult to obtain from direct application of the definition. For this reason we present some general theorems which will be of great assistance in this connection. Although the proofs of these theorems are not difficult, we shall omit them because of the limitations placed on this book.

THEOREM 4–1. Given the two sequences $\{r_n\}$ and $\{s_n\}$ which converge to the limits r and s, respectively.

(a) If c is a constant, $\lim\limits_{n\to\infty} cr_n = cr$ (b) $\lim\limits_{n\to\infty} (r_n + s_n) = r + s$

(c) $\lim\limits_{n\to\infty} (r_n - s_n) = r - s$ (d) $\lim\limits_{n\to\infty} r_n s_n = rs$

(e) $\lim\limits_{n\to\infty} \dfrac{r_n}{s_n} = \dfrac{r}{s}, \quad s_n \neq 0, s \neq 0$

THEOREM 4–2. Given three sequences $\{r_n\}, \{s_n\}, \{t_n\}$. If

$$\lim_{n\to\infty} r_n = \lim_{n\to\infty} s_n = l,$$

and if $r_n \leqslant t_n \leqslant s_n$ for all n, then

$$\lim_{n\to\infty} t_n = l.$$

We do not have immediate use for Theorem 4-2 but we list it here for future reference. However, we shall put Theorem 4-1 to work without delay.

We have already seen that

$$\lim_{n\to\infty} \frac{1}{n} = 0.$$

Therefore, from (d) in Theorem 4-1,

$$\lim_{n\to\infty} \frac{1}{n^2} = \lim_{n\to\infty} \frac{1}{n} \cdot \frac{1}{n} = \lim_{n\to\infty} \frac{1}{n} \cdot \lim_{n\to\infty} \frac{1}{n} = 0 \cdot 0 = 0.$$

Similarly,

$$\lim_{n\to\infty} \frac{1}{n^k} = 0,$$

where k is any positive integer. Also, using (a) of Theorem 4-1,

$$\lim_{n\to\infty} \frac{c}{n^k} = 0,$$

where c is any constant.

Example 4–13. Determine the limit of the sequence in Example 4-10 by the use of Theorem 4-1.

We write $(n + 1)/n$ in the form $1 + (1/n)$ and, applying (b), we have

$$\lim_{n\to\infty} \frac{n+1}{n} = \lim_{n\to\infty} \left(1 + \frac{1}{n}\right) = \lim_{n\to\infty} 1 + \lim_{n\to\infty} \frac{1}{n}.$$

What do we mean by the symbol $\lim_{n \to \infty} 1$? In conformity with the definition of our notation this means the limit of the sequence

$$1, 1, 1, \ldots, 1, \ldots.$$

Obviously, 1 satisfies the limit definition for this sequence since the difference between any term, regardless of its position, and 1 is 0. More generally

$$\lim_{n \to \infty} c = c,$$

where c is any constant. Thus

$$\lim_{n \to \infty} \frac{n+1}{n} = 1 + 0 = 1,$$

as we had already determined from the definition.

Example 4–14. Determine the limit of $\{s_n\}$ where

$$s_n = \frac{n^2 - 4}{2n^2 + 3}.$$

We write

$$s_n = \frac{1 - \dfrac{4}{n^2}}{2 + \dfrac{3}{n^2}}$$

by dividing both numerator and denominator by n^2. Applying (b) of Theorem 4-1, we have

$$\lim_{n \to \infty} \left(1 - \frac{4}{n^2}\right) = 1 - 0 = 1$$

and

$$\lim_{n \to \infty} \left(2 + \frac{3}{n^2}\right) = 2 + 0 = 2.$$

Then, from (e), we have

$$\lim_{n \to \infty} \frac{n^2 - 4}{2n^2 + 3} = \frac{1}{2}.$$

It may be that a sequence $\{s_n\}$ has the property that s_n increases without bound as n increases. That is to say, given any positive number M, however large, it is possible to find an N such that $s_n > M$ when $n > N$.

Example 4–15. Consider the behavior of the terms in the sequence 1, 2, 3, $\ldots, n, \ldots$, that is, $\{n\}$, as n increases.

It is clear that for every M, however large, it is possible to choose an N such that $s_n = n > M$ when $n > N$. Indeed we have only to take $N = M$ for this to be true. Thus, if $M = 10^3$, every term in the sequence beyond the 1000th term is greater than 10^3. Hence the sequence $\{n\}$ behaves in the manner described above.

Such a sequence does not have a limit as previously defined. However, it is convenient to have a language and notation to describe this situation. Since there is an essential characteristic of a limit present, namely, the fact that a property exists for all $n > N$, we make the following definition.

DEFINITION 4–8. The sequence $\{s_n\}$ is said to have the limit ∞ if for any positive number M, however large, it is possible to find an N such that $s_n > M$ when $n > N$. This property is indicated by writing

$$\lim_{n \to \infty} s_n = \infty.$$

It may be that a sequence $\{s_n\}$ has all its terms negative beyond a certain point and $|s_n|$ increases without bound as n increases. This situation is distinguished from the preceding one in the following definition.

DEFINITION 4–9. The sequence $\{s_n\}$ is said to have the limit $-\infty$ if the sequence $\{-s_n\}$ has the limit ∞. This property is indicated by writing

$$\lim_{n \to \infty} s_n = -\infty.$$

An alternate notation, referring to Definition 4-9, is

$$s_n \to -\infty \quad \text{as} \quad n \to \infty,$$

which is read s_n *approaches* (or *tends to*) *minus infinity as n approaches* (or *tends to*) *infinity*.

The following theorem is a useful collection of limits of some common combinations of sequences, some of which have infinite limits. These limits are presented without proof but the student should have no difficulty in convincing himself of their truth if he understands the definitions.

THEOREM 4–3. Given: the sequences $\{r_n\}$ and $\{s_n\}$, the limits of which are both ∞ and $\{t_n\}$, which has the finite limit $t \neq 0$; and the constant $c \neq 0$.

(a) $\displaystyle\lim_{n \to \infty} cr_n = \pm\infty$

(b) $\displaystyle\lim_{n \to \infty} (r_n + s_n) = \infty$

(c) $\displaystyle\lim_{n \to \infty} (r_n + t_n) = \infty$

(d) $\displaystyle\lim_{n \to \infty} (r_n - t_n) = \infty$

(e) $\displaystyle\lim_{n \to \infty} (t_n - r_n) = -\infty$

(f) $\displaystyle\lim_{n \to \infty} r_n t_n = \pm\infty$

(g) $\displaystyle\lim_{n \to \infty} r_n s_n = \infty$

(h) $\displaystyle\lim_{n \to \infty} \frac{r_n}{t_n} = \pm\infty, \quad t_n \neq 0$

(i) $\displaystyle\lim_{n \to \infty} \frac{1}{r_n} = 0, \quad r_n \neq 0$

(j) $\displaystyle\lim_{n \to \infty} \frac{t_n}{r_n} = 0, \quad r_n \neq 0$

If c and t are positive, the positive sign in the double sign $\pm$ prevails; otherwise the negative sign holds.

It is to be noted that nothing is said in the preceding theorem about sequences such as $\{r_n - s_n\}$ or $\{r_n/s_n\}$. Nothing general can be said about the limits of such sequences. They may, or may not, have a limit depending upon the particular sequence, and if a limit does exist, the value can be determined only by study of the individual case.

Next let us consider the sequence $\{1/s_n\}$ where $\{s_n\}$ is a sequence of positive terms such that $s_n \to 0$ as $n \to \infty$. We propose to show that

$$\lim_{n \to \infty} \frac{1}{s_n} = \infty,$$

that is, for any positive M, however large, it is possible to find an N such that $1/s_n > M$ when $n > N$. This will be true if it is always possible to find an N such that $s_n < 1/M$ when $n > N$. Since $s_n \to 0$ as $n \to \infty$, this last statement is true and we have proved the following theorem.

THEOREM 4-4. Let $\{s_n\}$ be a sequence of positive terms such that $s_n \to 0$ as $n \to \infty$. Then

$$\lim_{n \to \infty} \frac{1}{s_n} = \infty.$$

There is an obvious companion theorem for a sequence $\{s_n\}$ of negative terms.

Exercises 4-2 — approach infinity

1. Use the preceding theorems and examples to find the limits, if they exist, of the sequences whose general terms are:

(a) $1 - \dfrac{2}{n} + \dfrac{3}{n^2}$

(b) $1 + \dfrac{2}{n} - \dfrac{3}{n^2}$

(c) $\dfrac{3n^2 - 2n + 1}{n^2}$

(d) $\dfrac{2n^2 + n - 7}{3n^2 + 1}$

(e) $\dfrac{2n + 1}{n^2}$

(f) $\dfrac{3n^3}{5n^5 - 2n + 1}$

(g) $\dfrac{n - 1}{n + 1}$

(h) $\dfrac{n^2 - 1}{n + 1}$

(i) $\dfrac{(n + 2)(2n - 1)}{n^2}$

(j) $\dfrac{1}{(n - 1)(n + 2)}$

(k) $\dfrac{2n}{n - 1} - \dfrac{n + 1}{n}$

(l) $\dfrac{n^2 + 1}{n} - \dfrac{n^2}{n + 1}$

(m) $\dfrac{2n^2 + 1}{n} - \dfrac{n^2}{n + 1}$

(n) $\dfrac{n^2 + 1}{n} - \dfrac{2n^2}{n + 1}$

(o) $\left(\dfrac{3n}{n - 1}\right)^2$

(p) $1 + \dfrac{(-1)^n}{n}$

2. Given the sequences $\{s_n\}$, $\{t_n\}$ of positive terms such that $s_n \to s > 0$ and $t_n \to 0$ as $n \to \infty$. Evaluate the limits of the sequences whose general terms are

(a) $s_n^2 - s_n$ (b) $\dfrac{s_n^2 - 1}{s_n + 1}$ (c) $\dfrac{s_n}{t_n}$

(d) $1 - \dfrac{s_n}{t_n}$ (e) $1 + \dfrac{t_n}{s_n}$ (f) $\dfrac{t_n - 1}{t_n}$

3. Does the sequence $\{(1 + (-1)^n)/n\}$ have a limit? If so, what is it? Check your answer by use of the definition.

4–5. Limit of a Function

In the preceding section we defined and discussed limits of functions defining sequences. All such functions have the set of positive integers as their domain. Let us now consider a more general function whose domain is the set of all real numbers greater than some real number, that is, $x > a$. Then we may generalize Definition 4-7 as follows.

DEFINITION 4-10. Let f be a function whose domain is the set of real numbers $x > a$. Then the function values $f(x)$ are said to have the limit l as x approaches infinity if for any positive number ε, however small, it is possible to find a number N such that

$$|f(x) - l| < \varepsilon$$

when $x > N$. We write

$$\lim_{x \to \infty} f(x) = l.$$

This is read the *limit of $f(x)$ as x approaches infinity is l.* An alternate notation is

$$f(x) \to l \quad \text{as} \quad x \to \infty$$

and is read *$f(x)$ approaches (or tends to) l as x approaches (or tends to) infinity.*

This is completely analogous to Definition 4-7. The only change is from a *discrete* independent variable in the case of sequences to a *continuous* variable in the situation under consideration. The function value $f(x)$ may be made to approximate l to any required degree of accuracy by taking x sufficiently large. However, in the case of a continuous independent variable, we can define a more general and a more useful limit.

DEFINITION 4-11. Let f be a function whose domain includes $a \leqslant x \leqslant c$ with the possible exception of $x = b$, $a < b < c$. Then the function values

$f(x)$ are said to have the limit l as x approaches b if for any positive number ε, however small, it is possible to find a positive number δ such that

$$|f(x) - l| < \varepsilon$$

when

$$0 < |x - b| < \delta.$$

We write

$$\lim_{x \to b} f(x) = l.$$

This is read *limit of $f(x)$ as x approaches b is l*. The alternate notation

$$f(x) \to l \quad \text{as} \quad x \to b$$

is also used frequently.

This definition says that $f(x)$ can be made arbitrarily close to l by taking x sufficiently close to b ($x \neq b$), and that this approximation holds throughout an interval ($x = b$ excepted) which has b as its center. The exclusion of $x = b$ is very important because some of the most important limits we will meet will not have b within the domain of definition of the function.

Let us look at the graphical implications of this definition. We draw two horizontal lines $y = l - \varepsilon$, $y = l + \varepsilon$ where ε is chosen at will (Fig. 4-7). Then,

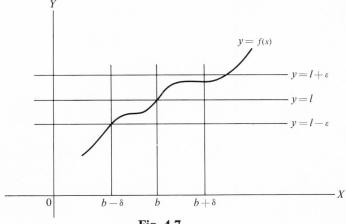

Fig. 4-7

if $f(x) \to l$ as $x \to b$, it is possible to determine two vertical lines $x = b - \delta$, $x = b + \delta$ such that the segment of the graph of $y = f(x)$ contained between the two vertical lines also lies completely between the two horizontal lines, except possibly for $x = b$, at which point f may not even be defined. It is to be emphasized that the two horizontal lines are chosen first and then the vertical lines are determined (and always can be determined) in accordance with that choice. In other words, we first specify how close $f(x)$ is to be to l, and then an interval for x about b as a center can be determined which will bring this about.

Example 4–16. Consider $\lim\limits_{x \to 1} \dfrac{x^2 - 1}{x - 1}$.

We have

$$f(x) = \frac{x^2 - 1}{x - 1} = x + 1$$

provided $x \neq 1$. But the definition of limit excludes this value for x, so we have

$$\lim_{x \to 1} \frac{x^2 - 1}{x - 1} = \lim_{x \to 1} (x + 1).$$

Now our intuition tells us that any reasonable definition of limit would give

$$\lim_{x \to 1} (x + 1) = 2.$$

Let us find whether our definition does indeed verify this result. We need to show that for any positive ε it is possible to determine a positive number δ such that

$$|(x + 1) - 2| < \varepsilon$$

when

$$0 < |x - 1| < \delta.$$

Since the first of these inequalities simplifies to $|x - 1| < \varepsilon$, it is clear that we may choose $\delta = \varepsilon$; that is, if

$$0 < |x - 1| < \delta = \varepsilon,$$

it follows that

$$|(x + 1) - 2| = |x - 1| < \varepsilon.$$

That is to say, for example, if $\varepsilon = 0.01$, then we take $\delta = 0.01$ and we have

$$|(x + 1) - 2| < 0.01$$

when

$$0 < |x - 1| < 0.01.$$

Hence, we have

$$\lim_{x \to 1} \frac{x^2 - 1}{x - 1} = 2,$$

as our intuition suggested.

Graphically, if we draw the lines $y = 2 \pm \varepsilon$, $x = 1 \pm \varepsilon$, we find that portion of the graph of

$$y = \frac{x^2 - 1}{x - 1}$$

included between the vertical lines also lies between the horizontal lines (Fig. 4-8). Although the function is undefined for $x = 1$, its value may be made

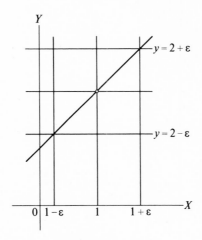

Fig. 4-8

as close to 2 as we please by confining x to a sufficiently small interval centered on $x = 1$.

If we define f to be the function such that

$$f(x) = \frac{x^2 - 1}{x - 1}, \qquad x \neq 1;$$

$$f(1) = 5,$$

then

$$\lim_{x \to 1} f(x) = 2$$

as before. However, the graph of $y = f(x)$ is that of Fig. 4-8 plus the isolated point $(1, 5)$. The fact that $(1, 5)$ does not fall between the lines $y = 2 \pm \varepsilon$ for $\varepsilon < 3$ is of no consequence as far as the limit is concerned, for $f(x)$ can still be made as close to 2 as we please by taking x sufficiently close to 1. This point $(1, 5)$ is of significance in another connection, as we shall see in the next section.

As in the case of sequences this general definition of limit is not adapted to the calculation of unknown limits. It is useful for checking suspected limits and for obtaining general theorems which prove helpful in finding limits. The fundamental theorem for operating with limits follows.

THEOREM 4–5. Let f and g be two functions such that

$$\lim_{x \to b} f(x) = k, \qquad \lim_{x \to b} g(x) = l.$$

Then (a)
$$\lim_{x \to b} [f(x) + g(x)] = k + l$$

(b)
$$\lim_{x \to b} [f(x) - g(x)] = k - l$$

(c)
$$\lim_{x \to b} [f(x) \cdot g(x)] = kl$$

(d)
$$\lim_{x \to b} \left[\frac{f(x)}{g(x)} \right] = \frac{k}{l}, \qquad l \neq 0$$

It is not within the scope of this book to prove this theorem. However, the student should consider these conclusions carefully and realize that they are reasonable in the light of the definition. The uses of the theorem are obvious, and it greatly simplifies the general problem of calculating limits.

THEOREM 4–6. If c is a constant,

$$\lim_{x \to b} c = c \qquad \text{for any } b.$$

In this case $f(x) = c$ and consequently $|f(x) - c| = 0$ for all x. Hence c satisfies the limit definition and the theorem is established.

THEOREM 4–7. $\lim_{x \to b} x = b$.

Here $f(x) = x$ so that the condition $|f(x) - b| < \varepsilon$ reduces to $|x - b| < \varepsilon$. Hence, once ε is chosen, we may take $\delta = \varepsilon$ and the conditions of the definition are satisfied. The theorem is true also with b replaced by ∞.

Application of Theorems 4-5(c) and 4-7 give

$$\lim_{x \to b} x^2 = b^2,$$

and it may be shown easily that

$$\lim_{x \to b} x^n = b^n, \tag{4-6}$$

where n is a positive integer. If $b \neq 0$, (4-6) may be established also for n a negative integer by using Theorems 4-5(d), 4-6, and 4-7.

A more general theorem on powers follows which we state without proof.

THEOREM 4–8. Let p and q be integers and let $[f(x)]^{p/q}$ be defined for $a \leqslant x \leqslant c$ except possibly for $x = b$, $a < b < c$. Then, if $\lim\limits_{x \to b} f(x)$ exists,

$$\lim_{x \to b} [f(x)]^{p/q} = \left[\lim_{x \to b} f(x)\right]^{p/q}.$$

Example 4–17. Find $\lim\limits_{x \to 2} (x^2 - 3x + 1)$.

According to Theorem 4-5(a) and (b) we may write

$$\lim_{x \to 2} (x^2 - 3x + 1) = \lim_{x \to 2} x^2 - \lim_{x \to 2} 3x + \lim_{x \to 2} 1.$$

These individual limits are covered in Theorems 4-5(c), 4-6, 4-7, and 4-8. We have, then,

$$\lim_{x \to 2} (x^2 - 3x + 1) = 2^2 - 3(2) + 1 = -1.$$

In this case we note that $\lim\limits_{x \to b} f(x) = f(b)$. That this is not always the case is indicated in the next example.

Example 4–18. Find

$$\lim_{x \to -1} \frac{x^3 - 7x - 6}{x^2 - 1}.$$

When we try to use Theorem 4-5 immediately to evaluate this limit we find that the hypothesis of the theorem is not satisfied, since

$$\lim_{x \to -1} (x^2 - 1) = 0.$$

However, we recall that the definition of the limit as $x \to -1$ does not involve $x = -1$, and since

$$\frac{x^3 - 7x - 6}{x^2 - 1} = \frac{x^2 - x - 6}{x - 1}$$

for $x \neq -1$, we may write

$$\lim_{x \to -1} \frac{x^3 - 7x - 6}{x^2 - 1} = \lim_{x \to -1} \frac{x^2 - x - 6}{x - 1}$$

$$= \frac{\lim\limits_{x \to -1} (x^2 - x - 6)}{\lim\limits_{x \to -1} (x - 1)} = \frac{-4}{-2} = 2.$$

Example 4–19. Find $\lim\limits_{x \to 1} \sqrt{(4 - x^2)^3}$.

Applying Theorem 4-8 we have

$$\lim_{x \to 1} \sqrt{(4 - x^2)^3} = \lim_{x \to 1} (4 - x^2)^{3/2}$$

$$= \left[\lim_{x \to 1} (4 - x^2)\right]^{3/2} = (3)^{3/2} = 3\sqrt{3}.$$

When the values $f(x)$ become indefinitely great as $x \to b$, no limit exists in the sense of Definition 4-11. However, in order to have a concise notation for describing such a situation, we make the following definition.

DEFINITION 4-12. The function values $f(x)$ are said to have the limit ∞ as x approaches b if for any $M > 0$, however large, it is possible to find a positive number δ such that $f(x) > M$ when $0 < |x - b| < \delta$. We write

$$\lim_{x \to b} f(x) = \infty.$$

DEFINITION 4-13. The function values $f(x)$ are said to have the limit $-\infty$ as $x \to b$ if

$$\lim_{x \to b} [-f(x)] = \infty.$$

We write

$$\lim_{x \to b} f(x) = -\infty.$$

Example 4–20. Find $\lim\limits_{x \to 1} \dfrac{1}{(x - 1)^2}$.

We note that the values of $1/(x - 1)^2$ increase as $x \to 1$. This causes us to suspect an "infinite limit," as described in the preceding definitions. This will be true if for an arbitrarily chosen $M > 0$, we can find a $\delta > 0$ such that

$$\frac{1}{(x - 1)^2} > M \qquad \text{when} \qquad 0 < |x - 1| < \delta.$$

This inequality implies that

$$\frac{1}{|x - 1|} > \sqrt{M} \qquad \text{or} \qquad |x - 1| < \frac{1}{\sqrt{M}}.$$

Hence, for any M, if we choose $\delta = 1/\sqrt{M}$, Definition 4-8 will be satisfied and we can write

$$\lim_{x \to 1} \frac{1}{(x - 1)^2} = \infty.$$

Example 4–21. Find $\lim\limits_{x \to \infty} \dfrac{x^2 - 1}{4 - x^2}$.

We divide numerator and denominator by x^2 and write

$$\lim_{x \to \infty} \frac{x^2 - 1}{4 - x^2} = \lim_{x \to \infty} \frac{1 - (1/x^2)}{(4/x^2) - 1} = \frac{\lim\limits_{x \to \infty}(1 - (1/x^2))}{\lim\limits_{x \to \infty}((4/x^2) - 1)}.$$

It is intuitively clear, and indeed not difficult to prove, that if n is a positive integer and k is a constant,

$$\lim_{x \to \infty} \frac{k}{x^n} = 0.$$

A portion of this proof is left for an exercise. Let us assume this result. Then

$$\lim_{x \to \infty} \frac{x^2 - 1}{4 - x^2} = \frac{\lim\limits_{x \to \infty} 1 - \lim\limits_{x \to \infty}(1/x^2)}{\lim\limits_{x \to \infty}(4/x^2) - \lim\limits_{x \to \infty} 1} = \frac{1 - 0}{0 - 1} = -1.$$

In the general definition of limit we permit the independent variable x to approach a value b in any manner whatsoever; that is, x may be greater than b, less than b, or any combination of these possibilities. The only requirement is that once δ is determined, $0 < |x - b| < \delta$. Occasions arise when it is convenient and useful to compute a special limit where x is restricted to values greater than b.

DEFINITION 4–14. $f(x)$ is said to have the right-hand limit l as x approaches b if for any positive number ε, however small, it is possible to find a positive number δ such that $|f(x) - l| < \varepsilon$ when $0 < x - b < \delta$. We write

$$\lim_{x \to b^+} f(x) = l.$$

Similarly, we have the next definition.

DEFINITION 4–15. $f(x)$ is said to have the left-hand limit l as x approaches b if for any positive number ε, however small, it is possible to find a positive number δ such that $|f(x) - l| < \varepsilon$ when $0 < b - x < \delta$. We write

$$\lim_{x \to b^-} f(x) = l.$$

Example 4–22. Given

$$f(x) = x + 1, \qquad x \leqslant 1;$$
$$f(x) = x^2 - 1, \qquad x > 1.$$

Calculate: (a) $\lim\limits_{x \to 1^-} f(x)$; (b) $\lim\limits_{x \to 1^+} f(x)$.

(a) $\lim\limits_{x \to 1^-} f(x) = \lim\limits_{x \to 1^-} (x + 1) = 2.$

(b) $\lim\limits_{x \to 1^+} f(x) = \lim\limits_{x \to 1^+} (x^2 - 1) = 0.$

Note that $\lim\limits_{x \to 1} f(x)$ does not exist.

The following theorem is one which will prove very useful in later work and also provides an excellent example of how theorems on limits are proved.

THEOREM 4–9. Let $f(x) \leqslant g(x) \leqslant h(x)$ for $a \leqslant x \leqslant c$ where $a < b < c$, and let

$$\lim_{x \to b} f(x) = \lim_{x \to b} h(x) = l.$$

Then

$$\lim_{x \to b} g(x) = l.$$

The existence of the limits of $f(x)$ and $h(x)$ implies that for any ε it is possible to choose δ_1 and δ_2 such that

$$|f(x) - l| < \varepsilon, \qquad 0 < |x - b| < \delta_1,$$

and

$$|h(x) - l| < \varepsilon, \qquad 0 < |x - b| < \delta_2.$$

If we take δ to be the smaller of δ_1 and δ_2, we may state

$$|f(x) - l| < \varepsilon, \qquad |h(x) - l| < \varepsilon, \qquad 0 < |x - b| < \delta.$$

These inequalities may be written

$$l - \varepsilon < f(x) < l + \varepsilon, \qquad l - \varepsilon < h(x) < l + \varepsilon, \qquad 0 < |x - b| < \delta.$$

Thus, making use of the original hypothesis,

$$l - \varepsilon < f(x) \leqslant g(x) \leqslant h(x) < l + \varepsilon, \qquad 0 < |x - b| < \delta,$$

or

$$|g(x) - l| < \varepsilon, \qquad 0 < |x - b| < \delta.$$

Therefore

$$\lim_{x \to b} g(x) = l.$$

Many more theorems on limits need to be established if we are to operate on a completely rigorous basis. However, as stated previously, it is not our intention to attempt this. We have given some of the commonly used theorems on limits and have proved a few of them to give the student some idea of the methods and perhaps a better understanding of the definitions. Now we leave it to the student to fill in the gaps with his intuition.

Exercises 4-3

Evaluate the limits in Exercises 1–30.

1. $\lim_{x \to 1} (2x^2 - 3x)$

2. $\lim_{x \to 2} (x^2 - 3x + 2)$

3. $\lim_{x \to 0} (3x^3 + 3)$

4. $\lim_{t \to -3} (-5t^2 - 7t + 4)$

5. $\lim\limits_{r\to 1} [(r^2 - 1)(3r^2 + 7)]$

6. $\lim\limits_{w\to -2} [(w^2 + 3w - 1)(w^3 - 7w^2 + 3)]$

7. $\lim\limits_{y\to -2} \dfrac{y^2 - 4}{y - 2}$

8. $\lim\limits_{z\to 4} \dfrac{2z^2 - 10}{z^2 + 1}$

similar to ones on test

9. $\lim\limits_{x\to -2} \dfrac{x^2 - 4}{x + 2}$

10. $\lim\limits_{r\to 4} \dfrac{r^2 - 16}{r^2 - 3r - 4}$

11. $\lim\limits_{x\to -1} \dfrac{3x^2 + x - 2}{2x^2 + x - 1}$

12. $\lim\limits_{y\to \frac{1}{3}} \dfrac{3y^2 + 5y - 2}{6y^2 - 5y + 1}$

13. $\lim\limits_{x\to 0} \dfrac{2x^2 - x}{3x}$

14. $\lim\limits_{w\to 1} \dfrac{w^3 - 2w + 1}{2w^3 - w - 1}$

15. $\lim\limits_{r\to -2} (2r^2 - 3)^3$

16. $\lim\limits_{y\to 0} \sqrt{y^2 + 9}$

17. $\lim\limits_{x\to 1} \sqrt{2x^2 + x + 1}$

18. $\lim\limits_{x\to 3^+} \sqrt{x^2 - 9}$

19. $\lim\limits_{x\to 3^-} \sqrt{9 - x^2}$

20. $\lim\limits_{x\to 4^+} \dfrac{\sqrt{x - 4}}{\sqrt{x^2 - 16}}$

21. $\lim\limits_{x\to 1^-} \dfrac{\sqrt{1 - x^2}}{\sqrt{1 - x}}$

22. $\lim\limits_{x\to 2} \dfrac{\sqrt{3x^2 + 2x}}{(x^2 - 4)^2}$

23. $\lim\limits_{x\to 0^+} \dfrac{2}{x}$

24. $\lim\limits_{x\to 0^-} \dfrac{2}{x}$

25. $\lim\limits_{x\to \infty} \dfrac{2}{x}$

26. $\lim\limits_{x\to -\infty} \dfrac{2}{x}$

27. $\lim\limits_{x\to \infty} \dfrac{2x^2 - 3x + 1}{x^2 + 5}$

28. $\lim\limits_{x\to \infty} \dfrac{3x^3 + 7x}{5x^3 + 3}$

29. $\lim\limits_{x\to -\infty} \dfrac{2 - x^2}{x + 2}$

30. $\lim\limits_{x\to -\infty} \dfrac{2x^2 + 3}{5x^3 - 7x + 1}$

divide by highest power

31. If $f(x)$ is a polynomial, show by means of the theorems stated in this section that $\lim\limits_{x\to b} f(x) = f(b)$.

32. If $f_1(x), f_2(x)$ are polynomials, show by means of the theorems stated in this section that

$$\lim\limits_{x\to b} \frac{f_1(x)}{f_2(x)} = \frac{f_1(b)}{f_2(b)}, \qquad f_2(b) \neq 0.$$

33. Prove by means of Definition 4-10 that

$$\lim\limits_{x\to \infty} \frac{1}{x} = 0.$$

4–6. Continuous Functions

In the preceding section we discussed the behavior of $f(x)$ for values of x near a particular value $x = b$. The question of whether $x = b$ was a number in the domain of f was not pertinent to that discussion. Now we wish to define a property possessed by some functions for a particular $x = b$ which requires that the domain of f contains $x = b$.

DEFINITION 4–16. The function f is said to be continuous for $x = b$ if:
   (a) f is defined for $x = b$;
   (b) $\lim_{x \to b} f(x) = f(b)$.

This definition implies three things about f: f is defined for $x = b$; the limit of $f(x)$ as $x \to b$ exists; this limit is the value of the function for $x = b$. An intuitive view of this property is that as $x \to b$, $f(x)$ is tending toward a particular value and the function value $f(b)$ is consistent with this trend.
   Definition 4-16 may be expressed in equivalent form.

DEFINITION 4–17. The function f is continuous for $x = b$ if for every positive ε, however small, it is possible to determine a positive δ such that $|f(x) - f(b)| < \varepsilon$ when $|x - b| < \delta$.

This definition says precisely the same as Definition 4-16 because (1) the use of the symbol $f(b)$ implies that it has meaning, that is, $x = b$ is in the domain of f; and (2) the statement involving the inequalities includes the definition of the statement $\lim_{x \to b} f(x) = f(b)$. This form of the definition illustrates one important property of continuous functions more vividly than Definition 4-16. *If f is continuous at $x = b$, it is possible to produce an arbitrarily small change in the function value by making a sufficiently small change in x from $x = b$.* This is an important intuitive concept.

Example 4–23. Consider the continuity of f defined in Example 4-16.

We have

$$f(x) = \frac{x^2 - 1}{x - 1}.$$

If $x \neq 1$, $f(x) = x + 1$, and by the results of Sec. 4-5,

$$\lim_{x \to b} f(x) = \lim_{x \to b} (x + 1) = b + 1 = f(b).$$

Hence, by Definition 4-16, f is continuous for every x different from 1. If $x = 1$, f is undefined and therefore fails to satisfy the definition of continuity for that value of x.

Example 4–24. Consider the continuity of g where

$$g(x) = \frac{x^2 - 1}{x - 1}, \qquad x \neq 1;$$

$$g(1) = 5.$$

Here g is defined for all values of x but $\lim_{x \to 1} g(x) = 2$, whereas $g(1) = 5$. Therefore $\lim_{x \to 1} g(x) \neq g(1)$, and g fails to be continuous for $x = 1$. Note that an arbitrarily small change in $g(x)$ cannot be effected by a sufficiently small change in x from $x = 1$. As x changes from 1 to a nearby value, $g(x)$ jumps from $g(1) = 5$ to a value close to 2 since $\lim_{x \to 1} g(x) = 2$. As a matter of fact, the smaller the change in x the closer the jump in the value of $g(x)$ is to 3. Obviously a number close to 3 is not arbitrarily small!

For all other values of x, g is identical to f in Example 4-23 and is therefore continuous.

Example 4–25. Consider the continuity of h where

$$h(x) = \frac{x^2 - 1}{x - 1}, \qquad x \neq 1;$$

$$h(1) = 2.$$

This function is continuous for all values of x since it is defined for all values of x and, in all cases, $x = 1$ included,

$$\lim_{x \to b} h(x) = h(b).$$

Example 4–26. Consider the continuity of f where

$$f(x) = 1, \qquad x \leqslant 0;$$

$$f(x) = x, \qquad x > 0.$$

This function is defined for all values of x, and for $b \neq 0$,

$$\lim_{x \to b} f(x) = f(b).$$

Thus f is continuous for all values of x different from 0. At $x = 0$, we note that

$$\lim_{x \to 0^+} f(x) = 0, \qquad \lim_{x \to 0^-} f(x) = 1,$$

and $f(0) = 1$. Hence we might say that f is continuous at $x = 0$ from the left, but we cannot make the unqualified statement that f is continuous there because $\lim_{x \to 0} f(x)$ does not exist.

Example 4–27. Consider the continuity of f where

$$f(x) = \frac{1}{x - 1}.$$

Here f is defined for every value of x except $x = 1$, and for those values

$$\lim_{x \to b} f(x) = f(b).$$

Hence f is continuous except for $x = 1$. For $x = 1$ we have

$$\lim_{x \to 1^-} f(x) = -\infty, \qquad \lim_{x \to 1^+} f(x) = \infty.$$

Moreover, $f(1)$ is undefined. Thus f fails to be continuous for $x = 1$ for two reasons; $f(1)$ is undefined, and $\lim_{x \to 1} f(x)$ does not exist.

DEFINITION 4–18. If a function fails to be continuous for a value x, say $x = b$, it is said to be discontinuous for $x = b$.

There are many types of discontinuities and we shall make no attempt to define or classify them except to mention certain ones occurring in the examples already considered. In Examples 4-23 and 4-24 the functions f and g are said to have a _removable discontinuity_ at $x = 1$. That is to say, if we properly redefine f and g for $x = 1$, that is, $f(1) = g(1) = 2$, these redefined functions are continuous for $x = 1$. Note that this is not possible for the discontinuities of the functions in Examples 4-26 and 4-27. No redefinition of these functions for $x = 0$ and $x = 1$, respectively, will give continuous functions for these values of x. The function in Example 4-26 is said to have an _ordinary_ discontinuity at $x = 0$, and the function in Example 4-27 is said to have an _infinite_ discontinuity at $x = 1$.

An examination of the graphs of these functions (Fig. 4-9) shows the difference in these discontinuities. The graphical consideration also gives an intuitive notion of the idea of continuity. If we trace the graph of a function with a pencil, the graph can be traced continuously without lifting the point from the paper except at points whose abscissas are values of x for which the function is discontinuous. For this reason we often speak of a _point of discontinuity_ or a _point of continuity_. Similarly, it is common usage to say that a function is continuous or discontinuous at $x = b$.

If the domain of g is the closed interval $a \leq x \leq b$, we may include the end points in our definition of continuity by using right- and left-hand limits at these points. That is, we say that f is continuous at $x = b$ if $\lim_{x \to b^-} f(x) = f(b)$; similarly for $x = a$, f is continuous at $x = a$ if $\lim_{x \to a^+} f(x) = f(a)$.

DEFINITION 4–19. A function is said to be continuous over an interval if it is continuous for each value of the variable in the interval.

There are many theorems relating to continuous functions, some of which we shall find indispensable to our development of calculus. Two of these follow, which we present without proof.

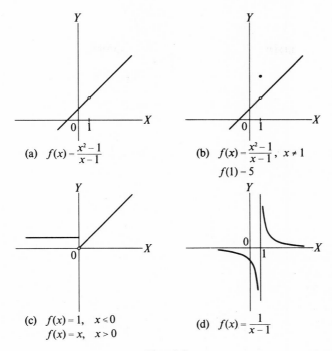

(a) $f(x) = \dfrac{x^2 - 1}{x - 1}$

(b) $f(x) = \dfrac{x^2 - 1}{x - 1}$, $x \neq 1$
  $f(1) = 5$

(c) $f(x) = 1$, $x \leqslant 0$
  $f(x) = x$, $x > 0$

(d) $f(x) = \dfrac{1}{x - 1}$

Fig. 4-9

THEOREM 4–10. Let the functions f and g be continuous at $x = b$. Then the functions $f \pm g$, $f \cdot g$, $c \cdot f$ (c a constant) are continuous at $x = b$. Moreover, f/g is continuous at $x = b$, provided $g(b) \neq 0$.

THEOREM 4–11. Let the function f be continuous on the closed interval $a \leqslant x \leqslant b$. Then

(a) f has a maximum and a minimum value on the interval; that is, there are numbers α and β in the interval such that $f(\alpha) = m$, $f(\beta) = M$ and

$$m \leqslant f(x) \leqslant M$$

for all x in the interval.

(b) If A is any number between $f(a)$ and $f(b)$, there exists a number c, $a < c < b$, such that $f(c) = A$.

Exercises 4-4

Discuss the continuity of the functions defined in Exercises 1–10.

1. $f(x) = 2x^2 + 3$
  continuous for all x

2. $f(x) = 3x^2 - 2x + 7$

3. $f(x) = \dfrac{x^2 + 2x}{x + 1}$ **4.** $f(x) = \dfrac{x^2 - 9}{x^2 + 9}$

5. $f(x) = \dfrac{x^2 + 9}{x^2 - 9}$ **6.** $f(x) = \dfrac{x - 3}{x^2 + 2x - 3}$

7. $f(x) = \dfrac{4x + 1}{1 - x^2}$ **8.** $f(x) = \dfrac{5x^2 + 7}{2x^3 + 5x^2 + 2x}$

9. $\begin{cases} f(x) = x + 1, & x \geqslant 0; \\ f(x) = x, & x < 0 \end{cases}$ **10.** $f(x) = x + |x|$

11. The total cost of producing x items is given by

$$C = 1500 + 4x, \qquad 0 \leqslant x \leqslant 500;$$
$$C = 2000 + 3x, \qquad 500 < x \leqslant 1000.$$

For what values of x is this function continuous? discontinuous? Draw its graph.

12. A corporation pays 30% tax on its profits up to and including $30,000 and 50% tax on any profits in excess of this amount. Write a statement (or statements) defining the taxes T as a function of profits P. What is the domain of this function? For what values of P is it continuous? Draw the graph of this function.

13. If $f(x)$ is a polynomial, prove that $f(x)$ is continuous for all values of x.

14. If $f_1(x)$, $f_2(x)$ are polynomials, prove that $f_1(x)/f_2(x)$ is continuous for all values of x except those for which $f_2(x) = 0$.

Chapter 5

THE DERIVATIVE

5-1. Introduction

In Sec. 4-3 we considered the problem of finding the equation of a line tangent to the curve $y = f(x)$ at a point (x_1, y_1). There we saw that a suitable definition of this tangent involved the behavior of the expression

$$\frac{\Delta y}{\Delta x} = \frac{f(x_1 + \Delta x) - f(x_1)}{\Delta x} \tag{5-1}$$

as $\Delta x \to 0$. We developed the concept of limit in Sec. 4-4 and 4-5 to describe this behavior and as the result we may define the slope m of the tangent line as follows.

DEFINITION 5-1. The slope of the tangent to the curve† $y = f(x)$ at the point (x_1, y_1) is defined to be

$$m = \lim_{\Delta x \to 0} \frac{f(x_1 + \Delta x) - f(x_1)}{\Delta x},$$

provided this limit exists.

Next let us consider the motion of a particle along a straight line where its distance s from a fixed point on the line at time t is given by

$$s = f(t).$$

The distance traveled by the particle from time t_1 to time $t_1 + \Delta t$ is given by

$$\Delta s = f(t_1 + \Delta t) - f(t_1)$$

and if we divide this distance by the elapsed time Δt, we obtain the average velocity of the particle during this time interval:

$$\frac{\Delta s}{\Delta t} = \frac{f(t_1 + \Delta t) - f(t_1)}{\Delta t}. \tag{5-2}$$

---

† We shall also refer to the slope of the tangent to a curve at a given point as the *slope of the curve* at that point.

This average velocity, however, does not give us precise information regarding the velocity of the particle at the instant t_1; it gives merely the average velocity over a time interval beginning with time t_1. We cannot obtain this velocity at time t_1 by taking $\Delta t = 0$ in (5-2) because the expression is undefined for this value. But if we allow the time interval to become shorter and shorter, that is, allow Δt to become smaller and smaller, it appears reasonable that each successive average velocity so obtained is a better representation of the velocity at time t_1 than are the preceding ones over longer time intervals. This line of reasoning leads us to the following definition of the *instantaneous velocity at t_1*.

DEFINITION 5-2. Let the path of a moving particle be a straight line and let its distance s from a fixed point on the line at time t be given by $s = f(t)$. Then the velocity of the particle at time t_1 is defined to be

$$v = \lim_{\Delta t \to 0} \frac{\Delta s}{\Delta t} = \lim_{\Delta t \to 0} \frac{f(t_1 + \Delta t) - f(t_1)}{\Delta t},$$

provided this limit exists.

The notion of instantaneous velocity, or *instantaneous rate of change of distance with respect to time,* can be extended to a more general situation. Let two variables x and y be related by the equation $y = f(x)$, and consider the change in y produced by a change in x from x_1 to $x_1 + \Delta x$. We have

$$\Delta y = f(x_1 + \Delta x) - f(x_1).$$

Now if we form the ratio

$$\frac{\Delta y}{\Delta x} = \frac{f(x_1 + \Delta x) - f(x_1)}{\Delta x},$$

we have the change in y per unit change in x as x varies from x_1 to $x_1 + \Delta x$, or the *average rate of change of y with respect to x over the interval Δx*. Again, if we allow Δx to become smaller and smaller, we get better and better approximations for what we conceive to be the *instantaneous rate of change of y with respect to x at $x = x_1$* and we are led to the next definition.

DEFINITION 5-3. If x and y are related by $y = f(x)$, the instantaneous rate of change of y with respect to x at $x = x_1$ is defined to be

$$r = \lim_{\Delta x \to 0} \frac{\Delta y}{\Delta x} = \lim_{\Delta x \to 0} \frac{f(x_1 + \Delta x) - f(x_1)}{\Delta x},$$

provided this limit exists.

5–2. The Derivative

All three definitions in the preceding section involve the same limit. This limit is one of the fundamental tools of the calculus and serves as the basis for one main branch of the subject "Differential Calculus." For this reason we give it a name and define a notation to apply to it.

DEFINITION 5–4. The derivative of f at x_1, indicated by $f'(x_1)$, is defined to be

$$f'(x_1) = \lim_{\Delta x \to 0} \frac{f(x_1 + \Delta x) - f(x_1)}{\Delta x},$$

provided this limit exists.

It is to be emphasized that this definition is stated in terms of a particular x and this has been underlined by using $x = x_1$. However, from this point on, for reasons of simplicity, unless there is opportunity for misinterpretation, we shall drop the subscript but keep in mind always that $f'(x)$ is the value of a limit for a *particular* x. Thus *we have associated with the function f a new function f' such that*

$$f'(x) = \lim_{\Delta x \to 0} \frac{f(x + \Delta x) - f(x)}{\Delta x},$$

when this limit exists.

Many notations for the derivative are used by different writers. We shall confine ourselves to the following, which we shall use interchangeably:

$$f'(x) = \frac{df(x)}{dx} = \frac{df}{dx} = y' = \frac{dy}{dx},$$

it being understood that $y = f(x)$. The student will have no difficulty in relating other notations, found in various places, to those above.

Definitions 5-1, 5-2, 5-3 may now be restated, using this notation.

DEFINITION 5–1a. The slope of the tangent to the curve $y = f(x)$ at the point (x_1, y_1) is defined to be $m = f'(x_1)$, provided $f'(x_1)$ exists.

DEFINITION 5–2a. Let the path of a moving particle be a straight line and let its distance from a fixed point on the line at time t be given by $s = f(t)$. Then the velocity of the particle at time t_1 is defined to be $v = f'(t_1)$, provided $f'(t_1)$ exists.

DEFINITION 5–3a. If x and y are related by $y = f(x)$, the instantaneous rate of change of y with respect to x at $x = x_1$ is defined to be $r = f'(x_1)$, provided $f'(x_1)$ exists.

Example 5–1. (a) Calculate $f'(x)$ where $f(x) = x^2$. (b) Find the equation of the tangent to $y = x^2$ at the point $(\frac{1}{2}, \frac{1}{4})$. (c) Find the instantaneous rate of change of y with respect to x at $x = -2$ and interpret the result.

$$(a)\ f'(x) = \lim_{\Delta x \to 0} \frac{(x + \Delta x)^2 - x^2}{\Delta x}$$

$$= \lim_{\Delta x \to 0} \frac{2x\,\Delta x + \overline{\Delta x}^2}{\Delta x}$$

$$= \lim_{\Delta x \to 0} (2x + \Delta x) = 2x.$$

(b) According to Definition 5-1a, the slope of the required tangent is given by

$$m = f'\left(\frac{1}{2}\right) = 2\left(\frac{1}{2}\right) = 1.$$

Therefore its equation is

$$y - \frac{1}{4} = 1\left(x - \frac{1}{2}\right), \quad \text{slope intercept formula}$$

or

$$4x - 4y - 1 = 0, \quad ?$$

which agrees with the result obtained in Sec. 4-3.

(c) From Definition 5-3a, we have

$$r = f'(-2) = 2(-2) = -4.$$

Thus, if we consider x as increasing, at the instant it assumes the value $x = -2$, y is decreasing four times as fast as x is increasing. Graphically this means that a point moving along the parabola from left to right, at the instant it passes through $(-2, 4)$, is falling† four units for each unit it moves to the right. This is true only for this particular instant. At a later instant, say when the point passes through $(-1, 1)$, the rate of falling is only two units for each unit of motion to the right. The only curve for which this rate remains a constant is a straight line.‡

Example 5–2. A ball is thrown vertically upward so that t seconds later it is s feet above the ground, where $s = -16t^2 + 96t$. (a) What is its velocity when

---

† Hereafter, in interpreting rates of change, we shall always assume the independent variable to be increasing. Thus, when we speak of a falling curve, we mean that as x increases, y decreases.

‡ For the remainder of this chapter we shall assume $f'(x) > 0$ means $f(x)$ increases as x increases; $f'(x) < 0$ means $f(x)$ decreases as x increases. We shall prove this in Chapter 6.

2 sec have elapsed? (b) When will it stop rising? (c) Why is its velocity negative when $t = 4$ sec?

(a) From Definition 5-2a, the velocity of the ball at any particular time t is given by

$$v = f'(t) = \lim_{\Delta t \to 0} \left\{ \frac{[-16(t + \Delta t)^2 + 96(t + \Delta t)] - [-16t^2 + 96t]}{\Delta t} \right\}$$

$$= \lim_{\Delta t \to 0} \left\{ \frac{-32t\,\Delta t - 16\overline{\Delta t}^2 + 96\Delta t}{\Delta t} \right\}$$

$$= \lim_{\Delta t \to 0} (-32t - 16\Delta t + 96) = -32t + 96.$$

At the particular instant that $t = 2$, we have

$$v = f'(2) = -32(2) + 96 = 32 \text{ ft/sec.}$$

(b) The ball will stop rising when its velocity is zero, that is, when

$$f'(t) = -32t + 96 = 0.$$

Thus the required time is $t = 3$.

(c) The velocity at $t = 4$ is

$$f'(4) = -32(4) + 96 = -32 \text{ ft/sec.}$$

When s is decreasing, its rate of change with respect to an increasing t will be negative. At time $t = 4$, the ball is falling.

Example 5–3. Calculate $f'(x)$ where $f(x) = (x^2 + 1)/2x$.

$$f'(x) = \lim_{\Delta x \to 0} \frac{1}{\Delta x} \left\{ \frac{(x + \Delta x)^2 + 1}{2(x + \Delta x)} - \frac{x^2 + 1}{2x} \right\}$$

$$= \lim_{\Delta x \to 0} \frac{1}{\Delta x} \left\{ \frac{x[(x + \Delta x)^2 + 1] - (x^2 + 1)(x + \Delta x)}{2x(x + \Delta x)} \right\}$$

$$= \lim_{\Delta x \to 0} \frac{1}{\Delta x} \left\{ \frac{x^2\,\Delta x - \Delta x + x\,\overline{\Delta x}^2}{2x(x + \Delta x)} \right\}$$

$$= \lim_{\Delta x \to 0} \frac{x^2 - 1 + x\,\Delta x}{2x(x + \Delta x)} = \frac{x^2 - 1}{2x^2}.$$

Example 5–4. Compute $f'(x)$ where $f(x) = 1/\sqrt{x^2 + 1}$.

$$f'(x) = \lim_{\Delta x \to 0} \frac{1}{\Delta x} \left\{ \frac{1}{\sqrt{(x + \Delta x)^2 + 1}} - \frac{1}{\sqrt{x^2 + 1}} \right\}$$

$$= \lim_{\Delta x \to 0} \frac{\sqrt{x^2 + 1} - \sqrt{(x + \Delta x)^2 + 1}}{\Delta x \sqrt{x^2 + 1}\sqrt{(x + \Delta x)^2 + 1}}.$$

If we attempt to evaluate the limit in this form, our theorems do not apply because the denominator has the limit zero. To surmount this difficulty, we alter the form by multiplying above and below by the nonzero expression $\sqrt{x^2 + 1} + \sqrt{(x + \Delta x)^2 + 1}$ and obtain

$$f'(x) = \lim_{\Delta x \to 0} \left\{ \frac{(x^2 + 1) - [(x + \Delta x)^2 + 1]}{\Delta x \sqrt{x^2 + 1}\sqrt{(x + \Delta x)^2 + 1}[\sqrt{x^2 + 1} + \sqrt{(x + \Delta x)^2 + 1}]} \right\}$$

$$= \lim_{\Delta x \to 0} \left\{ \frac{-2x - \Delta x}{\sqrt{x^2 + 1}\sqrt{(x + \Delta x)^2 + 1}[\sqrt{x^2 + 1} + \sqrt{(x + \Delta x)^2 + 1}]} \right\}$$

$$= \frac{-2x}{2(x^2 + 1)\sqrt{x^2 + 1}} = \frac{-x}{(x^2 + 1)^{3/2}}.$$

5–3. Some Additional Interpretations of the Derivative

The rate of change property of the derivative permits its use in many fields. One very fertile field is that of economics. For a quick look into this area, let us define some terminology.

DEFINITION 5–5. Marginal cost (marginal revenue; marginal profit) is the change produced in the total cost (total revenue; total profit) by increasing production (sales) by one unit.

Now suppose f is the cost function, that is, $f(x)$ is the total cost of producing x units, where $f(x)$ has valid economic interpretation only for nonnegative integral values of x. However, if we temporarily neglect this fact, f is usually defined over an interval. Then, if f is differentiable over this interval, $f'(x)$ represents the instantaneous rate of change of $f(x)$ with respect to x, that is, the instantaneous change in $f(x)$ per unit change in x. This is precisely the marginal cost for admissible values of x and we may rephrase the first part of Definition 5-5 as follows.

DEFINITION 5–6. If f is the cost function, $f'(x)$ is the marginal cost at the production level of x units.

Similar reasoning leads to a restatement of the last two parts of Definition 5-5.

DEFINITION 5–7. If f is the total revenue (total profit) function, $f'(x)$ is the marginal revenue (marginal profit) at the sales level of x units.

Example 5–5. A company produces and sells x items per week. If the selling price is \$3.00 each and the production costs are estimated to be $10,000 + 150x^{1/2}$ dollars, what is the marginal profit at the production level of 10,000 units weekly?

The total weekly revenue is $3x$ dollars. Hence the profit function is defined by

$$P(x) = 3x - (10{,}000 + 150x^{1/2}).$$

According to Definition 5-7, we need to find the value of $P'(x)$ for $x = 10{,}000$. We have

$$P'(x) = \lim_{\Delta x \to 0} \frac{1}{\Delta x}\{[3(x + \Delta x) - 10{,}000 - 150(x + \Delta x)^{1/2}]$$

$$- [3x - 10{,}000 - 150x^{1/2}]\}$$

$$= \lim_{\Delta x \to 0} \frac{1}{\Delta x}\{3\Delta x - 150[(x + \Delta x)^{1/2} - x^{1/2}]\}$$

$$= \lim_{\Delta x \to 0} \frac{1}{\Delta x}\left\{3\Delta x - 150\left[\frac{x + \Delta x - x}{(x + \Delta x)^{1/2} + x^{1/2}}\right]\right\}$$

$$= \lim_{\Delta x \to 0} \left\{3 - \frac{150}{(x + \Delta x)^{1/2} + x^{1/2}}\right\} = 3 - \frac{75}{x^{1/2}}.$$

Thus the marginal profit at the required level is

$$P'(10{,}000) = 3 - \frac{75}{100} = \$2.25.$$

We shall see many more uses of the derivative as we proceed with our study of calculus.

Exercises 5-1

Find $f'(x)$ in Exercises 1–12.

1. $f(x) = 2x + 3$

2. $f(x) = 4 - 5x$

3. $f(x) = x^2 - 2x + 7$

4. $f(x) = 6x^2 + 5$

5. $f(x) = \dfrac{1}{x}$

6. $f(x) = \dfrac{1}{x - 1}$

7. $f(x) = \dfrac{x}{x - 1}$

8. $f(x) = \dfrac{2x^2}{x + 1}$

9. $f(x) = \sqrt{x}$

10. $f(x) = \dfrac{1}{\sqrt{x}}$

11. $f(x) = x^{3/2}$

12. $f(x) = \sqrt{4 - x^2}$

Find the equation of the tangent to the curves at the indicated points in Exercises 13–18.

13. $y = 1 - 2x^2,\ (2, -7)$

14. $y = 4x^2 - 10x + 1,\ (3, 7)$

15. $y = \dfrac{3}{1-x}, (-2, 1)$ **16.** $y = x + \dfrac{1}{x}, \left(-2, -\dfrac{5}{2}\right)$

17. $y = x^3, (1, 1)$ **18.** $y = \dfrac{1}{x^2}, \left(\dfrac{1}{2}, 4\right)$

19. What is the total income function when the price is a constant $10 per unit sold? Find the derivative of the income with respect to the number of units sold. Interpret your result.

20. A retail outlet has a uniform policy of making a 25 percent markup on the cost. Find the derivative of the selling price with respect to the cost. Interpret your result.

21. The height of a ball thrown vertically upward from the ground with an initial velocity of 128 ft/sec is given by

$$s = 128t - 16t^2.$$

(a) What is its average velocity during the time interval from
(1) $t = 1$ to $t = 3$; (2) $t = 2$ to $t = 3$; (3) $t = 2.5$ to $t = 3$; (4) $t = 2.75$ to $t = 3$?
(b) What is its instantaneous velocity at $t = 3$?

22. The height s (in feet) of a ball above the ground at time t (in seconds) is given by
$$s = -16t^2 + 160t + 192.$$

Find (a) the velocity at time t; (b) the maximum value of s; (c) the initial velocity.

23. A ball is thrown vertically upward from the top of a building 80 ft high. In t sec it is s ft above the ground, where
$$s = -16t^2 + 64t + 80.$$

Find (a) the velocity at time t; (b) the initial velocity; (c) the velocity at $t = 3$; (d) the velocity when it strikes the ground; (e) the time when it ceases to rise.

24. A man finds that his bank account varies according to the law
$$A = 3t^2 - 30t - 72,$$

where t is time in months from the initial accounting and A is in hundreds of dollars. (a) Does he have any long-run financial problems? Why? (b) Is he ever in debt? If so, during what period? (c) At what time does his situation undergo a basic change?

25. A manufacturer finds that his total revenue y is expressed by the equation
$$y = 10,000 - (x - 100)^2,$$

where x is the number of units sold per week. Find the marginal revenue at the level of 75 sales per week; 125 sales per week. Interpret your results.

26. A company finds its costs for producing x items are approximated by the relation

$$C = 5000 + 5x + \frac{10}{x}.$$

Find the marginal cost at the production level of 1000 units.

27. A company finds that the cost of producing x articles is given by

$$C = 1000 + 5x + .03x^2$$

where C is the cost in dollars. Find by two different methods the cost of producing the eleventh article.

28. A firm has a cost function

$$C(x) = 2x^2 - 1000x + 10,000$$

for producing x items per week and a revenue function

$$R(x) = 2000x - x^2$$

for the sale of these items. Find the net profit when the level of production is such that the marginal cost is equal to the marginal revenue.†

5–4. Some Basic Theorems

As was observed in the preceding section, the process of calculating the derivative directly from the definition is often tedious and may be quite difficult even for apparently simple functions. We shall devote the rest of this chapter to developing formulas and methods which will eliminate much of this difficulty for certain types of functions.

DEFINITION 5–8. A function f is said to be differentiable at x if $f'(x)$ exists.

The process of calculating the derivative is called *differentiation,* and to *differentiate* a function is to calculate its derivative.

THEOREM 5–1. Let $f(x) = u(x) + v(x)$. Then f is differentiable for those values of x for which u and v are differentiable and

$$f'(x) = u'(x) + v'(x).‡$$

---

† It can be shown that this level of production gives the maximum net profit.
‡ The hypotheses of this theorem are a "sufficient" set of conditions for the differentiability of f. It may be that f is differentiable for values of x for which neither u nor v are differentiable.

To prove this theorem we write

$$f'(x) = \lim_{\Delta x \to 0} \frac{f(x + \Delta x) - f(x)}{\Delta x}$$

$$= \lim_{\Delta x \to 0} \left\{ \frac{u(x + \Delta x) + v(x + \Delta x) - [u(x) + v(x)]}{\Delta x} \right\}$$

$$= \lim_{\Delta x \to 0} \left\{ \frac{u(x + \Delta x) - u(x)}{\Delta x} + \frac{v(x + \Delta x) - v(x)}{\Delta x} \right\}$$

$$= \lim_{\Delta x \to 0} \frac{u(x + \Delta x) - u(x)}{\Delta x} + \lim_{\Delta x \to 0} \frac{v(x + \Delta x) - v(x)}{\Delta x}.$$

If u and v are differentiable, the two limits in the right member exist and are $u'(x)$ and $v'(x)$, respectively. Thus the theorem is extablished.

If we employ the notation

$$\Delta f = f(x + \Delta x) - f(x),$$

and a similar one for the functions u and v, the preceding statements may be expressed in the form

$$f'(x) = \lim_{\Delta x \to 0} \frac{\Delta f}{\Delta x} = \lim_{\Delta x \to 0} \frac{\Delta u}{\Delta x} + \lim_{\Delta x \to 0} \frac{\Delta v}{\Delta x}.$$

We may also express the conclusion of Theorem 5-1 in the notation

$$\frac{df}{dx} = \frac{du}{dx} + \frac{dv}{dx} \qquad \text{or} \qquad f' = u' + v'.$$

However we may choose to write it, *the derivative of the sum of differentiable functions is the sum of their derivatives.*

THEOREM 5–2. Let $f(x) = c$, where c is a constant. Then $f'(x) = 0$.

We have

$$f'(x) = \lim_{\Delta x \to 0} \frac{f(x + \Delta x) - f(x)}{\Delta x} = \lim_{\Delta x \to 0} \frac{c - c}{\Delta x} = 0.$$

THEOREM 5–3. Let $f(x) = cg(x)$, where c is a constant and g is a differentiable function. Then $f'(x) = cg'(x)$.

To prove this we write

$$f'(x) = \lim_{\Delta x \to 0} \frac{f(x + \Delta x) - f(x)}{\Delta x}$$

$$= \lim_{\Delta x \to 0} \frac{cg(x + \Delta x) - cg(x)}{\Delta x}$$

$$= c \lim_{\Delta x \to 0} \frac{g(x + \Delta x) - g(x)}{\Delta x}$$

$$= cg'(x).$$

THEOREM 5-4. Let $f(x) = x^n$ where n is an integer. Then $f'(x) = nx^{n-1}$; $x \neq 0$ if $n < 0$.

We consider three cases in the proof:

CASE I $(n = 0)$: Since $x^0 = 1$ this case is verified by Theorem 5-2.

CASE II $(n > 0)$: In this case and the next one too we make use of the algebraic identity

$$u^n - v^n = (u - v)(u^{n-1} + u^{n-2}v + u^{n-3}v^2 + \cdots + uv^{n-2} + v^{n-1})$$

which the student can readily verify by actual multiplication. We write

$$f'(x) = \lim_{\Delta x \to 0} \frac{f(x + \Delta x) - f(x)}{\Delta x}$$

$$= \lim_{\Delta x \to 0} \frac{(x + \Delta x)^n - x^n}{\Delta x}$$

$$= \lim_{\Delta x \to 0} \frac{1}{\Delta x} \{[(x + \Delta x) - x][(x + \Delta x)^{n-1} + (x + \Delta x)^{n-2}x$$

$$+ (x + \Delta x)^{n-3}x^2$$

$$+ \cdots + (x + \Delta x)x^{n-2} + x^{n-1}]\}$$

$$= \lim_{\Delta x \to 0} \frac{\Delta x}{\Delta x} [(x + \Delta x)^{n-1} + (x + \Delta x)^{n-2}x + \cdots + (x + \Delta x)x^{n-2} + x^{n-1}]$$

$$= [\overbrace{x^{n-1} + x^{n-1} + \cdots + x^{n-1}}^{n \text{ terms}}] = nx^{n-1}.$$

This completes Case II.

CASE III $(n < 0, x \neq 0)$: Let $m = -n$. Then m is a positive integer and

$$x^n = \frac{1}{x^{-n}} = \frac{1}{x^m}.$$

Thus

$$f'(x) = \lim_{\Delta x \to 0} \frac{f(x + \Delta x) - f(x)}{\Delta x}$$

$$= \lim_{\Delta x \to 0} \frac{1}{\Delta x} \left\{ \frac{1}{(x + \Delta x)^m} - \frac{1}{x^m} \right\}$$

$$= \lim_{\Delta x \to 0} \frac{1}{\Delta x} \left\{ \frac{x^m - (x + \Delta x)^m}{(x + \Delta x)^m x^m} \right\}$$

$$= \lim_{\Delta x \to 0} \frac{1}{\Delta x} \left\{ \frac{[x - (x + \Delta x)][x^{m-1} + x^{m-2}(x + \Delta x)}{(x + \Delta x)^m x^m} + x^{m-3}(x + \Delta x)^2 + \cdots + (x + \Delta x)^{m-1}] \right\}$$

$$= \lim_{\Delta x \to 0} \frac{-\Delta x}{\Delta x} \left\{ \frac{x^{m-1} + x^{m-2}(x + \Delta x) + x^{m-3}(x + \Delta x)^2}{(x + \Delta x)^m x^m} + \cdots + (x + \Delta x)^{m-1}} \right\}$$

$$= - \left\{ \frac{x^{m-1} + x^{m-1} + \cdots + x^{m-1}}{x^{2m}} \right\}$$

$$= - \frac{m x^{m-1}}{x^{2m}} = -m x^{-m-1} = n x^{n-1}.$$

This completes the proof of the theorem as stated. However, *it is also true if n is any rational number, that is, any number of the form p/q where p and q are integers, provided $x \neq 0$ if $n < 1$.* We shall assume the theorem in this form now, although we find it expedient to postpone the proof to Sec. 5-7. In a later chapter we shall prove it is also true for n any real number.

Example 5–6. Find $f'(3)$, where $f(x) = 4x^2 - 10x + 1$.

We have, from Theorem 5-1,

$$f'(x) = \frac{d(4x^2)}{dx} + \frac{d(-10x)}{dx} + \frac{d(1)}{dx},$$

and Theorems 5-2 and 5-3 allow us to write this in the form

$$f'(x) = 4 \frac{d(x^2)}{dx} - 10 \frac{d(x)}{dx} + 0,$$

from which, by Theorem 5-4, we obtain

$$f'(x) = 8x - 10.$$

Then
$$f'(3) = 8(3) - 10 = 14.$$

Example 5–7. Find $f'(x)$, where $f(x) = 5x^3 - 3/x^2$.

First we write
$$f(x) = 5x^3 - 3x^{-2}.$$

Then, applying Theorems 5-1, 5-3, 5-4 as in the preceding example, we obtain

$$f'(x) = \frac{d(5x^3)}{dx} + \frac{d(-3x^{-2})}{dx}$$

$$= 5\frac{d(x^3)}{dx} - 3\frac{d(x^{-2})}{dx}$$

$$= 15x^2 + 6x^{-3}$$

$$= 15x^2 + \frac{6}{x^3}.$$

Example 5–8. Find $f'(x)$, where $f(x) = 2\sqrt{x} + \dfrac{1}{2\sqrt{x}}$.

We write this in the form
$$f(x) = 2x^{1/2} + \frac{1}{2}x^{-1/2}.$$

Then, as before,

$$f'(x) = \frac{d(2x^{1/2})}{dx} + \frac{d(\frac{1}{2}x^{-1/2})}{dx}$$

$$= 2\frac{d(x^{1/2})}{dx} + \frac{1}{2}\frac{d(x^{-1/2})}{dx}.$$

Hence, assuming Theorem 5-4 for n rational, we have

$$f'(x) = 2\left(\frac{1}{2}\right)x^{-1/2} + \frac{1}{2}\left(-\frac{1}{2}\right)x^{-3/2}$$

$$= x^{-1/2} - \frac{1}{4}x^{-3/2},$$

which may be written as

$$f'(x) = \frac{1}{\sqrt{x}} - \frac{1}{4\sqrt{x^3}},$$

or in a number of other forms such as

$$f'(x) = \frac{4x - 1}{4\sqrt{x^3}}.$$

Exercises 5-2

In Exercises 1–20 find f' where f is defined by the following equations:

1. $f(x) = 3x^2 - 5x + 7$ **2.** $f(t) = t^3 + 4t^2 - 7t + 5$

3. $f(r) = 1 - 5r - r^2 - 10r^3$ **4.** $f(z) = z^4 - 2z^{10}$

5. $f(x) = \dfrac{2}{3} + \dfrac{4x}{5} - \dfrac{3x^2}{17}$ **6.** $f(w) = 2(1 - w^2) + w^5$

7. $f(x) = 2x^2 - \dfrac{3}{x} + \dfrac{11}{x^2}$ **8.** $f(x) = -\dfrac{1}{x^3} - \dfrac{2}{x^2} + \dfrac{14}{x}$

9. $f(t) = \dfrac{2t^4 - 8t^3 + 3t^2}{4t^3}$ **10.** $f(z) = (z^2 - 1)(2z + 3)$

11. $f(x) = (x^2 - 1)^3$ **12.** $f(x) = 7(1 - x^2) + 4(x^3 + 1)$

13. $f(r) = \dfrac{2}{r^3} - \dfrac{3 - r}{r^2}$ **14.** $f(t) = t^{10} - \dfrac{1}{t^{10}}$

15. $f(x) = x^{1/2}$ **16.** $f(x) = \sqrt{x^3}$

17. $f(t) = t\sqrt{t} + \dfrac{2}{\sqrt{t}}$ **18.** $f(w) = w^{3/2} + \dfrac{1}{w^{3/2}}$

19. $f(x) = \dfrac{(x^2 + 1)(x - 1)}{\sqrt{x}}$ **20.** $f(z) = \dfrac{5}{\sqrt[3]{z^5}}$

21. A bullet fired vertically upward with a muzzle velocity of 1600 ft/sec will reach a height $s = -16t^2 + 1600t$ at the end of t sec. (a) How high does it rise? (b) What is its velocity when it reaches a height of 7600 ft?

22. A sphere is expanding. What is the rate of change in its volume with respect to its radius at the instant the radius is three units long?

23. Find the equation of the tangent to the curve $y = \sqrt{x} + x^{-1}$ at the point $(4, 9/4)$.

24. Find the points on the curve $y = \sqrt{x^3} + (3/\sqrt{x})$ at which the tangents are parallel to the x-axis.

5–5. Derivatives of Products and Quotients

THEOREM 5–5. Let $f(x) = u(x)v(x)$. Then f is differentiable for those values of x for which u and v are differentiable, and

$$f'(x) = u(x)v'(x) + v(x)u'(x).\dagger$$

---

† See footnote to Theorem 5-1.

We have
$$f(x + \Delta x) - f(x) = u(x + \Delta x)v(x + \Delta x) - u(x)v(x).$$

In order to write this in a form in which the differences Δu and Δv appear, we add and subtract the quantity $u(x + \Delta x)v(x)$ and obtain

$$f(x + \Delta x) - f(x) = u(x + \Delta x)v(x + \Delta x) - u(x + \Delta x)v(x)$$
$$+ u(x + \Delta x)v(x) - u(x)v(x)$$
$$= u(x + \Delta x)[v(x + \Delta x) - v(x)] + v(x)[u(x + \Delta x) - u(x)].$$

Thus

$$f'(x) = \lim_{\Delta x \to 0} \frac{f(x + \Delta x) - f(x)}{\Delta x}$$

$$= \lim_{\Delta x \to 0} u(x + \Delta x)\left[\frac{v(x + \Delta x) - v(x)}{\Delta x}\right]$$

$$+ \lim_{\Delta x \to 0} v(x)\left[\frac{u(x + \Delta x) - u(x)}{\Delta x}\right]$$

$$= u(x)v'(x) + v(x)u'(x)$$

for those values of x for which u and v are differentiable.†

THEOREM 5-6. Let $f(x) = u(x)/v(x)$. Then f is differentiable for those values of x for which u and v are differentiable and for which $v(x) \neq 0$, and

$$f'(x) = \frac{v(x)u'(x) - u(x)v'(x)}{[v(x)]^2}.$$

To prove this theorem we write

$$\frac{f(x + \Delta x) - f(x)}{\Delta x} = \frac{1}{\Delta x}\left\{\frac{u(x + \Delta x)}{v(x + \Delta x)} - \frac{u(x)}{v(x)}\right\}$$

$$= \frac{1}{\Delta x}\left\{\frac{u(x + \Delta x)v(x) - u(x)v(x + \Delta x)}{v(x + \Delta x)v(x)}\right\}.$$

----

† This assumes that
$$\lim_{\Delta x \to 0} u(x + \Delta x) = u(x).$$

This is true under the hypothesis that u is differentiable at x. We have

$$\lim_{\Delta x \to 0}\left[u(x + \Delta x) - u(x)\right] = \lim_{\Delta x \to 0} \Delta u = \lim_{\Delta x \to 0} \Delta x \cdot \frac{\Delta u}{\Delta x} = 0 \cdot \frac{du}{dx} = 0,$$

which establishes the assumption. Hence u satisfies all conditions of continuity at x. Thus *differentiability implies continuity*. The converse is not true, as may be verified by considering

$$u(x) = x^{1/3}$$

at $x = 0$.

Then, adding and subtracting $u(x)v(x)$ in the numerator of the right member, we have

$$f'(x) = \lim_{\Delta x \to 0} \frac{1}{\Delta x}\left\{\frac{u(x + \Delta x)v(x) - u(x)v(x) - u(x)v(x + \Delta x) + u(x)v(x)}{v(x + \Delta x)v(x)}\right\}$$

$$= \lim_{\Delta x \to 0}\left\{\frac{v(x)\left[\dfrac{u(x + \Delta x) - u(x)}{\Delta x}\right] - u(x)\left[\dfrac{v(x + \Delta x) - v(x)}{\Delta x}\right]}{v(x + \Delta x)v(x)}\right\}$$

$$= \frac{v(x)u'(x) - u(x)v'(x)}{[v(x)]^2}, \qquad v(x) \neq 0,$$

if u and v are differentiable.

Example 5–9. Find $f'(x)$, where $f(x) = (2x^2 - 1)(1 - 5x^3)$.

Let $u(x) = 2x^2 - 1$ and $v(x) = 1 - 5x^3$. Then $u'(x) = 4x$, $v'(x) = -15x^2$, and by Theorem 5-5 we have

$$f'(x) = (2x^2 - 1)(-15x^2) + (1 - 5x^3)(4x)$$
$$= -50x^4 + 15x^2 + 4x.$$

Example 5–10. Find $f'(x)$, where

$$f(x) = \frac{x^3 + x}{1 + x^2}.$$

Let $u(x) = x^3 + x$ and $v(x) = 1 + x^2$. Then $u'(x) = 3x^2 + 1$, $v'(x) = 2x$, and by Theorem 5-6 we have

$$f'(x) = \frac{(1 + x^2)(3x^2 + 1) - (x^3 + x)(2x)}{(1 + x^2)^2}$$
$$= \frac{x^4 + 2x^2 + 1}{(1 + x^2)^2}.$$

Since the student will need to perform these operations many times and under varying circumstances, he will find it useful to learn Theorems 5-5 and 5-6 as *rules* rather than formulas. These can be stated as follows:

Product Rule. The derivative of a product is the first factor times the derivative of the second plus the second factor times the derivative of the first.

Quotient Rule. The derivative of a quotient is the denominator times the derivative of the numerator minus the numerator times the derivative of the denominator, all divided by the square of the denominator.

5–6. The Chain Rule

There still remain many functions, closely related to the ones just considered, for which we have no adequate tool for differentiation. The differentiation of the function defined by $f(x) = \sqrt[3]{x^2 + 1}$ would, at this point, force us to go directly to the definition. To deal with this problem and countless others, we now develop the *chain rule*.

THEOREM 5–7. Let the variables x, y, and u be related by $y = f(u)$, $u = g(x)$, where f and g are differentiable for common values of x.† Then, for these common values of x,

$$\frac{dy}{dx} = \frac{dy}{du}\frac{du}{dx}.$$

We have shifted to a different notation from that used in the preceding theorems in the belief that in this theorem it is easier to understand and use. Using Δu to represent the change in u produced by a change Δx in x (that is, $\Delta u = g(x + \Delta x) - g(x)$), we write the algebraic identity

$$\frac{\Delta y}{\Delta x} = \frac{\Delta y}{\Delta u} \cdot \frac{\Delta u}{\Delta x}. \tag{5-3}$$

Thus

$$\frac{dy}{dx} = \lim_{\Delta x \to 0} \frac{\Delta y}{\Delta x} = \lim_{\Delta x \to 0} \frac{\Delta y}{\Delta u} \cdot \frac{\Delta u}{\Delta x}.$$

Since $\Delta u \to 0$ as $\Delta x \to 0$, and since, by the hypothesis of the differentiability of f and g,

$$\lim_{\Delta u \to 0} \frac{\Delta y}{\Delta u} = \frac{dy}{du}, \qquad \lim_{\Delta x \to 0} \frac{\Delta u}{\Delta x} = \frac{du}{dx}, \tag{5-4}$$

we may write

$$\frac{dy}{dx} = \frac{dy}{du} \cdot \frac{du}{dx}$$

except for one possibility. It may be that as $\Delta x \to 0$, $\Delta u = g(x + \Delta x) - g(x)$ may assume the value zero. In that case the factor $\Delta y/\Delta u$ in (5-3) is undefined and we cannot draw our conclusion so simply. In order to overcome this difficulty we define the function ε by

$$\varepsilon(\Delta u) = \frac{\Delta y}{\Delta u} - \frac{dy}{du}, \qquad \Delta u \neq 0,$$

$$\varepsilon(0) = 0.$$

---

† That is to say, some values of x for which g is differentiable produce values of u for which f is differentiable. In many cases, f and g are differentiable for all values of u and x. Example 5-12 illustrates this situation.

Then we may write (5-3) in the form

$$\frac{\Delta y}{\Delta x} = \left\{\frac{dy}{du} + \varepsilon(\Delta u)\right\}\frac{\Delta u}{\Delta x}$$

which is defined for $\Delta u = 0$, thus eliminating the difficulty experienced with (5-3) in this case. Moreover, from (5-4), $\varepsilon(\Delta u) \to 0$ as $\Delta u \to 0$, and therefore $\varepsilon(\Delta u) \to 0$ as $\Delta x \to 0$. Hence

$$\frac{dy}{dx} = \lim_{\Delta x \to 0}\left\{\frac{dy}{du} + \varepsilon(\Delta u)\right\}\frac{\Delta u}{\Delta x}$$

$$= \lim_{\Delta x \to 0}\frac{dy}{du}\frac{\Delta u}{\Delta x} + \lim_{\Delta x \to 0}\varepsilon(\Delta u)\frac{\Delta u}{\Delta x}$$

$$= \frac{dy}{du}\lim_{\Delta x \to 0}\frac{\Delta u}{\Delta x} + \lim_{\Delta x \to 0}\varepsilon(\Delta u)\lim_{\Delta x \to 0}\frac{\Delta u}{\Delta x}$$

$$= \frac{dy}{du}\frac{du}{dx} + 0 \cdot \frac{du}{dx}$$

$$= \frac{dy}{du}\frac{du}{dx},$$

which establishes the theorem.

Example 5–11. Find dy/dx, where $y = \sqrt[3]{x^2 + 1}$.

The relationship between y and x may be written

$$y = u^{1/3}, \qquad u = x^2 + 1.$$

Then

$$\frac{dy}{du} = \frac{1}{3}u^{-2/3}, \qquad \frac{du}{dx} = 2x,$$

and by Theorem 5-7,

$$\frac{dy}{dx} = \frac{1}{3}u^{-2/3} \cdot 2x = \frac{2x}{3\sqrt[3]{(x^2 + 1)^2}}.$$

Example 5–12. Find dy/dx, where $y = (x^3 + 2x - 1)^5$.

This result may be obtained, without appealing to Theorem 5-7, by expanding the right member, thereby obtaining a polynomial. However, Theorem 5-7 reduces the labor considerably. We write

$$y = u^5, \qquad u = x^3 + 2x - 1;$$

thus

$$\frac{dy}{du} = 5u^4, \qquad \frac{du}{dx} = 3x^2 + 2,$$

from which we obtain

$$\frac{dy}{dx} = 5u^4(3x^2 + 2) = 5(x^3 + 2x - 1)(3x^2 + 2).$$

We shall usually refer to Theorem 5-7 as the *chain rule*.

One immediate consequence of the chain rule is an extended form of Theorem 5-4.

THEOREM 5-8. Let $y = u^n$, $u = g(x)$, where n is an integer and g is differentiable. Then

$$\frac{dy}{dx} = nu^{n-1}\frac{du}{dx}, \qquad u \neq 0 \quad \text{if } n < 0.$$

This theorem, as in the case of Theorem 5-4, can be extended to include all real n. We shall assume it to be true for n rational pending the proof of this fact in Sec. 5-7. In this case we must place the restriction $u \neq 0$ if $n < 1$.

Many situations call for a combination of the various differentiation formulas. The next two examples illustrate this.

Example 5-13. Find dy/dx, where

$$y = (x^2 + 1)^3 \sqrt{9 - x^2}.$$

We first note that this is a product, so we have

$$\frac{dy}{dx} = (x^2 + 1)^3 \frac{d(\sqrt{9 - x^2})}{dx} + \sqrt{9 - x^2}\,\frac{d(x^2 + 1)^3}{dx}.$$

Then we apply the chain rule, or Theorem 5-8, to each of the indicated derivatives in the right member and obtain

$$\frac{d(\sqrt{9 - x^2})}{dx} = \frac{d(9 - x^2)^{1/2}}{dx} = \frac{1}{2}(9 - x^2)^{-1/2}(-2x)$$

$$= \frac{-x}{\sqrt{9 - x^2}}$$

and

$$\frac{d(x^2 + 1)^3}{dx} = 3(x^2 + 1)^2(2x) = 6x(x^2 + 1)^2.$$

Thus

$$\frac{dy}{dx} = (x^2 + 1)^3 \left[\frac{-x}{\sqrt{9 - x^2}} \right] + \sqrt{9 - x^2} [6x(x^2 + 1)^2]$$

$$= \frac{(x^2 + 1)^2 x}{\sqrt{9 - x^2}} [-(x^2 + 1) + 6(9 - x^2)]$$

$$= \frac{-x(x^2 + 1)^2 (7x^2 - 53)}{\sqrt{9 - x^2}}.$$

Example 5–14. Find $f'(x)$, where

$$f(x) = \frac{1 - x^2}{\sqrt{x^2 - 4}}.$$

This example may be approached in either of two ways. We may write it in the form $f(x) = (1 - x^2)(x^2 - 4)^{-1/2}$ and proceed as in Example 5-13, or we may write it as

$$f(x) = \frac{1 - x^2}{(x^2 - 4)^{1/2}}$$

and treat it as a quotient. We elect to do it in the latter fashion.

We have

$$f'(x) = \frac{(x^2 - 4)^{1/2} \dfrac{d(1 - x^2)}{dx} - (1 - x^2) \dfrac{d(x^2 - 4)^{1/2}}{dx}}{[(x^2 - 4)^{1/2}]^2}.$$

Since

$$\frac{d(1 - x^2)}{dx} = -2x,$$

and, by the chain rule,

$$\frac{d(x^2 - 4)^{1/2}}{dx} = \frac{1}{2}(x^2 - 4)^{-1/2}(2x) = x(x^2 - 4)^{-1/2},$$

we obtain

$$f'(x) = \frac{(x^2 - 4)^{1/2}(-2x) - (1 - x^2)x(x^2 - 4)^{-1/2}}{x^2 - 4}.$$

An effective means of simplifying this expression is to factor $x(x^2 - 4)^{-1/2}$ from the numerator. If we do this,

$$f'(x) = \frac{x(x^2 - 4)^{-1/2}[(-2)(x^2 - 4) - (1 - x^2)]}{x^2 - 4}$$

$$= \frac{x(-2x^2 + 8 - 1 + x^2)}{(x^2 - 4)^{3/2}}$$

$$= \frac{-x(x^2 - 7)}{(x^2 - 4)^{3/2}}.$$

Exercises 5-3

Find dy/dx in Exercises 1–28.

1. $y = (2 + 3x^2)^3$

2. $y = (1 - x^3)^4$

3. $y = 2(3x^2 - 5)^5$

4. $y = \dfrac{1}{2}(x^2 - 3x + 5)^{-2}$

5. $y = \sqrt{4 - x^2}$

6. $y = \sqrt[3]{(2x^2 - 5)^2}$

7. $y = \sqrt{x^3 + 9}$

8. $y = 3\sqrt{2x - x^2}$

9. $y = \dfrac{x}{x + 1}$

10. $y = x^3(x^2 - 1)$

11. $y = (x^3 + 3)(2x^2 - 3x + 1)$

12. $y = \dfrac{x^2}{x - 1}$

13. $y = \dfrac{2x}{x^2 + 1}$

14. $y = \dfrac{x^2 + 1}{x^2 - 1}$

15. $y = \dfrac{1 - 2x}{4 + x^2}$

16. $y = 3(4 - x^2)(x^3 - x - 1)$

17. $y = (x - 1)\sqrt{x^2 - 1}$

18. $y = \dfrac{x - 1}{\sqrt{x^2 - 1}}$

19. $y = (x^2 - 1)\sqrt{x - 1}$

20. $y = \dfrac{\sqrt{x^2 - 1}}{x - 1}$

21. $y = x\sqrt{(9 - x^2)^3}$

22. $y = \dfrac{(x + 3)^2}{\sqrt{x^2 - 9}}$

23. $y = \dfrac{1}{(x - 2)^2\sqrt{x^2 - 4}}$

24. $y = \dfrac{x}{(x + 1)(2x - 3)^2}$

25. $y = \sqrt{\dfrac{x^3}{2x + 3}}$

26. $y = (2x - 3)^2(9 - x^2)^3(3x^2 - 2x)$

27. $y = (1 + \sqrt{1 + x})^2$

28. $y = \sqrt{1 + \sqrt{1 - x}}$

29. Find the points on the curve $y = x^2/(x^2 - 4)$ at which the tangents are horizontal. *horiz tan when 1st d = 0*

30. Find the points of intersection of the two curves $y = x/(2 - x^2)$ and $y = 1/x$. At what angles do their tangents intersect at these points?†

---

† We shall say that the angle of intersection of two curves at a common point is the angle between their tangents at that point.

31. Find the points of intersection, if any, of the two curves $y = x\sqrt{x-2}$ and $y = 3/\sqrt{x-2}$. At what angle do the curves intersect at these points?

32. Does the curve $y = (x-1)^{2/3}$ have a vertical tangent? If so, at what point? (derivative undefined)? Sketch the curve.

33. Find the points at which tangents to $y = x/\sqrt{9-x^2}$ are parallel to the line $x - 3y + 3 = 0$.

5-7. Functions Defined Implicitly and Their Derivatives

The functions we have encountered thus far have been defined by equations of the form $y = f(x)$, and we say that y is defined *explicitly* as a function of x. However, we often encounter equations of the form $F(x, y) = 0$, such as

$$x^2 - y^2 - 1 = 0, \qquad x^2 y^2 - 1 = 0,$$
$$x^2 + xy - 2y^2 - 2x + 5y - 3 = 0,$$

which define y as one or more functions of x but not in explicit form. Many equations of this kind, including those above, can be written in explicit form. Thus, $x^2 - y^2 - 1 = 0$ can be written in the form

$$y = \pm\sqrt{x^2 - 1},$$

so that this equation *implies* the two functions f_1 and f_2 defined by

$$f_1(x) = \sqrt{x^2 - 1}, \qquad f_2(x) = -\sqrt{x^2 - 1}.$$

Not all equations $F(x, y) = 0$ imply that y is a real function of x. The equation

$$x^2 + 2y^2 + 1 = 0$$

does not define y as a function of x since there is no pair of real values (x_1, y_1) which satisfy this equation. It is beyond the scope of this book to deal with the problem of when $F(x, y) = 0$ defines a functional relationship and when this function possesses special properties such as continuity and differentiability. We shall, from this point on, unless confronted with evidence to the contrary, assume that the equations encountered define y as a function of x in some domain and that this function is differentiable in at least a portion of that domain.

Many equations that define functions implicitly either cannot be converted into explicit form or are of such nature as to make such a conversion undesirable. Therefore it is important to develop a means of calculating derivatives of these functions without defining them in explicit form. We accomplish this by assuming that the equation $F(x, y) = 0$ implies an unknown function f such that $y = f(x)$. Thus, if we substitute the unknown function value $f(x)$ for y, we obtain the identity

$$F(x, f(x)) \equiv 0,$$

where the symbol $\equiv$ means that the statement is true for all values of x in the domain of f. Then the derivatives of the two identical functions will be equal and we have

$$\frac{dF}{dx} = \frac{d(0)}{dx} = 0,$$

where, in the calculation of dF/dx, we consider y as an unknown function of x wherever it occurs.

Example 5–15. Calculate dy/dx, where $x^3 + 2xy^2 + y^3 = 0$.

As indicated in the preceding discussion, we have

$$\frac{d}{dx}(x^3 + 2xy^2 + y^3) = \frac{d(0)}{dx} = 0, \quad \checkmark$$

or

$$\frac{d(x^3)}{dx} + \frac{d(2xy^2)}{dx} + \frac{d(y^3)}{dx} = 0, \quad \checkmark$$

or

$$3x^2 + 2\left(x\frac{d(y^2)}{dx} + y^2\right) + 3y^2\frac{dy}{dx} = 0,$$

or

$$3x^2 + 2\left(x \cdot 2y\frac{dy}{dx} + y^2\right) + 3y^2\frac{dy}{dx} = 0,$$

or

$$(4xy + 3y^2)\frac{dy}{dx} + 3x^2 + 2y^2 = 0,$$

or, solving for dy/dx,

$$\frac{dy}{dx} = \frac{-(3x^2 + 2y^2)}{4xy + 3y^2}, \qquad 4xy + 3y^2 \neq 0.$$

Example 5–16. Find the equation of the tangent to $x^3 + xy^3 = 9$ at the point $(1, 2)$.

Note first that the basis of *implicit differentiation* does not require that the equation always have zero as the right member. We merely take the equation in a convenient form and *differentiate both sides with respect to x, assuming y to be a function of x.*

To calculate the slope of the curve at a general point (x, y) we have

$$\frac{d(x^3 + xy^3)}{dx} = \frac{d(9)}{dx}$$

or

$$\frac{d(x^3)}{dx} + \frac{d(xy^3)}{dx} = 0,$$

from which we obtain

$$3x^2 + x\frac{d(y^3)}{dx} + y^3 = 0$$

or

$$3x^2 + 3xy^2\frac{dy}{dx} + y^3 = 0,$$

or, solving for dy/dx,

$$\frac{dy}{dx} = \frac{-3x^2 - y^3}{3xy^2}.$$

Then the slope of the required tangent is

$$\frac{dy}{dx}\bigg]_{\substack{x=1 \\ y=2}} = \frac{-3(1)^2 - (2)^3}{3(1)(2)^2} = -\frac{11}{12},$$

and its equation is

$$y - 2 = -\frac{11}{12}(x - 1)$$

or

$$11x + 12y = 35.$$

The process of implicit differentiation enables us to extend Theorems 5-4 and 5-8 to include the case where n is any rational number. We have already assumed the truth of the following theorem, which we now prove:

THEOREM 5–9. Let $y = u^n$, $u = g(x)$ where n is a rational number and g is differentiable. Then

$$\frac{dy}{dx} = nu^{n-1}\frac{du}{dx}, \qquad u \neq 0 \quad \text{if } n < 1.$$

If n is rational, we may write

$$y = u^{p/q},$$

where p and q are integers. If x and y satisfy this equation, they also satisfy

$$y^q = u^p.$$

We differentiate both sides of this equation with respect to x and obtain

$$q y^{q-1} \frac{dy}{dx} = p u^{p-1} \frac{du}{dx}$$

or

$$\frac{dy}{dx} = \frac{p u^{p-1}(du/dx)}{q y^{q-1}} .$$

Now we make use of the original equation and write

$$\frac{dy}{dx} = \frac{p}{q} \frac{u^{p-1}(du/dx)}{(u^{p/q})^{q-1}}$$

$$= \frac{p}{q} \frac{u^{p-1}(du/dx)}{u^{p-p/q}}$$

$$= \frac{p}{q} u^{p/q-1} \frac{du}{dx}$$

$$= n u^{n-1} \frac{du}{dx} .$$

If the equation $y = f(x)$ implicitly defines x as a function of y, that is, if it implies the existence of a function h such that $x = h(y)$, the process of implicit differentiation yields an important formula. It is not our intention at this point to discuss what hypotheses are needed on f to imply the existence of h, but rather to present another example of the usefulness of implicit differentiation.

We differentiate both members of

$$y = f(x)$$

with respect to y, assuming x to be a function of y, and obtain

$$1 = \frac{df}{dx} \frac{dx}{dy} = f'(x) \frac{dx}{dy}, \qquad (5\text{-}5)$$

where the chain rule was used on the right member. If $f'(x) \neq 0$, we have

$$\frac{dx}{dy} = \frac{1}{f'(x)}, \qquad (5\text{-}6)$$

or, writing $dy/dx = f'(x)$,

$$\frac{dx}{dy} = \frac{1}{dy/dx}, \qquad \frac{dy}{dx} \neq 0. \qquad (5\text{-}7)$$

Equation (5-5) can be used equally well to obtain

$$\frac{dy}{dx} = \frac{1}{dx/dy}, \qquad \frac{dx}{dy} \neq 0. \tag{5-8}$$

The usefulness of these formulas will be evident in various connections later on.

5–8. Higher Derivatives

The process of differentiation provides us with a means of determining a function f' from a given differentiable function f. If f' is a differentiable function, we may relate to it a function f'', defined by

$$f''(x) = \lim_{\Delta x \to 0} \frac{f'(x + \Delta x) - f'(x)}{\Delta x},$$

which we call the *second derivative*. We denote the second derivative by the various notations

$$f''(x) \equiv \frac{d^2 f(x)}{dx^2} \equiv \frac{d^2 f}{dx^2} \equiv y'' \equiv \frac{d^2 y}{dx^2}.$$

In the same way, if f'' is differentiable, we may define the third derivative, and so on.

As we proceed with our study of calculus we shall see various uses for higher derivatives. One will occur in Example 5-20. Another will appear in the next chapter, where we use the second derivative to determine special points on the graph of a function. Others will present themselves from time to time.

Example 5–17. Calculate

$$\frac{dy}{dx}, \qquad \frac{d^2 y}{dx^2}, \dots, \frac{d^5 y}{dx^5}$$

where $y = 2x^4 - 3x^2 + 5x + 7$.

We have

$$\frac{dy}{dx} = 8x^3 - 6x + 5, \qquad \frac{d^2 y}{dx^2} = 24x^2 - 6, \qquad \frac{d^3 y}{dx^3} = 48x,$$

$$\frac{d^4 y}{dx^4} = 48, \qquad \frac{d^5 y}{dx^5} = 0.$$

Example 5–18. Find $f''(x)$, where

$$f(x) = \frac{x}{\sqrt{1 - x^2}}.$$

We have

$$f'(x) = \frac{(1-x^2)^{1/2}(1) - x\,\dfrac{d(1-x^2)^{1/2}}{dx}}{[(1-x^2)^{1/2}]^2}$$

$$= \frac{(1-x^2)^{1/2} - x(\tfrac{1}{2})(1-x^2)^{-1/2}(-2x)}{1-x^2},$$

and if we factor $(1-x^2)^{-1/2}$ from the numerator, we obtain

$$f'(x) = \frac{(1-x^2)^{-1/2}[(1-x^2) + x^2]}{1-x^2}$$

$$= (1-x^2)^{-3/2}.$$

Then

$$f''(x) = -\frac{3}{2}(1-x^2)^{-5/2}(-2x)$$

$$= \frac{3x}{(1-x^2)^{5/2}}.$$

The methods of implicit differentiation can be used to obtain higher derivatives, as the following example illustrates.

Example 5–19. Find d^2y/dx^2, where $x^{1/2} + y^{1/2} = 1$.

Differentiating implicitly, we have

$$\frac{d(x^{1/2})}{dx} + \frac{d(y^{1/2})}{dx} = \frac{d(1)}{dx}$$

or

$$\frac{1}{2}x^{-1/2} + \frac{1}{2}y^{-1/2}\frac{dy}{dx} = 0,$$

from which we obtain

$$\frac{dy}{dx} = -\frac{x^{-1/2}}{y^{-1/2}} = -\frac{y^{1/2}}{x^{1/2}}. \tag{5-9}$$

Then, differentiating implicitly again, we have

$$\frac{d}{dx}\left(\frac{dy}{dx}\right) = \frac{d}{dx}\left(-\frac{y^{1/2}}{x^{1/2}}\right)$$

or

$$\frac{d^2y}{dx^2} = -\frac{x^{1/2}\,\dfrac{d(y^{1/2})}{dx} - y^{1/2}\,\dfrac{d(x^{1/2})}{dx}}{(x^{1/2})^2}$$

$$= -\left[\frac{x^{1/2}(\tfrac{1}{2})y^{-1/2}(dy/dx) - y^{1/2}(\tfrac{1}{2})x^{-1/2}}{x}\right].$$

We use (5-9) to eliminate dy/dx from the right member of this equation and get

$$\frac{d^2y}{dx^2} = -\frac{1}{2}\left[\frac{x^{1/2}y^{-1/2}(-y^{1/2}/x^{1/2}) - y^{1/2}x^{-1/2}}{x}\right]$$

$$= \frac{1}{2}\left[\frac{1 + y^{1/2}x^{-1/2}}{x}\right]$$

$$= \frac{1}{2}\left[\frac{1 + (y^{1/2}/x^{1/2})}{x}\right] = \frac{1}{2}\left[\frac{x^{1/2} + y^{1/2}}{x^{3/2}}\right].$$

Since $x^{1/2} + y^{1/2} = 1$, this result may be simplified to

$$\frac{d^2y}{dx^2} = \frac{1}{2}\left(\frac{1}{x^{3/2}}\right) = \frac{1}{2x^{3/2}}.$$

Example 5-20. A particle moves along a straight line in such a manner that its distance s in feet from a fixed point on the line at the end of t seconds is given by

$$s = t^3 - 3t^2 - 24t.$$

(a) Find its velocity when its acceleration is zero, and (b) its acceleration when its velocity is zero.

Earlier in this chapter we defined (Definition 5-2a) the instantaneous velocity of such a moving particle to be

$$v = \frac{ds}{dt}.$$

Then the *acceleration a* (for straight-line motion) is given by

$$a = \frac{dv}{dt} = \frac{d^2s}{dt^2}.$$

Hence

$$v = \frac{ds}{dt} = 3t^2 - 6t - 24,$$

and

$$a = \frac{d^2s}{dt^2} = 6t - 6.$$

Thus $v = 0$ when

$$3t^2 - 6t - 24 = 3(t^2 - 2t - 8)$$
$$= 3(t - 4)(t + 2) = 0,$$

that is, when

$$t = 4, -2.$$

Likewise, $a = 0$ when

$$6t - 6 = 0,$$

or

$$t = 1.$$

Then, for part (a), we have

$$v = 3t^2 - 6t - 24]_{t=1} = -27 \text{ ft/sec};$$

and for part (b),

$$a = 6t - 6]_{t=4} = 18 \text{ ft/sec/sec};$$

and

$$a = 6t - 6]_{t=-2} = -18 \text{ ft/sec/sec}.$$

The interpretations to be placed on the signs of these results are:
(a) The motion is in the direction of decreasing s;
(b) The velocity is increasing at $t = 4$ and decreasing at $t = -2$.

There is a distinction to be made between *velocity* and *speed*. Velocity takes into account the direction of the motion as well as the rate of change of distance with respect to time. We define

$$\text{Speed} = |\text{velocity}|.$$

Thus the speed may be increasing when the velocity is decreasing. As a matter of fact, the speed will always be increasing when the velocity and acceleration have the same sign.

The value $t = -2$ would not be admissible in many physical applications because the domain of the distance function is often $t \geqslant 0$.

Exercises 5-4

Find the indicated derivatives in Exercises 1–4.

1. $y = 2x^2 + 3x + 5$; $\dfrac{d^2y}{dx^2}$

2. $y = 3x^6 - 4x^5 - 6x^2$; $\dfrac{d^4y}{dx^4}$

3. $y = x^k$; $\dfrac{d^k y}{dx^k}$; k a positive integer

4. $y = x^p$; $\dfrac{d^k y}{dx^k}$; p, k positive integers; $p > k$

Find dy/dx in Exercises 5–14.

5. $x^2 + y^2 = 25$ 6. $x^2 - y^2 = 1$

7. $x^2 + xy + y^2 = 1$ 8. $x^2 + 3xy - y^2 = 5$

9. $x^3 + xy^3 = 1$ **10.** $x^3 + xy + y^3 = 12$

11. $y^2(2x^2 + y) = a^2$ **12.** $x^2 + y^2 = 2y^3$

13. $x^2 y + xy^2 = x^3$ **14.** $x^2 y^2 = x^2 + y^2$

Find $d^2 y/dx^2$ in Exercises 15–20.

15. $x^2 + 4y^2 = 4$ **16.** $x^3 + y^3 = 1$

17. $x^{2/3} + y^{2/3} = a^{2/3}$ **18.** $y^3 = \dfrac{5}{x^2} + 1$

19. $y = x\sqrt{1 - x^2}$ **20.** $y = \dfrac{\sqrt{x^2 - a^2}}{x}$

21. Find the angles of intersection of the two curves $2y^2 = x^3$ and $xy = 4$.

22. Find the slope of the curve $x^2 y^2 = 25 - x^2$ at the points where the line $x = 4$ intersects it.

In the following exercises it is to be understood that the motion is along a straight line, the distance s is in feet, and the time t is in seconds.

In Exercises 23–26, find the velocity and acceleration at the specified time for the motion described by the given equation.

23. $s = t^3 + 3t^2 - t + 3,\ t = 2$

24. $s = 3t^3 - 5t^2 + 2t - 14,\ t = 3$

25. $s = t + \dfrac{1}{t},\ t = \dfrac{1}{2}$

26. $s = t\sqrt{t^2 + 9},\ t = 4$

27. If $s = t^3 - 9t^2 + 15t + 4$, find the acceleration when the velocity is zero.

28. If $s = 16(t + (1/t^3))$, find the velocity when the acceleration is 6 ft/sec/sec.

29. If $s = 2t^3 - 15t^2 + 24t + 5$, find the values of t for which the speed is increasing.

30. The costs involved in manufacturing x units of a product are:
(a) Development costs, \$10,000.
(b) Cost of materials, $100x$ dollars.
(c) Manufacturing costs, $15x^{2/3}$ dollars.
Find the marginal cost if the production level is 125 units. Use the second derivative to show that the marginal cost decreases as x increases.

Chapter 6

SOME APPLICATIONS

6-1. The Mean Value Theorem

We begin this chapter with one of the most important theorems in calculus.

THEOREM 6-1 (Mean Value Theorem). Let f be continuous on $a \leqslant x \leqslant b$ and let f' exist for each x on $a < x < b$. Then there exists a number η, $a < \eta < b$, such that

$$f(b) - f(a) = (b - a)f'(\eta) \tag{6-1}$$

We omit the proof of this theorem. However, a look at the geometric implications will convince the student of its plausibility. If we write (6-1) in the form

$$\frac{f(b) - f(a)}{b - a} = f'(\eta),$$

we see, from Fig. 6-1, that the left member is the slope of the chord PQ. Hence the theorem states that there is a point $(\eta, f(\eta))$ on the graph of $y = f(x)$ between $P(a, f(a))$ and $Q(b, f(b))$ at which the tangent is parallel to PQ. If the curve is as shown in Fig. 6-1, there are clearly two values of η

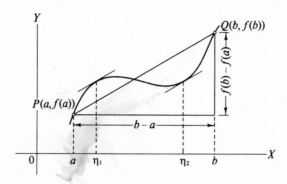

Fig. 6-1

satisfying the requirements. If there is a cusp, as in Fig. 6-2(a), or a multiple tangent as in Fig. 6-2(b), it is evident that such a parallel tangent may not exist. However, these cases are ruled out by requiring the existence of f' for $a < x < b$. Thus the theorem seems geometrically plausible. We shall, without further argument, *assume* the truth of the theorem.

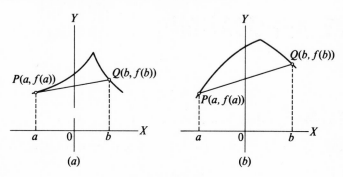

Fig. 6-2

Example 6–1. Can the mean value theorem be applied to

$$f(x) = x^3 - x$$ *Yes, polynomial, continuous for all x*

on the interval $-1 \leqslant x \leqslant 2$? If so, determine the value, or values, of η.

The theorem can be applied in this case. As a matter of fact, it can be applied to this function on any interval whatever, since f is continuous and f' exists for all x.

We have

$$f(b) = f(2) = 6, \qquad f(a) = f(-1) = 0,$$

so

$$\frac{f(b) - f(a)}{b - a} = \frac{6}{3} = 2.$$

Also

$$f'(x) = 3x^2 - 1,$$

whence

$$f'(\eta) = 3\eta^2 - 1.$$

Therefore

$$3\eta^2 - 1 = 2,$$

or

$$\eta = \pm 1.$$

The value $\eta = -1$ does not lie in the interior of the specified interval, but $\eta = 1$ satisfies all requirements. The tangent to the graph of $y = f(x)$ at $x = 1$ (Fig. 6-3) is parallel to the chord joining the points $(-1, 0)$, $(2, 6)$.

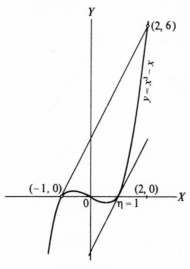

Fig. 6-3

Example 6–2. Can the mean value theorem be applied to

$$f(x) = (x - 1)^{2/3}$$

on the interval $0 \leqslant x \leqslant 2$? If so, determine the value, or values, of η.

f is continuous on the given interval; in fact, it is continuous for all x. However,

$$f'(x) = \frac{2}{3}(x - 1)^{-1/3}$$

does not exist for $x = 1$. Hence the hypotheses of the theorem are not satisfied and we cannot be sure a suitable η exists. Indeed, it is quite clear from Fig. 6-4

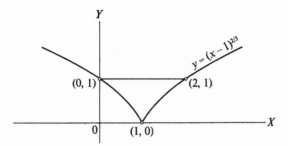

Fig. 6-4

that there is no tangent to the graph of $y = f(x)$ in this interval which is parallel to the chord joining the points (0, 1) and (2, 1).

However, for any interval not containing $x = 1$ in its interior, the theorem may be applied. Consider the interval $1 \leqslant x \leqslant 4$. We have

$$f(1) = 0, \qquad f(4) = 3^{2/3},$$

whence

$$\frac{f(b) - f(a)}{b - a} = \frac{3^{2/3}}{3} = 3^{-1/3}.$$

Moreover

$$f'(\eta) = \left. \frac{2}{3(x - 1)^{1/3}} \right]_{x=\eta} = \frac{2}{3(\eta - 1)^{1/3}} \cdot$$

We set these two results equal and obtain

$$\frac{2}{3(\eta - 1)^{1/3}} = 3^{-1/3},$$

or, cubing both members,

$$\frac{8}{27(\eta - 1)} = \frac{1}{3} \cdot$$

We solve this equation for η and get

$$\eta = \frac{17}{9} \cdot$$

The tangent to the graph of $y = (x - 1)^{2/3}$ at $(17/9, 4/3 \sqrt[3]{3})$ is parallel to the chord joining the points (1, 0) and (4, $\sqrt[3]{9}$), shown in Fig. 6-5.

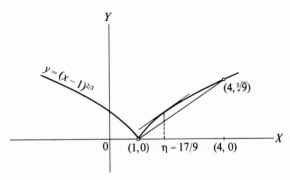

Fig. 6-5

The student should note that the theorem applies to the interval $1 \leqslant x \leqslant 4$, although f' is undefined at $x = 1$. It is required only that f' be defined at *interior points* of the interval.

The conclusion of the mean value theorem may be stated in various forms for special purposes. Three of the common forms of (6-1) are

$$f(x) - f(a) = (x - a)f'(\eta), \qquad a < \eta < x; \qquad (6\text{-}2)$$

$$f(x + h) - f(x) = hf'(\eta), \qquad x < \eta < x + h; \qquad (6\text{-}3)$$

$$f(x + h) - f(x) = hf'(x + \theta h), \qquad 0 < \theta < 1. \qquad (6\text{-}4)$$

The theorem, in the form originally stated, asserts the existence of *at least one* η between a and b, but does not provide any means of determining its specific value or values. However, its principal usefulness hinges on the existence of η rather than the actual value of it. The following theorem (assumed in Chapter 5) illustrates the manner in which it may be used to obtain useful results.

THEOREM 6–2. Let f be a function continuous on $a \leqslant x \leqslant b$ and possessing a derivative on $a < x < b$. Then

(a) If $f'(x) > 0$ for $a < x < b$, $f(x)$ is increasing on $a \leqslant x \leqslant b$;
(b) If $f'(x) < 0$ for $a < x < b$, $f(x)$ is decreasing on $a \leqslant x \leqslant b$;
(c) If $f'(x) = 0$ for $a < x < b$, $f(x)$ is constant on $a \leqslant x \leqslant b$.

To prove (a) we need to show that $f(x_2) > f(x_1)$, or $f(x_2) - f(x_1) > 0$ if $a \leqslant x_1 < x_2 \leqslant b$. The conditions of the mean value theorem are satisfied for the entire interval $a \leqslant x \leqslant b$ and therefore are satisfied for the interval $x_1 \leqslant x \leqslant x_2$. Hence we may use it to write

$$f(x_2) - f(x_1) = (x_2 - x_1)f'(\eta), \qquad x_1 < \eta < x_2.$$

Under the stated hypotheses both factors in the right member are positive; hence the left member is positive and the proof of (a) is complete.

We leave the similar proof of (b) as an exercise.

Part (c) is the converse of Theorem 5-2. For its proof we write, using the mean value theorem,

$$f(x) = f(a) + (x - a)f'(\eta)$$

where $a < x \leqslant b$ and $a < \eta < x$. By hypothesis, $f'(\eta) = 0$, $a < \eta < b$, and therefore

$$f(x) = f(a), \qquad a < x \leqslant b.$$

For $x = a$, we also have

$$f(x) = f(a),$$

so the theorem is true for $a \leqslant x \leqslant b$.

Example 6-3. Find the range of values of x for which $f(x)$ is increasing if $f(x) = 2x^3 + 3x^2 - 12x + 3$.

The solution of this problem requires that we determine the values of x for which $f'(x) > 0$ (Theorem 6-2(a)). We have

$$f'(x) = 6x^2 + 6x - 12$$
$$= 6(x^2 + x - 2)$$
$$= 6(x + 2)(x - 1).$$

If the two factors $x + 2$ and $x - 1$ have the same sign, we will have $f'(x) > 0$ as required. We note that

$$x + 2 > 0 \qquad \text{if} \quad x > -2,$$

and

$$x - 1 > 0 \qquad \text{if} \quad x > 1.$$

Both are satisfied if $x > 1$. Hence $f'(x) > 0$ for $x > 1$. On the other hand

$$x + 2 < 0 \qquad \text{if} \quad x < -2,$$

and

$$x - 1 < 0 \qquad \text{if} \quad x < 1.$$

So $f'(x) > 0$ also when $x < -2$. Thus the complete range of x for which $f(x)$ is increasing is

$$x > 1 \qquad \text{and} \qquad x < -2.$$

An immediate result of Theorem 6-2(c) follows.

THEOREM 6-3. Let f and g be two functions continuous on $a \leqslant x \leqslant b$ and $f'(x) = g'(x)$ for $a < x < b$. Then $f(x)$ and $g(x)$ differ by a constant for $a \leqslant x \leqslant b$.

To prove this we write

$$F(x) = f(x) - g(x).$$

Then F is continuous for $a \leqslant x \leqslant b$ and

$$F'(x) = f'(x) - g'(x) = 0$$

for $a < x < b$. Therefore, by Theorem 6-2(c), $F(x)$ is a constant, that is,

$$f(x) - g(x) = c$$

for $a \leqslant x \leqslant b$.

We shall find this theorem very useful in Chapter 7.

Exercises 6-1

In each of Exercises 1–6 determine (a) if the mean value theorem can be applied to the interval stated; (b) if not, why not; (c) if the theorem applies, find a value for η.

1. $f(x) = x^2 + 1, 0 \leqslant x \leqslant 1$ 2. $f(x) = \dfrac{x^2 + 1}{x^2}, 1 \leqslant x \leqslant 2$

3. $f(x) = \dfrac{x}{x + 1}, -2 \leqslant x \leqslant 2$ 4. $f(x) = \dfrac{x}{x + 1}, 0 \leqslant x \leqslant 2$

5. $f(x) = x^{2/3}, -1 \leqslant x \leqslant 1$ 6. $f(x) = x^{2/3}, 0 \leqslant x \leqslant 1$

7. Find the range of values of x for which $2x^3 + 3x^2 - 36x$ is increasing; decreasing.

8. Find the range of values of x for which $x + (1/x)$ is increasing; decreasing.

9. Find the range of values of x for which $2x/\sqrt{(2 + x^2)^3}$ is increasing; decreasing.

10. Prove Theorem 6-2(b).

11. A manufacturer finds that his total revenue y is expressed by the equation

$$y = 10{,}000 - (x - 100)^2,$$

where x is the number of units sold per week. (a) Find the rate at which the total revenue is changing with respect to units sold when $x = 50, 100, 150$. Interpret your results. (b) At what rate is the average revenue changing when $x = 50$? Interpret this result.

6–2. Sign of the First Derivative

Theorem 6-2 has important uses in sketching the graph of $y = f(x)$. If f is continuous, parts (a) and (b) give us conditions for a *rising* or *falling* graph.†

Example 6–4. Find the range of x for which the graph of $y = x^3 - 9x^2 + 24x - 17$ is (a) rising; (b) falling.

The function defined by this equation is continuous and possesses a derivative for all values of x. Therefore Theorem 6-2 applies and the problem reduces to the determination of the ranges of x for which y' is (a) positive; (b) negative. We write

$$y' = 3x^2 - 18x + 24$$
$$= 3(x - 2)(x - 4).$$

---

† We remark again that the terms *increasing, decreasing, rising,* and *falling* used in connection with functions and graphs always imply that the independent variable is increasing.

Thus y' will be positive when the factors in the right member have the same sign and will be negative when they have opposite signs. Since $x - 2 > 0$ when $x > 2$; $x - 2 < 0$ when $x < 2$; and $x - 4 > 0$ when $x > 4$; $x - 4 < 0$ when $x < 4$; we obtain

$$y' > 0 \qquad \text{when } x \geq 4 \text{ or } x \leq 2,$$

and

$$y' < 0 \qquad \text{when } 2 < x < 4.$$

Thus we may conclude that the graph is

(a) Rising when $x > 4$ or $x < 2$;
(b) Falling when $2 < x < 4$.

These results are verified by the graph shown in Fig. 6-6. A fact to note in connection with this graph is that the points at which it changes from rising to falling, or vice versa, are points on the curve at which $y' = 0$. This is usually the case, although not always, as we shall see in the next example.

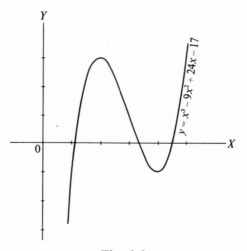

$$y = x^3 - 9x^2 + 24x - 17$$

Fig. 6-6

Example 6-5. Find the range of x for which the graph of $y = x^{2/3}(x - 5)$ is (a) rising; (b) falling.

The function defined by this equation is continuous for all x. Moreover

$$y' = \frac{5}{3}\left(\frac{x - 2}{x^{1/3}}\right)$$

exists for all values of x except $x = 0$. Hence we may apply Theorem 6-2 to any interval not containing $x = 0$ in its interior. However, we may use an interval which has $x = 0$ as one of its end points.

Let us consider first the interval $0 < x < \infty$. On this interval $x^{1/3} > 0$. Also $x - 2 > 0$ for $x > 2$, but $x - 2 < 0$ for $0 < x < 2$. Therefore

$$y' < 0 \qquad \text{when } 0 < x < 2,$$

$$y' > 0 \qquad \text{when } x > 2,$$

and we may say that the graph is

(a) Rising for $x > 2$;
(b) Falling for $0 < x < 2$.

Now let us consider the interval $-\infty < x < 0$. On this interval, $x^{1/3} < 0$ and likewise $x - 2 < 0$. Therefore $y' > 0$ when $x < 0$, so the graph is always rising on this interval. Figure 6-7 shows these results. We note in this case that the curve changes from rising to falling, or vice versa, not only at $x = 2$ where $y' = 0$, but also at $x = 0$ where y' does not exist.

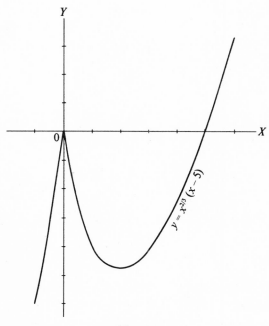

$$y = x^{2/3}(x - 5)$$

Fig. 6-7

6–3. Sign of the Second Derivative

DEFINITION 6–1. The graph of a function f is said to be concave upward at the point $(x_1, f(x_1))$ if there exists an interval containing x_1 in its interior throughout which the graph of f lies above the tangent line at $(x_1, f(x_1))$.

Similarly, changing "above" to "below," we define *concavity downward*. The obvious meaning will be attached to the statements "the graph of f is concave upward (downward) on the interval $a < x < b$."

The sign of the second derivative, if it exists, gives us a test for the direction of concavity. We have the following theorem.

THEOREM 6-4. Let f be twice differentiable on $a < x < b$. Then
(a) If $f''(x) > 0$ for $a < x < b$, the graph of f is concave upward on that interval;
(b) If $f''(x) < 0$ for $a < x < b$, the graph of f is concave downward on that interval.

We shall not give a formal proof of this theorem but content ourselves with a geometric argument supporting (a). Since $f''(x)$ exists and is positive, $f'(x)$, by Theorem 6-2, is increasing on $a < x < b$. Moreover, since f' is differentiable, it is continuous on the interval. These two facts imply that tangents to the graph of f are rotating continuously in a counterclockwise direction as the point of tangency moves along the curve from left to right (Fig. 6-8). This is the geometric situation produced by a graph being concave upward over an interval. A similar argument could be given in support of (b).

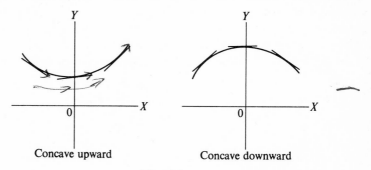

Concave upward   Concave downward

Fig. 6-8

Example 6-6. Find the range of x for which the graph of $y = x^3 - 9x^2 + 24x - 17$ is (a) concave upward; (b) concave downward.

The function f defined by this equation satisfies all hypotheses of Theorem 6-4. Hence our problem is to find the ranges of x for which (a) $y'' > 0$ and (b) $y'' < 0$. We have

$$y'' = 6x - 18 = 6(x - 3).$$

Thus $y'' > 0$ if $x > 3$, and $y'' < 0$ if $x < 3$; or we have (a) concavity upward for $x > 3$ and (b) concavity downward for $x < 3$. This is verified by Fig. 6-6.

We note that $y'' = 0$ at the point (3, 1) at which the direction of concavity changes from downward to upward. Also we note that the tangent crosses the curve at this point.

DEFINITION 6-2. A point on the graph of f at which a unique tangent exists and at which the direction of concavity changes is called a point of inflection.

As noted in Example 6-5, a point of inflection of the curve discussed in that example occurs at (3, 1) where $y'' = 0$. This is in accord with the following theorem.

THEOREM 6-5. If f'' is continuous on $a < x < b$, the only points of inflection of the graph of f on that interval occur at points for which $f''(x) = 0$.

Since f'' is continuous on $a < x < b$, the required change in sign in $f''(x)$ for a point of inflection can occur only at points where $f''(x) = 0$. This does not mean, however, that points of inflection may not occur at points at which f'' is discontinuous or even undefined.

Example 6-7. Find the points of inflection of the graph of $y = x^{1/3}$.

This equation defines a continuous function f for all values of x, but $f''(x) = -(2/9)x^{-5/3}$ is undefined at $x = 0$. Clearly, $f''(x)$ changes sign when x changes sign, so by Definition 6-2, the point (0, 0) is a point of inflection. f'' is continuous for all other values of x; hence Theorem 6-5 applies everywhere except at $x = 0$. Since there are no values of x for which $f''(x) = 0$, we conclude that the point (0, 0) is the only point of inflection of this curve (Fig. 6-9).

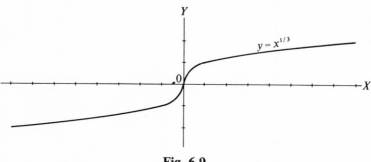

Fig. 6-9

Thus, when we are looking for points of inflection, we need to examine points at which

(a) f'' is continuous and $f''(x) = 0$;
(b) f'' is not continuous, or fails to exist.

However, after these points are located, it is still necessary to check whether $f''(x)$ changes sign at these points. Theorem 6-5 does not guarantee the existence of points of inflection when $f''(x) = 0$. It merely states the conditions under which they *may* occur.

Example 6–8. Find the points of inflection of the graph of $y = x^4$.

Here the conditions of Theorem 6-5 are satisfied. We have

$$y'' = f''(x) = 12x^2.$$

f'' is continuous everywhere and $f''(x) = 0$ when $x = 0$. Therefore the only possible point of inflection is $(0, 0)$. But $f''(x)$ is never negative, so the direction of concavity is always upward. Hence, by definition, this graph has no points of inflection (Fig. 6-10).

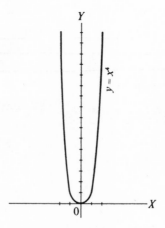

Fig. 6-10

Exercises 6-2

In Exercises 1–8 find the range of x over which the curve is (a) rising, (b) falling, (c) concave upward, (d) concave downward. Also (e) find the points of inflection and (*f*) sketch the curve together with the tangents at the points of inflection.

1. $y = x^2 + x - 2$ 2. $y = x^3 + 2x^2 + x$

3. $y = \dfrac{1}{6}(2x^3 + 9x^2)$ 4. $y = \dfrac{x^3}{3} - 3x^2 + 8x - 4$

5. $y = \dfrac{1}{2}(x^3 + 3x^2 - 9x - 7)$ 6. $y = \dfrac{1}{4}(x^4 + 2x^3 - 2x - 4)$

7. $y = x + \dfrac{1}{x}$ 8. $y = \dfrac{1}{1 + x^2}$

9. Prove that the graph of the cubic equation $y = ax^3 + bx^2 + cx + d$, $a \neq 0$, always has a point of inflection.

6–4. Maxima and Minima

Another important feature of the graph of a function f is the location of its high and low points. The ordinates of these points are called extremes of f. We make the following definitions.

DEFINITION 6-3. A function f is said to have a relative maximum (relative minimum) at x_0 if there is an interval containing x_0 in its interior such that $f(x_0) > f(x)$, $(f(x_0) < f(x))$, for all $x \neq x_0$ in the interval.

The term *relative* is used to distinguish between the case where the inequalities are satisfied only for values of x sufficiently close to x_0 as contrasted with the case where the inequalities are satisfied for all values of x. In the latter case we say we have an *absolute maximum* (*absolute minimum*). The function f defined by $f(x) = x^3 - 9x^2 + 24x - 17$ (see Fig. 6-6) has a relative maximum of 3 at $x = 2$ and a relative minimum of -1 at $x = 4$, but it has no absolute maximum or minimum.

The function f defined by $f(x) = x^4$ (see Fig. 6-10) has an absolute minimum of 0 at $x = 0$ and has no maximum, relative or otherwise.

According to Definition 6-3, relative maxima and minima occur at points where the graph of f changes from a rising to a falling curve. Therefore the general problem of determining values of x for which a function has extremes is that of determining values of x which separate the rising and falling portions of its graph. Values of x which *may possibly* do this are called *critical values*.

Quite clearly, values of x for which $f'(x) = 0$ are critical values.

THEOREM 6-6. Let f be defined on $a < x < b$ and have a relative maximum or minimum at $x = x_0$ where $a < x_0 < b$. Then if $f'(x_0)$ exists, $f'(x_0) = 0$.

Let us consider the case of a relative maximum at $x = x_0$. We have, by definition,

$$f'(x_0) = \lim_{\Delta x \to 0} \frac{f(x_0 + \Delta x) - f(x_0)}{\Delta x}$$

independent of the manner in which $\Delta x \to 0$. We have

$$\frac{f(x_0 + \Delta x) - f(x_0)}{\Delta x} < 0, \qquad \Delta x > 0,$$

and

$$\frac{f(x_0 + \Delta x) - f(x_0)}{\Delta x} > 0, \qquad \Delta x < 0,$$

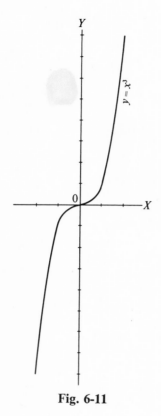

Fig. 6-11

since the numerator in both cases is negative for sufficiently small Δx. Therefore

$$f'(x_0) = \lim_{\substack{\Delta x \to 0 \\ \Delta x > 0}} \frac{f(x_0 + \Delta x) - f(x_0)}{\Delta x} \leqslant 0,$$

and

$$f'(x_0) = \lim_{\substack{\Delta x \to 0 \\ \Delta x < 0}} \frac{f(x_0 + \Delta x) - f(x_0)}{\Delta x} \geqslant 0.$$

These two numbers must be the same. Thus $f'(x_0) = 0$.

Obvious modifications in the argument will take care of the case of a relative minimum at $x = x_0$.

However, the converse of this theorem is not true; that is, we may have $f'(x_0) = 0$, but $f(x_0)$ is neither a relative maximum nor minimum. For example, consider $f(x) = x^3$ (Fig. 6-11). We have $f'(0) = 0$ but $f(0)$ is neither a relative maximum nor a relative minimum.

Also the theorem fails to cover the situations where a maximum or a minimum occurs at

(a) A point x_0 for which $f'(x_0)$ fails to exist;
(b) An end point of the interval of definition of f.

Since the existence of f' is part of the hypothesis of Theorem 6-6, it is clear why it does not cover (a). We illustrate two cases by the following examples.

Example 6–9. Discuss the extremes of $f(x) = 1 + (x - 2)^{2/3}$.

f is defined for all x, and from

$$f'(x) = \frac{2}{3}(x - 2)^{-1/3}$$

we note that $f'(x)$ is defined for all x except $x = 2$, and $f'(x) \neq 0$ for any x. Hence Theorem 6-6 says that f has no maxima or minima except possibly at $x = 2$. To investigate the situation there, we observe that

$$f'(x) < 0, \qquad x < 2;$$
$$f'(x) > 0, \qquad x > 2.$$

Therefore $(2, 1)$ is a point on the graph of f at which it changes from a falling to a rising curve (Fig. 6-12) and consequently $f(2) = 1$ is a relative minimum.

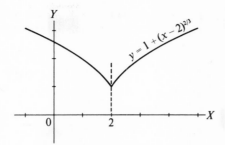

Fig. 6-12

Clearly this is an absolute minimum.

Example 6–10. Discuss the extremes of f where

$$f(x) = x^2, \qquad x \leqslant 1;$$
$$f(x) = 2 - x, \qquad x > 1.$$

We consider this example in three parts.

Part (a): $x < 1$. For these values of x, $f'(x) = 2x$, which satisfies Theorem 6-6, and we obtain the single critical value $x = 0$. Moreover

$$f'(x) < 0, \qquad x < 0;$$
$$f'(x) > 0, \qquad 0 < x < 1.$$

Hence the graph of f changes from a falling to a rising curve at $x = 0$ and $f(0) = 0$ is a relative minimum (Fig. 6-13). There are no other extremes on this part of the graph.

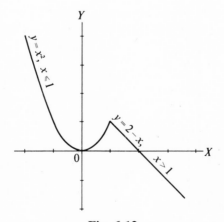

Fig. 6-13

Part (b): $x > 1$. For these values of x, $f'(x) = -1$, which satisfies Theorem 6-6, but $f'(x) \neq 0$ for any x. Hence there are no extremes on this part of the graph.

Part (*c*): $x = 1$. At this point there is not a unique tangent to the graph and consequently $f'(1)$ does not exist. Hence Theorem 6-6 does not apply. But from the preceding discussion we see that

$$f'(x) > 0, \qquad 0 < x < 1;$$
$$f'(x) < 0, \qquad x > 1.$$

Therefore $f(1) = 1$ satisfies the requirements of a relative maximum.

If the interval of definition of f is restricted, as it often is in applications, situation (b) may occur. A modification of the use of the term *relative maximum* (*relative minimum*) is necessary when talking about end points.

DEFINITION 6-4. $f(x_0)$, where x_0 is an end point of the interval of definition of f, is a relative maximum (relative minimum) if there is a subinterval with x_0 as an end point such that $f(x_0)$ is an absolute maximum (absolute minimum) over the subinterval.

It is clear that this definition will not require $f'(x_0) = 0$ in order that $f(x_0)$ be a relative maximum or a relative minimum if x_0 is an end point.

Example 6–11. Discuss the extremes of f where $f(x) = x^2 - 6x + 5, 1 \leqslant x \leqslant 6$.

We have $f'(x) = 2x - 6$ from which we obtain the critical value $x = 3$. The procedure used in the preceding examples shows that $f(3) = -4$ is a relative minimum (Fig. 6-14).

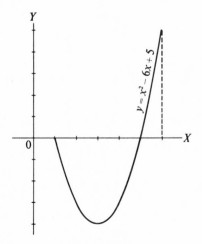

$$y = x^2 - 6x + 5$$

Fig. 6-14

Next we consider the end points $x = 1$ and $x = 6$. On the subinterval $1 \leqslant x \leqslant 3$, $f(1)$ is an absolute maximum due to the falling property of the curve. Hence $f(1) = 0$ is a relative maximum for $1 \leqslant x \leqslant 6$. Similarly, $f(6) = 5$ is a relative maximum since it is an absolute maximum for the subinterval

$3 \leqslant x \leqslant 6$ due to the rising property of the curve. As a matter of fact, it is an absolute maximum for any subinterval of $1 \leqslant x \leqslant 6$ having $x = 6$ as its right end point.

Thus this function has an absolute maximum of 5 and an absolute minimum of -4. This is in accordance with Theorem 4-11 which says that a function continuous on a closed interval has a greatest and a least value on that interval.

We are ready now to list the categories of critical values (values of the variable which *may* lead to extreme values of the function). They are

I Roots of the equation $f'(x) = 0$.
II Values of x for which $f'(x)$ does not exist.
III End points of the interval of definition of f.

The first step in the determination of extreme values of a function is to list the critical values. The second step is to check each one to see whether it produces a maximum, a minimum, or neither. The final step is to calculate the extreme values. We have already dealt with the first step and the final one is obvious. Then there remains the problem of the second step. We shall indicate two methods of approach to this problem.

Method 1 (*Sign of the first derivative*). This method has been used repeatedly in the preceding illustrative examples and we shall merely restate the basic results.

(a) For an interior critical value x_0:
    (1) Relative maximum if $f'(x) > 0$, $x < x_0$; $f'(x) < 0$, $x > x_0$.
    (2) Relative minimum if $f'(x) < 0$, $x < x_0$; $f'(x) > 0$, $x > x_0$.
    (3) Neither if $f'(x)$ has the same sign for $x < x_0$ and $x > x_0$.
(b) For an end-point critical value x_0:
    (1) Relative maximum for x_0, a right end point, if there is an interval $x_1 < x < x_0$ over which $f'(x) > 0$.
    (2) Relative minimum for x_0, a right end point, if there is an interval $x_1 < x < x_0$ over which $f'(x) < 0$.

For x_0 a left end point, the preceding rules have the intervals and the signs on $f'(x)$ reversed.

Method 2 (*Sign of the second derivative*). The sign of the second derivative may be used effectively in some instances for determining the nature of a critical value. Obviously, a second derivative test cannot be used for critical values of type II. Also, this method is not well suited for the study of end-point critical values. Therefore we shall confine our remarks to interior critical values x_0 for which $f'(x_0) = 0$.

Let x_0 be a critical value of type I. Then the graph of f has a horizontal tangent at $(x_0, f(x_0))$. If the graph of f is concave downward at this point, it is clear from Definition 6-1 that $f(x_0)$ satisfies the requirements of a relative maximum. Similarly, if the graph of f is concave upward at $(x_0, f(x_0))$, $f(x_0)$ is a relative minimum. If we combine these facts with Theorem 6-4, we obtain the following theorem.

THEOREM 6–7. Let x_0 be an interior point of the interval of definition of f, such that $f'(x_0) = 0$ and let $f''(x_0)$ exist. Then,
    (a) $f(x_0)$ is a relative maximum if $f''(x_0) < 0$;
    (b) $f(x_0)$ is a relative minimum if $f''(x_0) > 0$.

This theorem has limited application. For example, it offers no information if $f''(x_0) = 0$. Also, it may be that $f''(x_0)$ is so troublesome to calculate as to make the method impractical. But for those cases in which the second derivative is reasonable to calculate and different from zero, Theorem 6-7 provides a convenient test for interior critical values of type I.

Exercises 6-3

In Exercises 1–8 find the maximum and minimum points and sketch the curves.

1. $y = x^3 - 2$ **2.** $y = x^4 - 8x^2$

3. $y = 2x^3 - 3x^2 - 12x + 2$ **4.** $y = 2x^3 - 3x^2 - 36x + 20$

5. $y = x^3 - 3x + 1$ **6.** $y = x^4 - 6x^2 + 8x + 2$

In Exercises 7–10 determine any maxima and minima that the functions defined by the following equations may have. Also determine any points of inflection their graphs may have.

7. $f(x) = \dfrac{1 + x^2}{x}$ **8.** $f(x) = \dfrac{6x}{x^2 + 1}$

9. $f(x) = x^2 + \dfrac{1}{x^2}$ **10.** $f(x) = \dfrac{1}{5} x^{5/3} - 3x$

11. Find the maximum and minimum points on the curve

$$y = (2x + 1)(x - 2)^{2/3}$$

    and sketch it.

12. Find the extremes of the function defined by $f(x) = (x - 1)^{2/3}(x + 2)^2$ on the interval $-3 \leqslant x \leqslant 2$ and identify them.

13. Show that the parabola $y = ax^2 + bx + c$, $a \neq 0$, always has an absolute maximum or an absolute minimum point.

14. What conditions must be imposed on the coefficients of the cubic $ax^3 + bx^2 + cx + d$, $a \neq 0$, if it is to have one relative maximum and one relative minimum?

15. Prove the "relative minimum" case of Theorem 6-6.

6–5. Applications of Maxima and Minima

The methods developed in the preceding section for determining extreme values of a function find many important uses in applied fields. For example, it is important to a manufacturer to use the minimum amount of material in the construction of a particular type of container with a specified capacity. Or again, it is important to the owner of a public carrier to know the speed at which to operate it in order to obtain the maximum economy of operation. The numbers and nature of such problems are endless and the study of their solutions constitutes a major field in mathematics. We shall consider some of the simpler ones.

The principal distinctions between the problems encountered here and those of Sec. 6-4 are:

(a) It will be necessary to set up the function for which an extreme is to be determined.

(b) The domain of the function will often be restricted by the physical circumstances of the problem.

(c) Generally we shall be seeking an absolute maximum or an absolute minimum.

The general procedure for solving applied problems in maxima and minima may be outlined as follows:

Step 1. Determine by careful reading of the problem the quantity to be made an extreme.

Step 2. Express this quantity in terms of the other quantities in the problem. Sometimes it will be feasible to do this explicitly; other times it will be more convenient to obtain an implicit relationship.

Step 3. Use the special relations in the problem to express the function of Step 2 as a function of a single independent variable.† Be particularly careful to note the limitations imposed on the independent variable by the physical aspects of the problem.

Step 4. Determine the critical values of the independent variable by the methods of Sec. 6-4. Do not neglect the end points.

Step 5. Examine these critical values for one which will produce the required extreme.

It will prove desirable to modify this procedure somewhat in certain situations, particularly in the case where Step 2 leads to an implicit formula not readily convertible to explicit form. We shall illustrate this modification by means of examples.

Example 6–12. A rectangular plot 5000 sq ft in area is to be laid out so that a long straight wall serves as one side of it. What dimensions should be used so that the amount of fencing needed to enclose the remaining sides is a minimum?

Referring to the steps just outlined, the solution of this problem may be accomplished in the following manner.

---

† We shall consider only problems for which this is possible. Problems for which this is not possible require more complicated procedures.

Let the amount of fencing needed (the quantity to be minimized) be designated by L. Then, from Fig. 6-15,

$$L = 2x + y$$

and

(a) $$xy = 5000.$$

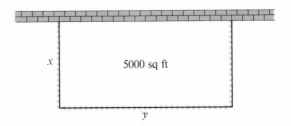

Fig. 6-15

Making use of this second relation, we may write

(b) $$L = 2x + \frac{5000}{x}.$$

If we ignore the physical character of L, we note that it is defined for all $x \neq 0$. However, the physical implications require that x be positive.

To determine the critical values of x we differentiate (b) and obtain

(c) $$\frac{dL}{dx} = 2 - \frac{5000}{x^2},$$

which we set equal to zero. We have

$$2 - \frac{5000}{x^2} = 0 \qquad \text{or} \qquad x = \pm 50.$$

We reject $x = -50$ for reasons already explained. From III, $x = 0$ would also be a critical value except that it is not within the domain of definition of L. Since the domain of L is $0 < x < \infty$, we do not have any end points to consider. Therefore the only admissible critical value is $x = 50$. Does this yield a minimum value of L? Either of our methods for testing critical values readily gives an affirmative answer to this question.

By Method 1:

$$\frac{dL}{dx} < 0, \quad x < 50; \qquad \frac{dL}{dx} > 0, \quad x > 50;$$

and the conclusion follows.

By Method 2:

$$\frac{d^2L}{dx^2} = \frac{10,000}{x^3},$$

so that

$$\frac{d^2L}{dx^2}\bigg]_{x=50} = \frac{10,000}{125,000} > 0,$$

and the conclusion that $x = 50$ yields a minimum follows from Theorem 6-7.

Thus $x = 50$ ft is one dimension and, from the equation (a), the other dimension is $y = 100$ ft.

Example 6–13. Determine the ratio of the radius to the height of a closed right-circular cylindrical container of capacity C cubic units if the surface area is to be a minimum.

Let r and h represent the radius and height, respectively, and let S represent the surface area. Then   lateral
surface

(a) $$S = 2\pi rh + 2\pi r^2, \qquad r > 0, h > 0,$$

and

(b) $$C = \pi r^2 h.$$

Thus we may write

$$S = \frac{2C}{r} + 2\pi r^2, \qquad \text{and} \qquad \frac{dS}{dr} = -\frac{2C}{r^2} + 4\pi r.$$

We set this equal to zero and obtain

$$-\frac{2C}{r^2} + 4\pi r = 0 \qquad \text{or} \qquad r = \left(\frac{C}{2\pi}\right)^{1/3},$$

the only critical value, since there are no values $r > 0$ for which dS/dr is undefined and the end points do not have to be considered because $0 < r < \infty$. This lone critical value gives a minimum because

$$\frac{d^2S}{dr^2}\bigg]_{r=(C/2\pi)^{1/3}} = \frac{4C}{r^3} + 4\pi\bigg]_{r=(C/2\pi)^{1/3}} > 0.$$

Then from (b),

$$h = \frac{C}{\pi r^2}\bigg]_{r=(C/2\pi)^{1/3}} = \left(\frac{4C}{\pi}\right)^{1/3},$$

and the required ratio is $r/h = 1/2$. Therefore the surface area of a closed cylindrical container of fixed volume will be a minimum when the height is equal to the diameter.

Now let us look at a modification of this method which will eliminate some of the unpleasant algebra.

Equation (b) defines h as a function of r implicitly and therefore the combination of (a) and (b) defines S as a function of r implicitly. Now we differentiate (a) implicitly with respect to r and obtain

(c) $$\frac{dS}{dr} = 2\pi\left(h + r\frac{dh}{dr} + 2r\right).$$

Next we calculate dh/dr from (b) by differentiating it implicitly with respect to r and get

$$0 = \pi\left(2rh + r^2\frac{dh}{dr}\right), \quad \text{or} \quad \frac{dh}{dr} = -\frac{2h}{r}.$$

We substitute this in (c) and set the result equal to zero. We have

$$\frac{dS}{dr} = 2\pi\left[h + r\left(\frac{-2h}{r}\right) + 2r\right] = 0,$$

or

$$2r - h = 0, \quad \text{or} \quad \frac{r}{h} = \frac{1}{2},$$

the result obtained by the previous method.

To demonstrate that this produces a minimum surface area, we could solve this equation with (b) and proceed as before. However, this loses most of the ground gained by this modified method. In many cases, as in this one, the following line of reasoning will answer the question.

The surface area S can be made as large as we please by choosing r sufficiently close to zero or sufficiently large. Since S is continuous and possesses a derivative at all points between these extremes, it must attain a minimum value where $dS/dr = 0$. This follows from Theorems 4-11 and 6-6. Since this occurs only when $r/h = 1/2$, the conclusion follows.

The method of implicit differentiation will often greatly simplify the work involved. Experience will teach the student when it is likely to be helpful.

Example 6–14. It is required to lay a cable from a point A on one bank of a river 1000 ft wide to a point B on the opposite bank 1200 ft downstream from the point C directly opposite A. (Assume that the river banks are parallel). If it costs $4.00 per foot to lay cable on land and $5.00 per foot to lay it in water, what route should be followed to minimize the cost?

Clearly the route to be taken is from A along a straight line to some point P on CB (Fig. 6-16) and then from P to B. If we call $CP = x$ we can write the formula for the cost of laying the cable:

$$K(x) = 5\sqrt{1000^2 + x^2} + 4(1200 - x)$$

where $0 \leqslant x \leqslant 1200$. Then

$$\frac{dK(x)}{dx} = 5x(1000^2 + x^2)^{-1/2} - 4$$

and when we set this equal to zero and solve for x, we obtain

$$x = \pm \frac{4000}{3} \cong \pm 1333.$$

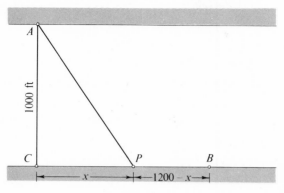

Fig. 6-16

But neither of these values of x fall within the interval $0 \leqslant x \leqslant 1200$. Moreover, $K(x)$ is continuous and differentiable in the interval. Consequently it assumes neither a maximum nor a minimum at any interior point. Therefore both maximum and minimum values of $K(x)$ occur at the end points. We examine $K(x)$ at these points and find

$$K(0) = \$9800, \qquad K(1200) \cong \$7810.$$

Therefore the maximum occurs when $x = 0$ and the minimum when $x = 1200$, and thus the most economical route is to go directly from A to B, laying all the cable in the water.

As a final illustrative example let us consider a problem in which the independent variable may assume integral values only. Many such cases arise, for example, in the field of economics.

Example 6-15. A dealer finds that if he prices an article at $\sqrt{300 - x^2}$ dollars, he can sell x articles per week. What value of x will bring the greatest total revenue per week?

The total weekly revenue may be expressed by the formula

$$R(x) = x(300 - x^2)^{1/2}$$

where, clearly, x takes on integral values in the interval $0 \leqslant x \leqslant \sqrt{300}$.

Strictly speaking, we cannot find the maximum value of this function by the methods just developed because the calculation of derivatives requires a continuous variable. However, if we consider x to be continuous, we can determine the maximum point on the graph of R, and all values of $R(x)$ for integral values of x will lie on this curve. Therefore we have but to examine

$R(x)$ for integral values of x in the neighborhood of the maximum point in order to determine the required value of x.

Then, considering x as a continuous variable, we have

$$R'(x) = (300 - x^2)^{1/2} - x^2(300 - x^2)^{-1/2},$$

which, when set equal to zero, gives the critical values

$$x = \pm\sqrt{150} \cong \pm 12.25.$$

Thus, taking into account the interval of definition and the integral character of x, we have

$$x = 0, 12, 13, 17$$

as the total list of possibilities which might give a maximum value of $R(x)$. Calculating $R(x)$ for each of these, we obtain

$$R(0) = 0; \quad R(12) = 149.9; \quad R(13) = 148.7; \quad R(17) = 56.1.$$

Therefore the greatest total revenue will result from selling 12 articles per week.

Exercises 6-4

1. Find two numbers whose sum is 14 and whose product is a maximum.

2. Find two numbers whose sum is 32 and the sum of whose squares is a minimum.

3. Find two numbers whose sum is 15 and the product of one by the square of the other is a maximum.

4. A closed rectangular box twice as long as it is wide is to have 192 sq in. of surface area including the top. Find its dimensions if the volume is a maximum.

5. A company estimates that its profit P is related to its advertizing costs x by $P = 100 + 108x - x^3$, where x and P are measured in units of 1000. Find the advertizing expenditure which will produce the maximum profit.

6. A rectangular plot is to be laid out so that a long straight wall serves as one side of it. One thousand yards of fencing are available to enclose the remaining three sides. What is the greatest area that can be enclosed?

7. Prove that the rectangle with a fixed perimeter having the greatest area is a square.

8. An open-top box with a square base has a total surface area of 75 sq ft. Find the dimensions that will make its volume a maximum.

9. A piece of sheet aluminum 18 in. square is to be made into an open-top box by cutting squares out of the corners and turning up the sides. What is the greatest volume that can be obtained?

10. Solve Exercise 4 with the added condition that the box must be stored in a space 10 in. square.

11. A rectangular swimming pool is to be surrounded by a concrete apron 4 ft wide along two parallel sides and 3 ft wide along the other two sides. If the swimming pool is to have an area of 600 sq ft, what should be its dimensions so that the least amount of concrete is used in building the apron?

12. Ship B is 150 miles due east of ship A. If B sails west at 12 mph and A sails south at 9 mph, when will the two ships be nearest each other?†

13. Find the dimensions of a rectangle of constant area 32 sq in. so that the distance from one corner to the midpoint of a nonadjacent side will be a minimum.

14. A man in a row boat is 4 miles from the nearest point A of a straight shore line. He wishes to reach point B, 5 miles from A, on the shoreline in the shortest possible time. If he can row 3 mph and walk 5 mph, how far should he land from A?

15. An open-top box with a square base, of volume 54 cu ft, is to be made. The material for the sides and bottom costs $10 and $5/sq ft, respectively. What dimensions should be used for a minimum cost?

16. A V-shaped trough of maximum capacity is to be made from a piece of sheet metal 24 in. wide by bending it down the middle. How wide should the trough be at the top?

17. On the parabola $y^2 = 2x$, find the point which is nearest the point $(0, 6)$.

18. Find the dimensions of the rectangle of perimeter 36 in. that will sweep out the greatest volume when rotated about one side.

19. Solve the problem in Example 6-12 if one end of the container is omitted.

20. Find the minimum first quadrant area bounded by the coordinate axes and a line through the point $(8, 1)$.

21. Find the length of the shortest ladder that will reach over an 8 ft fence to a house 1 ft beyond the fence.

22. A wire L feet long is to be cut into two pieces, one of which is to be bent into a circle and the other into a square. Where should it be cut if the combined area is to be a minimum? A maximum?

23. Two towns A and B are to be supplied with water from a common pumping station located on the bank of a straight river. A and B are 2 and 3 miles, respectively, from the nearest points C and D on the river. If C and D are 10 miles apart, where should the pumping station be located to make the total length of pipelines a minimum?

---

† Note that a positive quantity S may be maximized or minimized by maximizing or minimizing S^2. This device will often simplify the computation in a problem.

24. The cost of fuel for running a river steamer at V mph in still water is $V^3/32$ dollars per hour. Other costs per hour total \$160, independent of speed. Find the most economical speed (in still water) for a trip against a 4 mph current.

25. A printer contracts to print 2000 handbills at \$1.00 per hundred. If the number ordered exceeds 2000, he agrees to reduce the price of the whole lot by 2 cents per hundred for each 100 in excess of 2000. For what number of handbills would his receipts be the greatest?

26. A transportation company agrees to transport a minimum of 200 people for \$10 each. For each 100 tickets sold in excess of 200, a reduction of \$0.50 per ticket will be allowed for everyone. The capacity of the train is 800 people. What number of tickets will produce the greatest revenue for the company?

27. A distributor sells a lot of x articles at a price of $(700/(x + 2)) - (5)$ dollars each. Find the number of articles in a lot that will produce the maximum revenue for him.

28. A manufacturer finds that if he makes x articles per day, the costs involved are: (a) fixed costs, \$100 per day; (b) materials and labor, \$1.06 per article; (c) maintenance and replacement, $0.0002x^2$ dollars per day. Find the number he should produce per day to keep the cost per article at a minimum.

29. The costs involved in erecting an office building consist of: \$500,000 site purchase and development; \$75,000 construction cost for the first floor; \$100,000 for the second floor; \$125,000 for the third floor; and so on. Annual rental is \$15,000 per floor. Find the number of floors that will maximize the ratio of possible income to total investment.

30. To produce x tons of a material costs $50 + 0.002x$ dollars per ton and the number of tons which can be sold per year is $8000 - 50y$, where y is the selling price per ton. What price per ton will give the maximum profit?

31. A plant manager observes that a work force of 40 individuals produces an average of 150 articles daily per worker. He also notes that the average productivity drops 0.5 percent for each worker added above this number. How many people should he employ to obtain maximum daily production?

32. The probability $P(n, k)$ of getting exactly k successes in n trials, $0 \leqslant k \leqslant n$, is given by

$$P(n, k) = \frac{n!}{k!(n - k)!} \, p^k(1 - p)^{n-k},$$

where p is the probability of success in each trial. Find the value of p which maximizes $P(n, k)$.

6-6. Related Rates

The interpretation of the derivative as a rate of change leads to many useful applications. One is that of *related rates*.

If we have an equation expressing a relationship among a number of variables, each one of which is a function of a variable t (usually time), then the equation expresses an identity in t and we may differentiate with respect to t and obtain a relationship involving these variables and their derivatives with respect to t (rates of change with respect to t). Thus we have *related rates*.

Example 6-16. Ship A leaves a port at noon and sails west at 9 mph. Ship B leaves the same port at 1 P.M. and sails south at 12 mph. At what rate are the two ships separating at 3 P.M.?

Referring to Fig. 6-17, we see that we have the three variables s, x, y related by the equation

(a) $$s^2 = x^2 + y^2..$$

The question raised by this problem is simply: what is ds/dt when $x = 27$ and $y = 24$? The additional facts we have at our disposal are

$$\frac{dx}{dt} = 9, \qquad \frac{dy}{dt} = 12,$$

since these are the known speeds of ships A and B, respectively.

We can proceed in either of two ways at this point. First we arbitrarily choose a time origin, say $t = 0$, when the first ship leaves the port. Then

$$x = 9t, \qquad y = 12t - 12$$

and (a) takes the form

(b) $$s(t) = \sqrt{81t^2 + 144(t - 1)^2}.$$

Now we need to know the value of ds/dt when $t = 3$. We have, differentiating (b),

$$\frac{ds}{dt} = \frac{81t + 144(t - 1)}{\sqrt{81t^2 + 144(t - 1)^2}}.$$

Thus

$$\frac{ds}{dt}\bigg]_{t=3} = \frac{177\sqrt{145}}{145} \cong 14.7 \text{ mph.}$$

Another way in which we may proceed is to differentiate (a) with respect to t implicitly and obtain

$$2s\frac{ds}{dt} = 2x\frac{dx}{dt} + 2y\frac{dy}{dt}.$$

The specific value needed for s may be computed directly from Fig. 6-17. We have

$$s = \sqrt{(27)^2 + (24)^2} = 3\sqrt{145}.$$

Then, first dividing by 2,

$$3\sqrt{145}\,\frac{ds}{dt} = 27(9) + 24(12)$$

or

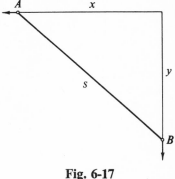

$$\frac{ds}{dt} = \frac{177}{\sqrt{145}} \cong 14.7 \text{ mph.}$$

Fig. 6-17

An important fact to note is that *particular values were substituted for the variables only after the differentiation had been performed.*

Example 6–17. An inverted conical container with a base radius of 4 in. and a height of 10 in. is losing water at the constant rate of 3 cu in. per minute. At what rate is the surface of the water dropping at the instant the water is 6 in. deep?

If we denote the volume and depth of water by v and h, respectively, we are given $dv/dt = -3$. We are required to find dh/dt when $h = 6$. Note that dv/dt is given a negative sign since v is a decreasing quantity with increasing time.

The volume of water at any instant is given by (Fig. 6-18)

$$v = \frac{\pi}{3}\, r^2 h,$$

and from similar triangles,

$$\frac{r}{h} = \frac{2}{5}.$$

Then the relationship between volume and depth of water may be written

$$v = \frac{4\pi h^3}{75}.$$

We differentiate this with respect to t and obtain

$$\frac{dv}{dt} = \frac{4\pi h^2}{25}\frac{dh}{dt},$$

or substituting the numerical values of h and dv/dt and solving for dh/dt,

$$\frac{dh}{dt} = -\frac{25}{48\pi} \cong -0.17 \text{ in./min.}$$

The negative sign of this result is compatible with the fact that h is a decreasing function of time.

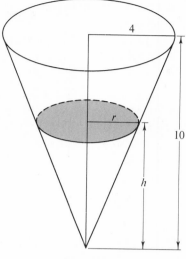

Fig. 6-18

Exercises 6-5

1. A stone is dropped in still water. The radius of the concentric ripples that result is increasing at the rate of 2 ft/sec. At what rate is the area of the outer circle increasing when the radius is 6 ft?

2. The edges of a cube are increasing at the rate of 2 in./min. At what rate is the volume increasing when an edge is 12 in. long?

3. At what rate is the surface area of the cube in the preceding problem increasing?

4. A boy is flying a kite at a vertical height of 75 ft. If the wind is blowing the kite in a horizontal direction at 8 ft/sec, at what rate is the boy releasing string when there are 125 ft of string out?

5. The area of a rectangle is increasing at the rate of 20 sq ft/min. At a particular instant, one side is 8 ft and increasing at the rate of 4 ft/min and the other side is 9 ft. How fast is the second side changing at that instant?

6. A ladder 15 ft long rests against the side of a house. The bottom of the ladder is sliding away from the house at the rate of 3 ft/sec. At what rate is the top of the ladder sliding down the house at the instant the bottom is 9 ft from the house?

7. The side of an equilateral triangle is increasing at the rate of 2 in./min. How fast is the area increasing at the instant the side is 20 in.?

8. A perfect gas confined within a container and kept at a constant temperature obeys the law $pv = $ constant, where p is the pressure and v is the volume. If at a particular instant the pressure is 30 lb/sq in. and increasing at the rate of 3 lb/sq in./min and the volume is 50 cu in., what is the rate of change in the volume at that time?

9. A man is approaching a tower 60 ft tall, standing on level ground, at the rate of 5 mph. At what rate is he approaching the top of the tower when he is 25 ft from the base? (Disregard the height of the man.)

10. A balloon filled with air is punctured in such a way that air escapes at the rate of 10 cu ft/min. Assuming that the balloon retains a spherical shape at all times, how fast is the diameter decreasing when it is 5 ft?

11. The volume of a sphere is increasing at a rate of cubic inches per minute equal to the numerical value of one-half its surface area. At what rate is the radius increasing?

12. The circumference of the great circle of a sphere is decreasing at the rate of 3 in./min. What is the rate of decrease of the volume at the instant the radius is 10 in?

13. A trough 20 ft long has as its cross section an isosceles triangle with its base uppermost. The base is 2 ft and the altitude is 1 ft. If a drain carries water away at such a rate that the depth of water decreases 2 in./min, what is the volume of water lost per min at the instant the water is 6 in. deep?

14. A man 6 ft tall is moving directly away from a light pole at the speed of 4 ft/sec. If the light is 24 ft above the ground, at what rate is his shadow lengthening? At what speed is the tip of his shadow moving along the ground?

15. In Exercise 14, how fast is the tip of the shadow moving away from the light at the instant the shadow is $4\frac{1}{2}$ ft long?

16. A man 6 ft tall is walking directly away from a spotlight on the ground toward a wall 50 ft distant from the light. If the man is walking at the rate of 240 ft/min, at what rate is his shadow on the wall shortening when he is 20 ft from the light?

17. In Example 6-17, how fast is the surface area of the water changing?

18. Car A is approaching an intersection at 60 mph and car B is approaching the same intersection on a road perpendicular to the first at 45 mph. If, when first observed, cars A and B were 1 mile and 3/4 mile, respectively, from the intersection, at what rate are they approaching each other when car A is $\frac{1}{2}$ mile from the intersection?

19. A man walks across a bridge as a barge passes directly under him. The barge is moving 3 ft/sec, the man 4 ft/sec, and the bridge is 24 ft above the stream. At what rate are the two separating at the end of 2 sec? Does this rate approach a limit as $t \to \infty$, and if so, what is it?

20. A man 6 ft tall walks at a speed of 5 ft/sec along a straight level path which at its closest point passes at a distance of 40 ft from a light on a 12-ft post. Find the speed at which his shadow is shortening when he is 30 ft short of the point on the path nearest the light.

6–7. The Differential

Consider a differentiable function f and let $y = f(x)$. Then

$$\lim_{\Delta x \to 0} \frac{\Delta y}{\Delta x} = f'(x),$$

from which we obtain

$$\frac{\Delta y}{\Delta x} = f'(x) + \eta,$$

where

$$\lim_{\Delta x \to 0} \eta = 0.$$

Thus an increment Δy produced by an increment Δx of the independent variable may be expressed in the form

$$\Delta y = f'(x)\, \Delta x + \eta\, \Delta x, \qquad \lim_{\Delta x \to 0} \eta = 0. \tag{6-5}$$

It is clear that $\eta\, \Delta x$ is a small quantity when Δx is small, and therefore, for such values of Δx, $f'(x)\, \Delta x$ is an approximation for Δy. Let us see what we can say about this approximation. We have

$$\lim_{\Delta x \to 0} \frac{\Delta y - f'(x)\, \Delta x}{\Delta x} = \lim_{\Delta x \to 0} \frac{\Delta y}{\Delta x} - f'(x) = f'(x) - f'(x) = 0.$$

That is,

$$\Delta y - f'(x)\, \Delta x \to 0$$

faster than $\Delta x \to 0$. Hence, for small Δx, $f'(x)\, \Delta x$ is indeed a good approximation for Δy.

These considerations, and others to be mentioned later, lead us to define a new function, which we call the *differential*.

DEFINITION 6–5. Let f be a differentiable function of x, and let $y = f(x)$. Then

(a) The differential dx of the independent variable x is defined by $dx = \Delta x$, where Δx is any real number whatever.

(b) The differential dy of the dependent variable y is defined by $dy = f'(x)\, dx$.

We use dy and df interchangeably.

Note first of all that $dx \equiv \Delta x$ is completely arbitrary and is therefore an independent variable whose range is the set of all real numbers. It is in no way dependent on x or the domain of f. If we had so desired, we could have used a completely unrelated symbol for it. However, the context in which it will be used is primarily that of an arbitrary increment of x and we have chosen the notation accordingly.

Secondly, since x and Δx are independent of each other and since there is a unique dy corresponding to each number pair $(x, \Delta x)$, we say *dy is a function of the two independent variables x and Δx*. We have not defined functions of two or more independent variables and have no intention of doing so at this point. The preceding statement is adequate for our purposes here. Such matters will be discussed in more detail in Chapter 13.

We have used the "approximation theme" as the basis of our definition of dy because this makes differentials immediately useful. In many cases it is tedious or even downright difficult to calculate Δy, whereas it may be relatively simple to determine dy.

A second and also important result of this definition is the fact that the ratio of dy to dx is $f'(x)$. This is our excuse for having introduced the notation

$$\frac{dy}{dx} = f'(x).$$

Thus, when we use the symbol dy/dx, we may think of it as the number $f'(x)$ resulting from

$$\lim_{\Delta x \to 0} \frac{\Delta y}{\Delta x} = \lim_{\Delta x \to 0} \frac{f(x + \Delta x) - f(x)}{\Delta x},$$

or we may think of it as an ordinary fraction whose numerator and denominator are dy and dx, respectively, and handle it accordingly. This latter interpretation proves useful in numerous situations.

Example 6–18. Let f be defined by $f(x) = 2x^2 + x - 1$, and let $y = f(x)$. Compare Δy and dy for (a) $x = 2$, $\Delta x = -1$; (b) $x = 2$, $\Delta x = 1/4$.

Since

$$f'(x) = 4x + 1,$$

we have, applying Definition 6-5,

$$dy = (4x + 1)\, dx.$$

Also,

$$\Delta y = [2(x + \Delta x)^2 + (x + \Delta x) - 1] - [2x^2 + x - 1].$$

(a) Setting $x = 2$ and $\Delta x = dx = -1$,

$$dy = [4(2) + 1](-1) = -9;$$
$$\Delta y = [2(2 - 1)^2 + (2 - 1) - 1] - [2(2)^2 + 2 - 1] = -7;$$
$$|\Delta y - dy| = 2.$$

(b) Setting $x = 2$ and $\Delta x = dx = 1/4$,

$$dy = [4(2) + 1]\frac{1}{4} = \frac{9}{4};$$

$$\Delta y = \left[2\left(2 + \frac{1}{4}\right)^2 + \left(2 + \frac{1}{4}\right) - 1\right] - [2(2)^2 + 2 - 1] = \frac{19}{8};$$

$$|\Delta y - dy| = \frac{1}{8}.$$

Observe that even in this simple case dy is easier to calculate than Δy. This is usually true. Note also how much better dy approximates Δy when $|\Delta x|$ is reduced from 1 to 1/4, x being held fixed at 2.

Let us look at the geometrical interpretations to be placed on differentials. Let $P(x, y)$, $Q(x + \Delta x, y + \Delta y)$ be two points on the graph (Fig. 6-19) of $y = f(x)$. Then the slope of the tangent at P is given by $f'(x) = dy/dx$. But the slope of this tangent is also given by the ratio RT/PR, and since $PR \equiv \Delta x \equiv dx$, it follows that $RT = dy$. The difference between Δy and dy is represented geometrically by the segment TQ.

If y is the dependent variable and x is the independent variable, the differentials of these variables have been defined so that their ratio is $f'(x)$. However, it may be that

$$y = f(x), \qquad x = h(t),$$

so that the independent variable is not x but t. Then both x and y are

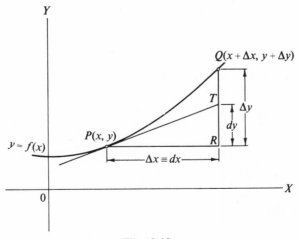

Fig. 6-19

dependent variables and dx and dy must be calculated accordingly. We have

$$y = f(h(t)), \qquad x = h(t),$$

so that

$$dx = h'(t)\, dt,$$

and using the chain rule to calculate $f'(h(t))$,

$$dy = f'(x)h'(t)\, dt.$$

Substituting from above, this reduces to

$$dy = f'(x)\, dx,$$

which is the same relation we had when x was an independent variable.
   This result is sufficiently important that we state it as a theorem.

THEOREM 6–8. If $y = f(x)$, then

$$dy = f'(x)\, dx$$

whether or not x is an independent variable.

   As long as we know y is a function of x, we can now freely manipulate the *fraction dy/dx* without bothering about the question of what the independent variable is.
   We could express all differentiation formulas of Chapter 5 in differential notation. For example, the formula

$$\frac{d}{dx}\left(\frac{u}{v}\right) = \frac{v\,\dfrac{du}{dx} - u\,\dfrac{dv}{dx}}{v^2}$$

becomes, in differential form,

$$d\left(\frac{u}{v}\right) = \frac{v\, du - u\, dv}{v^2}.$$

The transformation of other differentiation formulas into differential notation is left as an exercise for the student.

Exercises 6-6

In each of Exercises 1–6 find the general expression for dy.

1. $y = \sqrt{9 - x^2}$ **2.** $y = x\sqrt{1 + x^2}$

3. $y = \dfrac{x+1}{x^2-2}$ **4.** $y = \dfrac{\sqrt{x^2+4}}{x^2-1}$

5. $x^3 + xy + y^3 = 5$ **6.** $x^{2/3} + y^{2/3} = a^{2/3}$

7. Given f defined by $f(x) = x^2 + 2x + 2$. Find df for (a) $x = -2$, $\Delta x = 0.5$; (b) $x = 0$, $\Delta x = 0.5$; (c) $x = 2$, $\Delta x = -0.25$. Draw the graph of f and construct the tangent at each of the points where $x = -2, 0, 2$. Then show graphically dx, df, Δf for (a), (b), and (c).

8. Do the same as in Exercise 7 for f defined by $f(x) = 2/(x-1)$ and (a) $x = -1$, $\Delta x = 0.5$; (b) $x = 0$, $\Delta x = -0.5$; (c) $x = 1$, $\Delta x = 0.75$.

In each of Exercises 9–12 calculate the value of $dy - \Delta y$ to four decimal places.

9. $y = x^2 + 3x + 2$, $x = 2$, $\Delta x = 0.2$

10. $y = 2x^3 + x - 1$, $x = -2$, $\Delta x = 0.2$

11. $y = \dfrac{x+1}{x-2}$, $x = 3$, $\Delta x = -0.1$

12. $y = \dfrac{3}{x^2} + \dfrac{1}{x} - 1$, $x = 1$, $\Delta x = -0.03$

In each of Exercises 13–16 write the formula for the indicated differential, where u and v are differentiable functions of x.

13. $d(u+v)$ **14.** $d(u-v)$ **15.** $d(u^n)$ **16.** $d(uv)$

6-8. Applications of the Differential

Approximate solutions to many types of problems are obtainable by means of differentials. The following examples will illustrate some of these and the student will readily see how similar methods and interpretations will extend to others.

Example 6–19. Find an approximate value for $\sqrt[3]{28}$.

If we write $y = x^{1/3}$, we see that $y = 3$ when $x = 27$, so the problem reduces to finding the change produced in y by a change in x from 27 to 28. In other words, what is Δy when $x = 27$ and $\Delta x = 1$? We are in trouble when we try to calculate Δy, but the determination of dy is very simple. We have

$$dy = \frac{1}{3} x^{-2/3}\, dx,$$

and for $x = 27$, $\Delta x = 1$,

$$dy = \frac{1}{3}(27)^{-2/3}(1) = \frac{1}{27} \cong 0.037.$$

Therefore $\sqrt[3]{28} \cong 3 + 0.037 = 3.037$. A seven-figure table of cube roots gives $\sqrt[3]{28} = 3.036589$ which, when rounded off to three decimals, agrees with our approximation.

Example 6–20. The gas in a spherical balloon contracts, causing a reduction in the radius from 6 ft to 5.85 ft. What is the approximate change in the surface area of the balloon?

Let s represent the surface area. Then $s = 4\pi r^2$ and

$$\Delta s \cong ds = 8\pi r \, dr.$$

We have $r = 6$, $dr = -0.15$, and

$$\Delta s \cong 8\pi(6)(-0.15) = -7.2\pi \quad \text{sq ft.}$$

Example 6–21. Find an approximation for the volume of a cylindrical shell of height h, inner radius r, and wall thickness t.

We approach this problem from the standpoint of increasing the radius of a cylinder and finding the corresponding increase in the volume. For a cylinder

$$v = \pi r^2 h,$$

and

$$\Delta v \cong dv = 2\pi r h \, dr.$$

Now we take $dr = \Delta r = t$ and obtain the required approximation:

$$\Delta v \cong 2\pi r h t.$$

Example 6–22. The radius of the base and the altitude of a particular right-circular cone are equal and are measured as 10 in. with a possible error of 0.03 in. Find approximately the possible error in the calculated volume based on this measurement.

We have

$$v = \frac{\pi}{3} r^3$$

and

$$\Delta v \cong dv = \pi r^2 \, dr.$$

Now we take $r = 10$ and $dr = \pm 0.03$, and obtain

$$\Delta v \cong \pi(100)(\pm 0.03) = \pm 3\pi \quad \text{cu in.}$$

The magnitude of an error is more meaningful when compared to the magnitude of the quantity being measured. Hence we make the following definition.

DEFINITION 6–6. If Δv is the error in a quantity v, then $\Delta v/v$ is the relative error and $100(\Delta v/v)$ percent is the percentage error.

Applying this definition to the preceding example, the approximate relative error is given by

$$\frac{\Delta v}{v} \cong \frac{\pm 3\pi}{(\pi/3)(10)^3} = \pm 0.009,$$

and the approximate percentage error is

$$100\left(\frac{\Delta v}{v}\right)\% \cong (\pm 0.009)(100)\% = \pm 0.9\%.$$

Example 6-23. The radius of a sphere is found to be 12 in. with a possible error of 0.3 percent. (a) What is the approximate maximum possible error in the volume? (b) With what accuracy, approximately, must the radius be measured if the allowable error in the volume is 0.6 percent?

(a) We have

$$v = \frac{4}{3}\pi r^3,$$

whence

$$dv = 4\pi r^2 \, dr.$$

Since the possible error is 0.3 percent, we may write

$$\frac{dr}{r} = \frac{dr}{12} = \pm 0.003, \quad \text{or} \quad dr = \pm 0.036 \text{ in.}$$

Hence the approximate possible error in the volume is given by

$$dv = 4\pi(12)^2(\pm 0.036) \cong \pm 65 \text{ cu in.}$$

(b) If the allowable error in the volume is 0.6 percent, we have

$$\frac{dv}{v} = \frac{4\pi r^2 \, dr}{(4/3)\pi r^3} = 3\frac{dr}{r} = \pm 0.006.$$

Therefore the radius must be measured with an accuracy of

$$\frac{dr}{r} = \frac{1}{3}(\pm 0.006) = 0.2\%.$$

Exercises 6-7

Use differentials to compute an approximate value for each of Exercises 1–6.

1. $\sqrt{98}$ 2. $\sqrt[3]{123}$ 3. $\sqrt[4]{85}$ 4. $(220)^{4/3}$ 5. $\dfrac{1}{1005}$ 6. $(11.5)^{-3}$

7. Find the approximate value of $3x^4 - 7x^3 + 4x^2 - 5$ for $x = 1.98$.

8. Find the approximate value of $5x^5 - 3x^3 + 7$ for $x = 3.02$.

9. The area of a square is computed by using 12 ft as the length of a side. What is the approximate error, relative error, and percentage error in the computed area if the true length of a side is 11.98 ft?

10. Deposits in a bank follow the law $y = x^{3/2}$, where y is in millions of dollars and x is the number of years since the formation of the bank. Approximately, what will be the increase in deposits during the tenth year of the bank? Assuming that deposits follow this law exactly, what will be the exact amount of new deposits made during the tenth year?

11. The diameter of a circle is measured as 6 in. with a possible error of 0.02 in. Find the approximate possible error in the calculated area. What is the approximate relative error?

12. A closed cubical box of 6-ft inside dimensions is to be made from material 0.30 in. thick. Use differentials to find the approximate volume of material to be used.

13. The diameter of the base and altitude of a right-circular cylinder are the same and are measured as 6 in. with a possible error of 0.02 in. What is the approximate possible error in the area of the curved surface, the total area, and the volume?

14. A cylindrical tin can is to be made 4 in. in diameter and 4 in. high. What is the approximate relative error in the volume if an error of 0.4 percent is made in the dimensions?

15. Derive an approximate formula for the volume of a spherical shell of thickness t and internal radius r. How much error does your approximation introduce?

16. The edge of a cube is about 8 ft. Approximately how accurately must it be measured if the error in the volume is not to exceed 1 cu ft?

17. If the allowable error in the volume of a sphere is 0.9 percent, find approximately the allowable percentage error in the radius.

18. The cost of surfacing a hemispherical dome is $3.00/sq ft. Approximate the error in the estimated cost if an error of 6 in. is made in measuring the diameter as 120 ft. What is the approximate allowable error in measuring the diameter if the error in the cost estimate is not to exceed $75?

19. The intensity of illumination at a point is inversely proportional to the square of the distance of the light source from the point. If a light is 50 ft from a point, approximately how much nearer the point should it be placed to increase the illumination by 3 percent?

20. The period of a pendulum of length ℓ ft is given by

$$T = 2\pi \sqrt{\frac{\ell}{g}}$$

where T is in seconds and $g = 32.2$ ft/sec/sec. If an error of 0.03 ft is made in measuring ℓ as 4 ft, find the approximate relative error and percentage error in T as computed from this measurement.

21. If the period of the pendulum of a clock is supposed to be 2 sec, and if the clock loses 2 min/day, approximately how much should the pendulum be shortened to correct this error?

Chapter 7

THE INTEGRAL

7-1. Introduction

In Chapter 5 we introduced a limit which we called the derivative. As we have seen in the preceding pages, and will continue to see in those to come, the derivative provides us with a tool for solving many problems. In this chapter we shall introduce a second limit, which we shall call the definite integral, and which will also prove to be a powerful tool for solving certain types of problems. At first it will appear that there is no connection between these two limits—the derivative and the definite integral—but we shall soon see otherwise.

7-2. The Summation Notation

In order to simplify the writing of sums we define the summation, or sigma, notation.

DEFINITION 7-I. Let u_k be defined for $k = 1, 2, \ldots, n$. Then we write

$$u_1 + u_2 + u_3 + \cdots + u_n = \sum_{k=1}^{n} u_k.$$

The following examples illustrate this definition:

$$\sum_{k=1}^{5} \frac{1}{k} = \frac{1}{1} + \frac{1}{2} + \frac{1}{3} + \frac{1}{4} + \frac{1}{5};$$

$$\sum_{l=1}^{4} l(l + 2) = 1 \cdot 3 + 2 \cdot 4 + 3 \cdot 5 + 4 \cdot 6;$$

$$\sum_{m=2}^{n} \frac{m}{m^2 - 1} = \frac{2}{4 - 1} + \frac{3}{9 - 1} + \frac{4}{16 - 1} + \cdots + \frac{n}{n^2 - 1}.$$

Note that the summation variable may be changed at will, that is,

$$\sum_{k=1}^{5} \frac{1}{k} = \sum_{s=1}^{5} \frac{1}{s} = \sum_{p=1}^{5} \frac{1}{p} = \sum_{t=1}^{5} \frac{1}{t} = \cdots.$$

We shall interpret

$$\sum_{i=1}^{n} 1$$

to mean

$$\sum_{i=1}^{n} 1 = \underbrace{1 + 1 + 1 + 1 + \cdots + 1}_{n \text{ terms}} = n,$$

or more generally, if c is any constant,

$$\sum_{i=1}^{n} c = nc.$$

The following sums may be established by algebraic procedures. We present them here without proof for our future convenience.

$$\sum_{k=1}^{n} k = 1 + 2 + 3 + \cdots + n = \frac{n(n+1)}{2}. \qquad (7\text{-}1)$$

$$\sum_{k=1}^{n} k^2 = 1^2 + 2^2 + 3^2 + \cdots + n^2 = \frac{n(n+1)(2n+1)}{6}. \qquad (7\text{-}2)$$

$$\sum_{k=1}^{n} k^3 = 1^3 + 2^3 + 3^3 + \cdots + n^3 = \left[\frac{n(n+1)}{2} \right]^2. \qquad (7\text{-}3)$$

7–3. An Area Problem

In what follows we shall assume the *area* of rectangles and triangles to be that defined in plane geometry, that is, the area of a rectangle is the product of the length by the width and that of a triangle is one-half the product of the base by the altitude. Thus any plane figure that can be subdivided into rectangles and triangles may have its area defined by the sum of the areas of these subdivisions. But when such a subdivision is not possible (for example, when one boundary is not a straight line) such a definition of area is not available.

Let us see how we might give a different definition of the area of a plane figure which would be consistent with the one just given for special figures. In order that we may test this consistency, we shall consider first a plane area that can be subdivided into rectangles and triangles. Let us consider the plane area bounded by the lines $y = 0$, $y = x$, $x = 1$, $x = 2$. The student will readily determine that the area of this figure, as defined above, is 1.5 square units.

Let us subdivide into four equal parts the segment of the x-axis contained between $x = 1$ and $x = 2$. On each of these subdivisions as a base, let us construct a rectangle whose altitude is the ordinate of $y = x$ at the right end of the base (Fig. 7-1). If we indicate the sum of the areas of these four rectangles by S_4, we have

$$S_4 = (0.25)(1.25) + (0.25)(1.50) + (0.25)(1.75) + (0.25)(2) = 1.625.$$

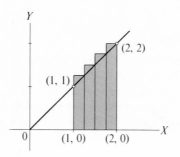

Fig. 7-1

It comes as no great surprise that $S_4 > 1.5$. It is clear from geometric considerations that this would be so. However, our intuition leads us to believe that such a process would lead us to a number nearer 1.5 if we used a greater number of rectangles. If we take ten subdivisions of the segment of the x-axis from $x = 1$ to $x = 2$, and construct rectangles as before, we have

$$S_{10} = (0.1)(1.1) + (0.1)(1.2) + (0.1)(1.3) + \cdots + (0.1)(2.0) = 1.55,$$

which confirms our intuition.

Now let us use n equal subdivisions and proceed as before. We have

$$S_n = \frac{1}{n}\left(1 + \frac{1}{n}\right) + \frac{1}{n}\left(1 + \frac{2}{n}\right) + \frac{1}{n}\left(1 + \frac{3}{n}\right) + \cdots + \frac{1}{n}\left(1 + \frac{n}{n}\right)$$

$$= \frac{1}{n}\left[n + \frac{1}{n}(1 + 2 + 3 + \cdots + n)\right].$$

Thus, making use of (7-1),

$$S_n = \frac{1}{n}\left\{n + \frac{1}{n}\left[\frac{n(n+1)}{2}\right]\right\} = 1 + \frac{1}{2}\left(\frac{n+1}{n}\right).$$

Since

$$\frac{n+1}{n} > 1 \qquad \text{for any } n,$$

we see that $S_n > 1.5$. However,

$$\lim_{n \to \infty} S_n = 1.5.$$

Therefore, if we define the area of our figure to be $\lim_{n \to \infty} S_n$, we shall have a definition that is consistent, at least in this special case, with our previous one.

The question immediately arises: Would the same result be obtained if the rectangles used in the definition of S_n had their altitudes taken as the ordinates of $y = x$ at the left end of the subdivisions, the midpoint, or any other point? The answer to this question is affirmative but we shall not prove it. Certain special cases will be presented as exercises for the student to verify.

7–4. The Definite Integral

The preceding example indicates that certain sums may be useful in defining area. For the moment, let us drop the discussion of areas and develop more generally the ideas involved in these sums.

Consider a function f defined and continuous on the interval $a \leqslant x \leqslant b$. Divide this interval into n subintervals by inserting the points x_k, $k = 1, 2, \ldots, n - 1$, ordered so that

$$a = x_0 < x_1 < x_2 < \cdots < x_{n-1} < x_n = b,$$

where we have renamed a and b for consistency of notation (Fig. 7-2). Let

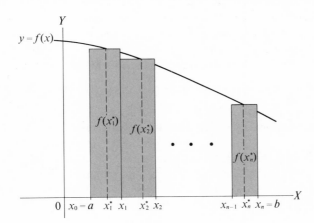

Fig. 7-2

$(\Delta x)_k$, $k = 1, 2, \ldots, n$, be defined by

$$(\Delta x)_k = x_k - x_{k-1},$$

and let x_k^*, $k = 1, 2, \ldots, n$, be any value satisfying

$$x_{k-1} \leqslant x_k^* \leqslant x_k.$$

Now form the sum

$$S_n = f(x_1^*)(\Delta x)_1 + f(x_2^*)(\Delta x)_2 + \cdots + f(x_n^*)(\Delta x)_n,$$

or, in summation notation,

$$S_n = \sum_{k=1}^{n} f(x_k^*)(\Delta x)_k.$$

Note that this sum is precisely the type used in the preceding section, although more general in nature. In the example f was defined by the equation $y = f(x) = x$, the x_k were uniformly spaced so that the $(\Delta x)_k$ were all equal, and the x_k^* were uniformly taken as x_k.

Now suppose an infinite sequence $\{S_n\}$ is defined by successive subdivisions of the interval and corresponding choices of x_k^* according to some law, subject only to the restriction that the length of the longest subdivision $(\Delta x)_k$ approaches zero. We express this important restriction by the notation

$$\max (\Delta x)_k \rightarrow 0.$$

If the limit of $\{S_n\}$ exists, independent of the manner of formation, subject to the one condition already mentioned, we say that

$$\lim_{n\to\infty} \sum_{k=1}^{n} f(x_k^*)(\Delta x)_k$$

exists and make the following definition.

DEFINITION 7–2. The definite integral of f from $x = a$ to $x = b$ is written $\int_a^b f(x)\, dx$ and is defined by

$$\int_a^b f(x)\, dx = \lim_{n\to\infty} \sum_{k=1}^{n} f(x_k^*)(\Delta x)_k$$

provided the limit exists in the sense indicated above.

It is not within the scope of this treatment of calculus to enter on a general discussion of functions for which the definite integral exists. We shall content ourselves with the statement, *without proof*, of a theorem which will cover all cases encountered in this book.

THEOREM 7–I. If a function f is continuous on the interval $a \leqslant x \leqslant b$, its definite integral over that interval exists.

We shall henceforth assume the truth of this theorem. Accordingly, we can say, since $f(x) = x$ is continuous for all values of x,

$$\int_1^2 x\, dx = 1.5.$$

This follows from the fact that the sequence $\{S_n\}$ formed by the scheme used in Sec. 7-3 has the limit 1.5. Theorem 7-1 states that for a continuous function this limit is independent of the manner in which $\{S_n\}$ is formed—subject always to the restriction that $\max(\Delta x)_k \rightarrow 0$. We shall understand from here on that this condition applies without specifically stating it.

Let us return to the function f defined by $f(x) = x$ on the interval $1 \leqslant x \leqslant 2$, and see what happens if we neglect the restriction $\max(\Delta x)_k \rightarrow 0$. For one manner of subdivision and choice of x_k^* (Sec. 7-3) we have

$$\lim_{n\to\infty} S_n = 1.5.$$

Now let us form another sequence $\{\bar{S}_n\}$ by subdividing the interval according to a different law. We subdivide the interval from $x = 1$ to $x = 1.75$ into $n - 1$ equal parts and leave the rightmost subdivision $(\Delta x)_n$ fixed at length 0.25 units (Fig. 7-3). Also, we choose x_k^* as the right end point of the corresponding

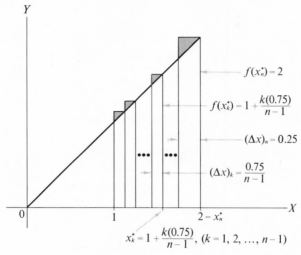

$$x_k^* = 1 + \frac{k(0.75)}{n-1}, \ (k = 1, 2, \dots, n-1)$$

Fig. 7-3

subinterval. Then

$$\bar{S}_n = \frac{0.75}{n-1}\left[1 + \frac{0.75}{n-1}\right] + \frac{0.75}{n-1}\left[1 + \frac{2(0.75)}{n-1}\right] + \frac{0.75}{n-1}\left[1 + \frac{3(0.75)}{n-1}\right]$$

$$+ \cdots + \frac{0.75}{n-1}\left[1 + \frac{(n-1)(0.75)}{n-1}\right] + (0.25)(2)$$

$$= \frac{0.75}{n-1}\left\{(n-1) + \frac{0.75}{n-1}[1 + 2 + \cdots + (n-1)]\right\} + 0.50$$

$$= 0.75 + \left[\frac{0.75}{n-1}\right]^2\left[\frac{(n-1)(n)}{2}\right] + 0.50$$

$$= 0.75 + \frac{(0.75)^2}{2}\left[\frac{(n-1)+1}{n-1}\right] + 0.50$$

$$= 0.75 + 0.28125\left[1 + \frac{1}{n-1}\right] + 0.50.$$

Hence

$$\lim_{n \to \infty} \bar{S}_n = 1.53125.$$

Thus

$$\lim_{n \to \infty} \bar{S}_n \neq \lim_{n \to \infty} S_n,$$

and on the surface it would appear that Theorem 7-1 is contradicted. This is not so, however, because in the formation of $\{\bar{S}_n\}$ we failed to enforce the restriction $\max(\Delta x)_k \to 0$. We have

$$(\Delta x)_n = 0.25$$

in every subdivision.

Geometrically, it is quite clear from Fig. 7-3 what happens. Considered as an area problem, as in Sec. 7-3, we see that

$$\lim_{n \to 0} \bar{S}_n$$

intuitively gives the area of the trapezoid bounded by $x = 1$, $x = 2$, $y = 0$, $y = x$, *plus the shaded triangle bounded by* $x = 1.75$, $y = x$, $y = 2$. All the other shaded triangles lose their effect in the limit because their bases $(\Delta x)_k$ (and altitudes) approach zero. Note that this geometrical consideration is verified numerically. We have

$$\lim_{n \to \infty} \bar{S}_n = 1.53125 = \lim_{n \to \infty} S_n + 0.03125,$$

and the area of the triangle described above is 0.03125.

Example 7–1. Find the value of $\int_0^2 x^2\,dx$.

The existence of this definite integral is assured by the continuity of x^2 on the interval $0 \le x \le 2$. Thus we have only to determine a suitable sequence $\{S_n\}$ and find its limit. We shall do this in two ways which will perhaps reinforce the student's confidence in Theorem 7-1.

First let us divide the interval into n equal parts and choose x_k^* at the left end of $(\Delta x)_k$. We have, then

$$(\Delta x)_1 = (\Delta x)_2 = (\Delta x)_3 = \cdots = (\Delta x)_n = \frac{2}{n}\,;$$

$$x_1^* = 0, \qquad x_2^* = \frac{2}{n}, \qquad x_3^* = \frac{4}{n}, \qquad x_4^* = \frac{6}{n}, \quad \cdots, x_n^* = \frac{2(n-1)}{n}\,;$$

$$f(x_1^*) = 0, \qquad f(x_2^*) = \left(\frac{2}{n}\right)^2, \qquad f(x_3^*) = \left(\frac{4}{n}\right)^2, \quad \cdots, f(x_n^*) = \left(\frac{2(n-1)}{n}\right)^2\,;$$

and

$$S_n = 0\left(\frac{2}{n}\right) + \left(\frac{2}{n}\right)^2\left(\frac{2}{n}\right) + \left(\frac{4}{n}\right)^2\left(\frac{2}{n}\right) + \cdots + \left(\frac{2(n-1)}{n}\right)^2\left(\frac{2}{n}\right).$$

Factoring out $(2/n)^3$, we obtain

$$S_n = \left(\frac{2}{n}\right)^3 [1^2 + 2^2 + 3^2 + \cdots + (n-1)^2].$$

Now if we apply Formula 7-2 with n replaced by $n - 1$, we get

$$S_n = \left(\frac{2}{n}\right)^3 \left[\frac{(n-1)(n)(2n-1)}{6}\right].$$

Hence

$$\int_0^2 x^2 \, dx = \lim_{n \to \infty} \left(\frac{2}{n}\right)^3 \left[\frac{(n-1)(n)(2n-1)}{6}\right] = \frac{8}{3}.$$

As a second way of obtaining a sequence $\{S_n\}$, let us again divide the interval into n equal parts, but this time let us choose x_k^* as the midpoint of $(\Delta x)_k$. Then

$$(\Delta x)_1 = (\Delta x)_2 = (\Delta x)_3 = \cdots = (\Delta x)_n = \frac{2}{n};$$

$$x_1^* = \frac{1}{n}, \qquad x_2^* = \frac{3}{n}, \ldots, x_n^* = \frac{2n-1}{n};$$

$$f(x_1^*) = \left(\frac{1}{n}\right)^2, \qquad f(x_2^*) = \left(\frac{3}{n}\right)^2, \ldots, f(x_n^*) = \left(\frac{2n-1}{n}\right)^2;$$

and

$$S_n = \left(\frac{1}{n}\right)^2 \left(\frac{2}{n}\right) + \left(\frac{3}{n}\right)^2 \left(\frac{2}{n}\right) + \left(\frac{5}{n}\right)^2 \left(\frac{2}{n}\right) + \cdots + \left(\frac{2n-1}{n}\right)^2 \left(\frac{2}{n}\right).$$

Factoring out $2/n^3$, we obtain

$$S_n = \frac{2}{n^3} [1^2 + 3^2 + 5^2 + \cdots + (2n-1)^2],$$

and if we add and subtract $2^2 + 4^2 + \cdots + (2n)^2$ inside the brackets, this may be written

$$S_n = \frac{2}{n^3} [(1^2 + 2^2 + 3^2 + \cdots + (2n)^2) - (2^2 + 4^2 + \cdots + (2n)^2)]$$

$$= \frac{2}{n^3} [(1^2 + 2^2 + 3^2 + \cdots + (2n)^2) - 2^2(1^2 + 2^2 + 3^2 + \cdots + n^2)].$$

If we apply (7-2) to each of the terms inside the brackets, we obtain

$$S_n = \frac{2}{n^3} \left[\frac{2n(2n+1)(4n+1)}{6} - \frac{4n(n+1)(2n+1)}{6}\right]$$

$$= \frac{1}{3} \left[2\left(2 + \frac{1}{n}\right)\left(4 + \frac{1}{n}\right) - 4\left(1 + \frac{1}{n}\right)\left(2 + \frac{1}{n}\right)\right],$$

or

$$\lim_{n \to \infty} S_n = \int_0^2 x^2 \, dx = \frac{8}{3}.$$

A still different choice of x_k^* is suggested to the student in one of the excercises.

Many geometrical and physical concepts may be defined as limits of sums. Among these are area of plane and curved surfaces, volume, arc length, mass, center of mass, moment of inertia, work, and numerous others. For this reason the definite integral finds wide application, and it is of the utmost importance that the student understand the basic ideas associated with its definition.

Exercises 7-1

The student will find equations (7-1), (7-2), and (7-3) useful in solving the following exercises.

1. Calculate $\int_1^2 x \, dx$ using equal subdivisions of the interval and choosing x_k^* as the left end point of $(\Delta x)_k$.

2. Calculate $\int_1^2 x \, dx$, using equal subdivisions of the interval and choosing x_k^* as the midpoint of $(\Delta x)_k$.

3. Calculate $\int_0^2 x^2 \, dx$, using equal subdivisions of the interval and choosing x_k^* as the right end point of $(\Delta x)_k$.

4. Evaluate $\int_a^b k \, dx$, where k is a constant.

5. Evaluate $\int_1^3 (2x - 4) \, dx$.

6. Evaluate $\int_0^1 x^3 \, dx$.

7-5. Properties and Definitions

The numbers a and b in the symbol $\int_a^b f(x) \, dx$ are referred to as the *lower* and *upper limits*, respectively, and $f(x)$ is called the integrand. The word "limit" here has the connotation of being an extreme of the interval of integration rather than the result of a limiting process.

We have defined the definite integral

$$\int_a^b f(x) \, dx$$

only for $a < b$. Now we wish to extend the definition to include the cases $a \geqslant b$.

Consider first $a > b$, and let us proceed as we did originally by subdividing the interval from $x = a$ to $x = b$. However, this subdivision is from right to left, whereas before it was from left to right. This, in accordance with our concept of directed line segments, produces a sign change (Fig. 7-4) in each

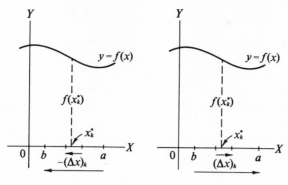

Fig. 7-4

subdivision $(\Delta x)_k$ but does not affect $f(x_k^*)$. Thus this reversal of direction changes the sign of S_n. Since

$$\int_b^a f(x)\, dx$$

is already defined, it is not necessary to go back to the summation notation to extend our definition.

DEFINITION 7–3

$$\int_a^b f(x)\, dx = -\int_b^a f(x)\, dx, \qquad a > b.$$

For $a = b$, we have an interval of zero length, and it follows that any subdivision of it would also be of zero length, that is, $(\Delta x)_k = 0$ for all k. Hence, for consistency with Definition 7-2, we define:

DEFINITION 7–4

$$\int_a^a f(x)\, dx = 0.$$

Referring back to Example 7-1, we have

$$\int_2^0 x^2\, dx = -\int_0^2 x^2\, dx = -\frac{8}{3}.$$

We may now state the following theorem.

THEOREM 7-2. If the upper and lower limits of a definite integral are interchanged, its sign is changed.

A very useful property of the definite integral is expressed in the next theorem.

THEOREM 7-3. If a function f is continuous over an interval containing a, b, and c,

$$\int_a^b f(x)\,dx = \int_a^c f(x)\,dx + \int_c^b f(x)\,dx.$$

We offer a sketch of the proof of this theorem. First consider the case $a < c < b$. The interval $a \leqslant x \leqslant b$ must be subdivided according to some law. We accept any law of subdivision as long as it always uses c as one of the points of subdivision (Fig. 7-5). Then

$$\int_a^b f(x)\,dx = \lim_{n\to\infty} \sum_{k=1}^n f(x_k^*)(\Delta x)_k$$

$$= \lim_{n\to\infty} S_n.$$

But

$$S_n = S_{n_1} + S_{n_2},$$

where S_{n_1} is the part of the sum over $a \leqslant x \leqslant c$, and S_{n_2} is the part of it over $c \leqslant x \leqslant b$. Hence,

$$\int_a^b f(x)\,dx = \lim_{n\to\infty} S_n = \lim_{n_1\to\infty} S_{n_1} + \lim_{n_2\to\infty} S_{n_2}$$

$$= \int_a^c f(x)\,dx + \int_c^b f(x)\,dx.$$

A second case is $a < b < c$. Here, the argument used above gives

$$\int_a^c f(x)\,dx = \int_a^b f(x)\,dx + \int_b^c f(x)\,dx$$

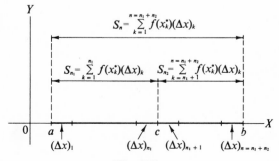

Fig. 7-5

or

$$\int_a^b f(x)\,dx = \int_a^c f(x)\,dx - \int_b^c f(x)\,dx.$$

This, making use of Theorem 7-2, may be written in the form stated in the theorem.

Other orderings of a, b, and c may be dealt with in a similar manner. We shall now establish *the theorem of the mean for integrals*.

THEOREM 7-4. If f is continuous for $a \leqslant x \leqslant b$, then there exists a number η, $a < \eta < b$, such that

$$\int_a^b f(x)\,dx = f(\eta)(b - a).$$

As in the mean value theorem (Theorem 6-1) this theorem does not provide a means of calculating η; it merely states its existence.

Since f is continuous for $a \leqslant x \leqslant b$, it has a maximum M and a minimum m on that interval (Theorem 4-11(a)). Therefore

$$\sum_{k=1}^n f(x_k^*)(\Delta x)_k \leqslant \sum_{k=1}^n M(\Delta x)_k = M \sum_{k=1}^n (\Delta x)_k = M(b - a)$$

for all subdivisions of the interval and corresponding choices of x_k^*. Therefore

$$\int_a^b f(x)\,dx = \lim_{n \to \infty} \sum_{k=1}^n f(x_k^*)(\Delta x)_k \leqslant M(b - a).$$

In a similar fashion,

$$\int_a^b f(x)\,dx \geqslant m(b - a),$$

so we have

$$m(b - a) \leqslant \int_a^b f(x)\,dx \leqslant M(b - a).$$

Hence we may write

$$\int_a^b f(x)\,dx = K(b - a), \qquad m \leqslant K \leqslant M.$$

The equality signs can hold only if f is a constant in which case the theorem is true. If neither equality sign holds, Theorem 4-11 asserts the existence of η, $a < \eta < b$, such that $f(\eta) = K$. Thus the theorem is established.

Example 7-2. Given $\int_2^4 (3x - 2)\,dx = 14$, find a value for η as guaranteed by Theorem 7-4.

We have

$$14 = (3\eta - 2)(4 - 2) = 2(3\eta - 2).$$

Hence

$$3\eta - 2 = 7 \quad \text{or} \quad \eta = 3.$$

Note that

$$2 < \eta < 4.$$

Example 7-3. Find the limits between which the value of $\int_1^6 (x^2 - 6x + 5)\, dx$ lies.

From the proof of Theorem 7-4 we know that

$$m(6 - 1) \leq \int_1^6 (x^2 - 6x + 5)\, dx \leq M(6 - 1),$$

where m and M are the minimum and maximum values of $x^2 - 6x + 5$ on the interval $1 \leq x \leq 6$. From Example 6-11 we have

$$m = -4, \qquad M = 5.$$

Hence

$$-20 \leq \int_1^6 (x^2 - 6x + 5)\, dx \leq 25.$$

Exercises 7-2

In Exercises 1–4 find the number η guaranteed by Theorem 7-4.

1. $\displaystyle\int_1^3 (2x - 4)\, dx = 0$ **2.** $\displaystyle\int_1^4 (x^2 - 1)\, dx = 18$

3. $\displaystyle\int_0^1 x^3\, dx = \tfrac{1}{4}$ **4.** $\displaystyle\int_1^3 (x^3 + 1)\, dx = 22$

5. Prove that the value of $\int_0^1 (dx/(x^2 + 1))$ is between $\tfrac{1}{2}$ and 1.

6. Prove that the value of $\int_{-1}^2 (x^2 + x + 1)\, dx$ is between 9/4 and 21.

7. Find the limits between which the value of $\int_0^3 \sqrt{x^2 + 9}\, dx$ lies.

8. Find the limits between which the value of $\int_{-2}^2 ((x^3/3) - 4x + 1)\, dx$ lies.

7-6. The Indefinite Integral

In the symbol $\int_a^b f(x)\, dx$, we refer to x as the *variable of integration*. It should be noted from the definition that the value of the integral depends only on the limits a and b and the function f. The character used to denote the variable of integration is of no importance. Thus

$$\int_a^b f(x)\,dx = \int_a^b f(t)\,dt = \int_a^b f(u)\,du = \int_a^b f(\oplus)\,d\oplus = \cdots.$$

For this reason we often refer to the variable of integration as a *dummy variable*.

Let f be continuous on $a \leqslant x \leqslant b$. Then the symbol $\int_a^x f(t)\,dt$ is defined for any x on that interval and thus defines a function of x. We write

$$A(x) = \int_a^x f(t)\,dt, \qquad a \leqslant x \leqslant b,$$

and call it an *indefinite integral of f*.

This indefinite integral will provide us with the link between the definite integral and the derivative. This, in turn, will give us a means of evaluating definite integrals without having to resort to the cumbersome limit of a sum, which for practical purposes is impossible to compute in all but the simplest cases. The idea that the definite integral is the limit of a sum is indispensable for formulating applications but essentially useless for computational purposes.

First we shall establish that $A(x)$ has a derivative by actually computing it. We have

$$A(x + \Delta x) - A(x) = \int_a^{x+\Delta x} f(t)\,dt - \int_a^x f(t)\,dt,$$

and by applying Theorem 7-3,

$$A(x + \Delta x) - A(x) = \int_a^x f(t)\,dt + \int_x^{x+\Delta x} f(t)\,dt - \int_a^x f(t)\,dt$$

$$= \int_x^{x+\Delta x} f(t)\,dt.$$

Now we apply the mean value theorem for integrals (Theorem 7-4) and obtain

$$\int_x^{x+\Delta x} f(t)\,dt = f(\eta)(x + \Delta x - x) = f(\eta)\,\Delta x, \qquad x < \eta < x + \Delta x.\dagger$$

Thus

$$\frac{A(x + \Delta x) - A(x)}{\Delta x} = f(\eta), \qquad x < \eta < x + \Delta x,$$

and

$$A'(x) = \lim_{\Delta x \to 0} \frac{A(x + \Delta x) - A(x)}{\Delta x} = \lim_{\Delta x \to 0} f(\eta) = f(x)$$

$\dagger$ This inequality takes the form $x + \Delta x < \eta < x$ if Δx is negative. However, regardless of this, $\eta \to x$ as $\Delta x \to 0$ and the same result follows.

since η must approach x as $\Delta x \to 0$. Thus we see that *the indefinite integral of f is a function whose derivative is f.*

THEOREM 7–5. If f is continuous on $a \leqslant x \leqslant b$, then

$$A'(x) = \frac{d}{dx} \int_a^x f(t)\,dt = f(x), \qquad a \leqslant x \leqslant b.$$

We have already seen (Theorem 6-3) that two functions which have the same derivative differ only by a constant. Therefore, if F is any function such that

$$F'(x) = f(x),$$

we may write

$$F(x) = \int_a^x f(t)\,dt + C.$$

The constant C may be evaluated by recalling that

$$\int_a^a f(t)\,dt = 0.$$

We have

$$F(a) = \int_a^a f(t)dt + C \qquad \text{or} \qquad F(a) = C.$$

Hence

$$\int_a^x f(t)\,dt = F(x) - F(a).$$

Now we can convert back to the definite integral by taking $x = b$ and obtain

$$\int_a^b f(t)\,dt = \int_a^b f(x)\,dx = F(b) - F(a).$$

This result is commonly known as the fundamental theorem of calculus.

THEOREM 7–6. Fundamental Theorem of Calculus. Let f be continuous on $a \leqslant x \leqslant b$,† and let F be any function such that $F' = f$. Then

$$\int_a^b f(x)\,dx = F(b) - F(a).$$

---

† We should note here what has been implied earlier. Continuous functions are not the only ones for which the definite integral exists. In more advanced courses the student will discover much more general functions for which this is true.

Any function F satisfying the relationship $F' = f$ is called an *antiderivative* of f. Thus the fundamental theorem reduces the problem of evaluating a definite integral essentially to that of finding an antiderivative. The notation

$$\int_a^b f(x)\,dx = F(x)\bigg]_a^b = F(b) - F(a)$$

is commonly used.

If we return now to Example 7-1, we find it much easier to calculate. We know that

$$\frac{d}{dx}\left(\frac{x^3}{3}\right) = x^2.$$

so, applying Theorem 7-6,

$$\int_0^2 x^2\,dx = \frac{x^3}{3}\bigg]_0^2 = \frac{(2)^3}{3} - \frac{(0)^3}{3} = \frac{8}{3}.$$

If F is a particular antiderivative of f, any other antiderivative of f may be obtained from

$$F(x) + C$$

by a proper choice of C (Theorem 6-3). Hence, if C is an *arbitrary constant*, that is, if it may be given any value whatever, the expression $F(x) + C$ represents *all* antiderivatives of f. All of the antiderivatives of f are usually denoted by the notation

$$\int f(x)\,dx = F(x) + C$$

and it is traditional to call this the *indefinite integral of f*.

When we recall the original definition we gave for the indefinite integral we see that this usage is not strictly correct. For example, the indefinite integral (as first defined) of the function f defined by $f(x) = 2x$ is

$$\int_a^x 2t\,dt = x^2 - a^2,$$

while in the sense just stated, it is

$$\int 2x\,dx = x^2 + C.$$

If we choose C any particular positive number, say $C = 1$, there is no real value of a which will make the right members of these two statements identical. However, this discrepancy will create no problems and we shall, as is customary, refer to either as the indefinite integral.

7-7. Evaluation of Indefinite Integrals

The evaluation of $\int f(x)\, dx$ requires, as seen in the preceding section, the determination of a function F from the equation

$$F'(x) = f(x),$$

or, what is equivalent, from the equation

$$dF(x) = f(x)\, dx.$$

For reasons that will be apparent shortly, we shall usually use the latter interpretation.

In Chapter 5 we were able to derive formulas and methods such that, given any algebraic function, we could follow a specific program and obtain the derivative of the function. Now we are confronted with the reverse process. This is more difficult and it is not possible to make such a general statement. In fact there are quite simple-appearing algebraic functions which our theory tells us have integrals, but these integrals cannot be written in terms of a finite number of algebraic combinations of elementary functions. However, we shall be able to evaluate many important integrals in simple form and apply them to various types of problems.

Each differentiation formula can be converted into a corresponding integration formula. Thus, Theorem 5-1, which states that the derivative of a sum is equal to the sum of the derivatives, may be reworded in integral language to say that the integral of a sum is equal to the sum of the integrals.

THEOREM 7-7. If f and g are continuous functions,

$$\int (f(x) + g(x))\, dx = \int f(x)\, dx + \int g(x)\, dx.$$

Any integration formula may be checked by differentiating the two members. For example, to check Theorem 7-7, we write

$$\frac{d}{dx} \int (f(x) + g(x))\, dx = \frac{d}{dx} \left\{ \int f(x)\, dx + \int g(x)\, dx \right\}$$

$$= \frac{d}{dx} \int f(x)\, dx + \frac{d}{dx} \int g(x)\, dx,$$

which, by Theorem 7-5, reduces to the identity

$$f(x) + g(x) = f(x) + g(x).$$

Similarly, we obtain the next theorem.

THEOREM 7-8. If f is continuous and C is a constant,

$$\int Cf(x)\,dx = C \int f(x)\,dx.$$

Also, since

$$\frac{d}{dx}\left(\frac{x^{n+1}}{n+1}\right) = x^n, \qquad n \neq -1,$$

we have the following theorem.

THEOREM 7-9

$$\int x^n\,dx = \frac{x^{n+1}}{n+1} + C, \qquad n \neq -1,$$

where C is an arbitrary constant.

The student should be careful never to omit the arbitrary constant; otherwise the result may be incorrect. Consider the function $f(x) = 2$. Then we should have

$$\int f'(x)\,dx = f(x) = 2.$$

But $f'(x) = 0$ and

$$\int f'(x)\,dx = \int (0)\,dx = 0,$$

if we neglect the constant. We must be sure that the indefinite integral represents all functions with the given derivative.

It will be greatly to our advantage in many cases to modify Theorem 7-9. If u is any differentiable function of x, we know by the chain rule that

$$d\left(\frac{u^{n+1}}{n+1}\right) = u^n \frac{du}{dx}\,dx = u^n\,du, \qquad n \neq -1.$$

THEOREM 7-10. If u is a differentiable function of x,

$$\int u^n\,du = \frac{u^{n+1}}{n+1} + C, \qquad n \neq -1.$$

   This is not a trivial modification of Theorem 7-9. It gives us the formula for integrating any power ($\neq -1$) of a differentiable function if we have the product of that power by the differential of the function.†

   The advantage of Theorem 7-10 over Theorem 7-9 will be apparent in the following examples.

Example 7–3. Evaluate $\int (1 + 2x)^5 \, dx$.

   If we rely on Theorem 7-9, we find it necessary to expand the integrand by the binomial theorem and obtain

$$\int (1 + 2x)^5 \, dx = \int (1 + 10x + 40x^2 + 80x^3 + 80x^4 + 32x^5) \, dx$$

$$= \int dx + 10 \int x \, dx + 40 \int x^2 \, dx + 80 \int x^3 \, dx$$

$$+ 80 \int x^4 \, dx + 32 \int x^5 \, dx$$

$$= x + 10\left(\frac{x^2}{2}\right) + 40\left(\frac{x^3}{3}\right) + 80\left(\frac{x^4}{4}\right) + 80\left(\frac{x^5}{5}\right) + 32\left(\frac{x^6}{6}\right) + C$$

$$= x + 5x^2 + \frac{40}{3} x^3 + 20x^4 + 16x^5 + \frac{16}{3} x^6 + C,$$

where the six arbitrary constants introduced by the six indefinite integrals have been combined into a single arbitrary constant.

   On the other hand, if we appeal to Theorem 7-10 and set

$$u = 1 + 2x,$$

we have

$$du = 2dx \quad \text{or} \quad dx = \tfrac{1}{2}du,$$

which enables us to write

$$\int (1 + 2x)^5 \, dx = \int u^5 \cdot \tfrac{1}{2} \, du = \tfrac{1}{2} \int u^5 \, du$$

$$= \frac{1}{2}\left(\frac{u^6}{6}\right) + C = \frac{1}{12}(1 + 2x)^6 + C.$$

It is left as an exercise to show that these two results differ only by a constant. This means that the two results can be made identical by appropriate choices of the arbitrary constants. However, the second method was far more painless.

---

† This statement should be accompanied by some exclusions, such as not permitting negative values of the function when $n = \tfrac{1}{2}$. However, what we have said is generally true and we prefer not to clutter up the statement of the theorem with excess words and notation.

The student will note that liberal use has been made of Theorem 7-8. It should be emphasized that *only constants* can be treated in this fashion.

In the example just concluded it was a matter of choice as to which method was to be used. There are many problems in which this choice is not available, at least at this point in the study of calculus. The following is an example of this sort.

Example 7-4. Evaluate $\int \sqrt{1 + 3x^2}\, x\, dx$.

The binomial expansion of $(1 + 3x^2)^{1/2}$ would not lead to a finite number of terms. Therefore this method cannot be used until we have studied sums with an infinite number of terms. But the method utilizing Theorem 7-10 works very well. We set

$$u = 1 + 3x^2$$

so

$$du = 6x\, dx \qquad \text{or} \qquad x\, dx = \frac{1}{6}\, du.$$

When we substitute these in our integral we have

$$\int (1 + 3x^2)^{1/2} x\, dx = \int u^{1/2} \cdot \frac{1}{6}\, du$$

$$= \frac{1}{6} \int u^{1/2}\, du = \frac{1}{6} \frac{u^{3/2}}{3/2} + C$$

$$= \frac{1}{9} u^{3/2} + C = \frac{1}{9} \sqrt{(1 + 3x^2)^3} + C.$$

The student should note that this method could not be applied to the evaluation of $\int (1 + 3x^2)^{1/2}\, dx$. When we set $u = 1 + 3x^2$, we have $du = 6x\, dx$, or $dx = du/6x$. Consequently this transformation does not give us an integral of the form $\int Cu^n\, du$. We shall have to devise new methods before this integral can be evaluated.

Example 7-5. Evaluate $\displaystyle\int_0^4 \frac{x\, dx}{\sqrt{x^2 + 9}}$.

First we evaluate

$$\int \frac{x\, dx}{\sqrt{x^2 + 9}}.$$

If we set $u = x^2 + 9$, $du = 2x\, dx$, or $x\, dx = du/2$, and

$$\int \frac{x\, dx}{\sqrt{x^2 + 9}} = \int u^{-1/2} \frac{du}{2} = \frac{1}{2} \frac{u^{1/2}}{1/2} + C$$

$$= (x^2 + 9)^{1/2} + C.$$

Then, applying Theorem 7-6,

$$\int_0^4 \frac{x \, dx}{\sqrt{x^2 + 9}} = (x^2 + 9)^{1/2} + C \Big]_0^4$$
$$= [(25)^{1/2} + C] - [(9)^{1/2} + C]$$
$$= 2.$$

It was a waste of time and space to write C into the formula when we applied Theorem 7-6 because it will always disappear in the algebra. Hence we commonly write

$$\int_0^4 \frac{x \, dx}{\sqrt{x^2 + 9}} = (x^2 + 9)^{1/2} \Big]_0^4 = (25)^{1/2} - (9)^{1/2} = 2.$$

This is not a contradiction of our earlier statement in which we cautioned never to omit the arbitrary constant. That statement was made with respect to the indefinite integral, which represents *all* functions having a given derivative. On the other hand, the fundamental theorem merely calls for *any* function having the given derivative.

A second, and usually simpler, method of evaluating this integral involves computing limits for u and abandoning x completely. When $x = 0$, $u = 9$ and when $x = 4$, $u = 25$. Then

$$\int_0^4 \frac{x \, dx}{\sqrt{x^2 + 9}} = \frac{1}{2} \int_9^{25} u^{-1/2} \, du = u^{1/2} \Big]_9^{25} = 5 - 3 = 2.$$

Exercises 7-3

In Exercises 1–18 find the indefinite integrals and check your result by differentiation.

1. $\int (2x^2 - 3x + 4) \, dx$

2. $\int (5x^3 + 2x - 7) \, dx$

3. $\int (u + 3)^2 \, du$

4. $\int (2t - 5)^3 \, dt$

5. $\int (2w^2 + 5)^2 \, dw$

6. $\int (s - 1)(s^2 + 2) \, ds$

7. $\int x(2x^2 + 5)^3 \, dx$

8. $\int x^2(3x^3 + 1)^2 \, dx$

9. $\int \sqrt{3r} \, dr$

10. $\int \frac{(x + 1)^2}{\sqrt{x}} \, dx$

11. $\int \sqrt{v}(2v + 3)^2 \, dv$

12. $\int u\sqrt{3 + u^2} \, du$

13. $\displaystyle \int \frac{x\,dx}{\sqrt[3]{x^2+5}}$

14. $\displaystyle \int \frac{dy}{\sqrt{10-3y}}$

15. $\displaystyle \int t^2\sqrt{t^3+10}\,dt$

16. $\displaystyle \int (x+2)\sqrt{x^2+4x+1}\,dx$

17. $\displaystyle \int x\sqrt{9-x^2}\,dx$

18. $\displaystyle \int \frac{(4w+10)\,dw}{\sqrt[3]{w^2+5w+3}}$

In Exercises 19–30 find the value of the definite integrals.

19. $\displaystyle \int_0^4 (x-\sqrt{x})\,dx$

20. $\displaystyle \int_{-1}^2 x(2x^2+3x)\,dx$

21. $\displaystyle \int_1^4 \frac{x+1}{x^4}\,dx$

22. $\displaystyle \int_0^4 x\sqrt{x^2+9}\,dx$

23. $\displaystyle \int_4^9 \frac{t+1}{\sqrt{t}}\,dt$

24. $\displaystyle \int_{-2}^2 x\sqrt{5+x^2}\,dx$

25. $\displaystyle \int_{\sqrt{6}}^3 \frac{u\,du}{\sqrt{u^2-5}}$

26. $\displaystyle \int_{2a}^{3a} \frac{x\,dx}{(x^2-a^2)^2}$

27. $\displaystyle \int_0^2 \frac{(2x+3)\,dx}{\sqrt{x^2+3x+1}}$

28. $\displaystyle \int_{\sqrt{7}/3}^{2/\sqrt{3}} \frac{s\,ds}{\sqrt{16-9s^2}}$

29. $\displaystyle \int_0^{1/2} \frac{x^3\,dx}{(1-x^4)^3}$

30. $\displaystyle \int_0^3 \sqrt[3]{(3x-1)^4}\,dx$

31. Show that the results of Example 7-2 differ only by a constant.

32. Prove Theorem 7-4 directly from Theorem 6-1. *Hint*: Let

$$F(x) = \int_a^x f(t)\,dt.$$

7–8. Areas of Plane Figures

In Sec. 7-3 we studied the problem of defining and calculating the area of a special plane figure. This figure had the *x*-axis as one boundary, the vertical lines $x=1$, $x=2$ as two others, and the final boundary was supplied by the line $y=x$ (Fig. 7-1). Such an *area* is often called the *area under the curve* $y = x$ *from* $x = 1$ *to* $x = 2$. A suitable definition of this area was found to be what we later defined (Sec. 7-4) as the definite integral

$$A = \int_1^2 x\,dx.$$

Following the same line of reasoning as that employed in Sec. 7-3, it seems logical to define the area under any continuous curve in a similar manner. This we do, and write the following definition.

DEFINITION 7-5. If f is continuous and nonnegative on $a \leqslant x \leqslant b$, the area under $y = f(x)$ from $x = a$ to $x = b$ is defined to be

$$A = \int_a^b y\, dx = \int_a^b f(x)\, dx.$$

In order to define area "under" a curve $y = f(x)$ over an interval in which f is nonpositive, we add the following definition.

DEFINITION 7-6. If f is continuous and nonpositive on $a \leqslant x \leqslant b$, the area under $y = f(x)$ from $x = a$ to $x = b$ is defined to be

$$A = -\int_a^b y\, dx = -\int_a^b f(x)\, dx.$$

These two cases have to be separated in order to avoid negative area, this being contrary to our intuitive concept of this quantity. Under the hypotheses of Definition 7-6 every term in $\sum_{k=1}^n f(x_k^*)(\Delta x)_k$ is negative or zero. Hence

$$\int_a^b f(x)\, dx \leqslant 0.$$

For this reason we introduce the negative sign in this case. If f is positive over a part of an interval and negative over the remainder, we can break it down into these two cases and use the additive property of area.

Example 7-6. Find the area under the curve $y = x^2 - 5x + 4$ (a) from $x = -1$ to $x = 1$; (b) from $x = 1$ to $x = 3$; (c) from $x = -1$ to $x = 3$.

From our sketch (Fig. 7-6)—an essential part of the solution of any such problem—we discover $y > 0$ when $x < 1$ or $x > 4$, and $y < 0$ where $1 < x < 4$. Actually, all we need is the sketch from $x = -1$ to $x = 3$. Sometimes the entire curve may be quite troublesome to sketch while that portion in which we are interested may offer little or no difficulty.

(a) This part is governed by Definition 7-5 and we have

$$A_1 = \int_{-1}^1 (x^2 - 5x + 4)\, dx = \left(\frac{x^3}{3} - \frac{5x^2}{2} + 4x\right)\Big]_{-1}^1$$
$$= 8\tfrac{2}{3} \text{ square units.}$$

(b) Applying Definition 7-6 to this part, we obtain

$$A_2 = -\int_1^3 (x^2 - 5x + 4)\, dx = -\left(\frac{x^3}{3} - \frac{5x^2}{2} + 4x\right)\Big]_1^3$$
$$= 3\tfrac{1}{3} \text{ square units.}$$

(c) We combine these results over the interval from $x = -1$ to $x = 3$ and get

$$A_3 = A_1 + A_2 = 12 \text{ square units.}$$

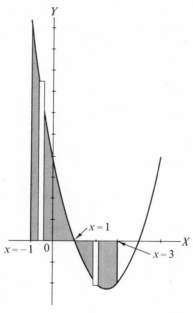

Fig. 7-6

If we had ignored the change in sign in f in the preceding solutions, we would have obtained

$$\int_{-1}^{3} (x^2 - 5x + 4) \, dx = \left(\frac{x^3}{3} - \frac{5x^2}{2} + 4x \right) \Big]_{-1}^{3} = 5\tfrac{1}{3}.$$

This is the excess of A_1 over A_2 and not at all the area desired.

The concept of an *element* of a definite integral is useful in many applications. By an "element" we mean a typical term in

$$\sum_{k=1}^{n} f(x_k^*)(\Delta x)_k,$$

thus

$$\text{Element} = f(x_k^*)(\Delta x)_k.$$

In the case of areas, we call the element an *element of area*. Geometrically it is a typical rectangle from the "approximating" sum. It usually will be quite helpful to make an element of area a part of the sketch, as in Fig. 7-6. For example, it is easy to see from this sketch why the total area from $x = -1$ to $x = 3$ must be broken into two parts. We shall find other uses for elements as we proceed.

Exercises 7-4

Find the area bounded by the *x*-axis and the following curves. Draw a sketch of the area showing an element of area.

1. $y = 3x + 6, x = 0$ **2.** $x + y - 9 = 0, x = 1, x = 5$

3. $y = 3 - 2x - x^2$ **4.** $y = x^2 - 4x + 3$

5. $y = x^2 - 4x + 4, x = 0$ **6.** $y = x + \sqrt{x}, x = 1$

7. $y = x^3 - 4x$ **8.** $y = x^3 + 4x, x = -1, x = 4$

9. $x^{1/2} + y^{1/2} = a^{1/2}, x = 0$ **10.** $y = x^3 - 2x^2 - 5x + 6$

11. $y = \dfrac{x + 1}{\sqrt{x^2 + 2x + 4}}, x = 0, x = 3$ **12.** $y = x + \dfrac{1}{\sqrt{x}}, x = 1, x = 9$

7-9. More General Plane Areas

In this section we wish to consider the area between two curves. Making use of the definitions of Sec. 7-8, we obtain the following basic theorem for this type of area.

THEOREM 7-11. Let f_1 and f_2 be continuous and $f_1(x) \geqslant f_2(x)$ on $a \leqslant x \leqslant b$. Then the area bounded by the curves $y = f_1(x)$, $y = f_2(x)$, $x = a$, and $x = b$ is given by

$$A = \int_a^b [f_1(x) - f_2(x)]\, dx.$$

This theorem holds true regardless of the signs of $f_1(x)$ and $f_2(x)$. This is obvious if both curves are completely above or completely below the *x*-axis. Consider the first of these (Fig. 7-7(a)). The preceding formula may be written

$$A = \int_a^b f_1(x)\, dx - \int_a^b f_2(x)\, dx,$$

the first term in the right member being the area under $y = f_1(x)$, the second being the area under $y = f_2(x)$, both from $x = a$ to $x = b$. The difference of

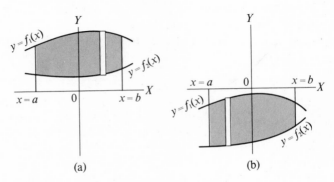

(a)   (b)

Fig. 7-7

these two areas leaves us with the area between the two curves. Note that the element of area for this integral is a rectangle, with its upper base in the curve $y = f_1(x)$ and its lower base in the curve $y = f_2(x)$. In actual practice we determine the element of area in this fashion and then write the integral which has this element.

Similar reasoning, keeping Definitions 7-5 and 7-6 in mind, will verify this theorem for the other possibilities, some of which are shown in Figs. 7-7(b) and 7-8.

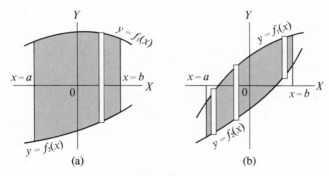

(a)   (b)

Fig. 7-8

If we consider this theorem from the element-of-area point of view we see that $f_1(x_k^*) - f_2(x_k^*)$ is always nonnegative and the algebraic combination gives the vertical distance between the two curves at $x = x_k^*$. Hence the rectangle we use to define the element of area always has its area given by

$$[f_1(x_k^*) - f_2(x_k^*)](\Delta x)_k$$

and the integral for the area is the one stated in the theorem. *The important thing here is not the signs of $f_1(x)$ and $f_2(x)$ but the relative positions of the two curves $y = f_1(x)$ and $y = f_2(x)$.* If they cross, thus changing their relative positions, we need to make a corresponding change in the integral or in the element of area in order to make the difference of the two functions nonnegative. This is analogous to the change necessary when finding the area under a curve if it crosses the x-axis.

These details, and certain others unmentioned so far, will be covered in the examples that follow.

Example 7-7. Find the area between $y = x$ and $4y = x^2$ from $x = 1$ to $x = 4$.

First we draw a sketch of the area (Fig. 7-9). This is an absolutely essential part of the solution of any such problem. From this sketch we observe that the curve $y = x$ is never below the curve $y = x^2/4$ in the range $1 \leqslant x \leqslant 4$. Then, by Theorem 7-10,

$$A = \int_1^4 \left(x - \frac{x^2}{4} \right) dx = \left(\frac{x^2}{2} - \frac{x^3}{12} \right) \Big]_1^4 = \frac{9}{4} \text{ square units.}$$

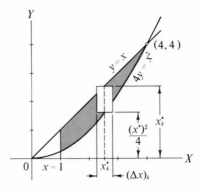

Fig. 7-9

If we approach this problem from the element-of-area point of view and draw one in our sketch (Fig. 7-9), we see that the ordinate of the upper base is x_k^* and that of the lower base is $(x_k^*)^2/4$. Hence the element of area is

$$\left[x_k^* - \frac{(x_k^*)^2}{4} \right](\Delta x)_k,$$

so the area is given by

$$\int_1^4 \left(x - \frac{x^2}{4} \right) dx$$

as before. In the element-of-area approach, the limits on the integral are determined by the extremes of x which may be used for x_k^*.

Example 7-8. Find the area bounded by $y = x + 1$ and $y = x^3 - x^2 - x + 1$.

When we draw a sketch of the area (Fig. 7-10) we discover that the two curves cross each other at the points $(-1, 0)$, $(0, 1)$, and $(2, 3)$. Hence the total area must be broken into two parts, A_1 for which $-1 \leqslant x \leqslant 0$, and A_2 for

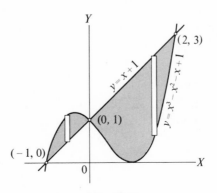

Fig. 7-10

which $0 \leqslant x \leqslant 2$. The element of area for A_1 has its upper base in $y = x^3 - x^2 - x + 1$ and its lower base in $y = x + 1$. Therefore its element of area is

$$[\{(x_k^*)^3 - (x_k^*)^2 - x_k^* + 1\} - \{x_k^* + 1\}](\Delta x)_k,$$

and consequently

$$A_1 = \int_{-1}^{0} [(x^3 - x^2 - x + 1) - (x + 1)] \, dx$$

$$= \int_{-1}^{0} (x^3 - x^2 - 2x) \, dx$$

$$= \left(\frac{x^4}{4} - \frac{x^3}{3} - x^2 \right) \Bigg]_{-1}^{0} = \frac{5}{12} \text{ square unit.}$$

Over A_2 the roles of the two curves are interchanged so that we have

$$A_2 = \int_{0}^{2} [(x + 1) - (x^3 - x^2 - x + 1)] \, dx$$

$$= \int_{0}^{2} (-x^3 + x^2 + 2x) \, dx$$

$$= \left(-\frac{x^4}{4} + \frac{x^3}{3} + x^2 \right) \Bigg]_{0}^{2} = \frac{8}{3} \text{ square units.}$$

Thus the total area is

$$A = A_1 + A_2 = \frac{5}{12} + \frac{8}{3} = \frac{37}{12} \text{ square units.}$$

Example 7-9. Find the area bounded by the curves $y^2 = x$ and $x + y = 2$.

The sketch of this area (Fig. 7-11) reveals that the element of area is not the same throughout the region. Those elements in the region for which

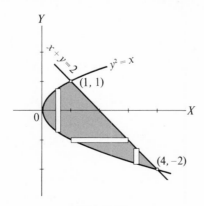

Fig. 7-11

$0 \leqslant x \leqslant 1$ have both the upper and lower bases in the parabola, while those in the region for which $1 \leqslant x \leqslant 4$ have the upper base in the line and the lower base in the parabola. This is brought about by the fact that the upper boundary is made up of two curves, $y = \sqrt{x}$, $0 \leqslant x \leqslant 1$, and $y = 2 - x$, $1 \leqslant x \leqslant 4$, and the lower boundary is $y = -\sqrt{x}$. The equation $y^2 = x$ defines two functions of x: $f_1(x) = \sqrt{x}$ and $f_2(x) = -\sqrt{x}$.

Therefore we find it necessary to break the area into two parts. Using either elements of area, or applying Theorem 7-10, we obtain

$$A = \int_0^1 [\sqrt{x} - (-\sqrt{x})]\, dx + \int_1^4 [(2 - x) - (-\sqrt{x})]\, dx$$

$$= 2 \int_0^1 x^{1/2}\, dx + \int_1^4 (2 - x + x^{1/2})\, dx$$

$$= \frac{4}{3} x^{3/2} \Big]_0^1 + \left(2x - \frac{x^2}{2} + \frac{2x^{3/2}}{3} \right) \Big]_1^4$$

$$= 4\tfrac{1}{2} \text{ square units.}$$

In this particular example, as in many others, it is possible to avoid the necessity of breaking the area into two or more parts if we reverse the roles of x and y. Then y becomes the independent variable and x the dependent one. This changeover does not involve any new concepts or assumptions regarding area. Reduced to elements of area, it amounts to taking *horizontal* instead of vertical elements. Thus we may find the area between a curve and the y-axis, or we may find the area between two curves $x = g_1(y)$ and $x = g_2(y)$.

We shall not formally state the definitions and theorems involved, but rather expect the student to make the necessary changes and interpretations based on his understanding of what has gone before.

Then, returning to Fig. 7-11, we take a horizontal element of area. The difficulty encountered with vertical elements is no longer present. Any horizontal element has its right end in $x = 2 - y$ and its left end in $x = y^2$, and therefore may be written as

$$[(2 - y_k^*) - (y_k^*)^2](\Delta y)_k .$$

Hence, since the extremes of y are -2 and 1, we may write

$$A = \int_{-2}^1 [(2 - y) - y^2]\, dy$$

$$= \left(2y - \frac{y^2}{2} - \frac{y^3}{3} \right) \Big]_{-2}^1 = 4\tfrac{1}{2} \text{ square units.}$$

Hereafter when the sketch is drawn, the first thing to check is whether horizontal or vertical elements of area are more appropriate for the particular problem at hand.

Exercises 7-5

Find the total area bounded by the following curves in Exercises 1–19.

1. $y = 9 - x^2$ and the x-axis (Calculate in two ways.)
2. $y = x^2$ and $x = y^2$ (Calculate in two ways.)
3. $x^2 + 4y + 8 = 0$ and $y = -5$
4. $y = x^3$, $x = 2$, and the x-axis (Calculate in two ways.)
5. $y^2 = 4x$ and $y^2 + 4x - 8 = 0$
6. $y^2 + 3x - 3 = 0$ and $x = -4$ (Calculate in two ways.)
7. $3y = x^3$, $y = 0$, $x = -2$, and $x = 3$
8. $y^2 = 2x$ and $y^2 = 4x - 1$
9. $x^2 = 4y$ and $x^2 = 8y - 4$
10. $x^2 - 5x + y = 0$ and $x^2 - 5x + 2y = 0$
11. $y = x^2 - 1$ and $y = x + 1$ (Calculate in two ways.)
12. $y^2 + 4x - 8 = 0$ and $y = 2x$
13. $x^2 - 4x + 2y = 0$ and $2y + x + 6 = 0$
14. $y^2 - 2x - 4 = 0$ and $x - y = 2$
15. $y = x^3 - 2x^2 - 3x$ and $y = 5x$
16. $y = x^3 + 2x^2 - 3x$ and $y = 12x$
17. $y = x^3 + 2x^2 - 7x + 1$ and $y = x + 1$
18. $y = x^3 - 9x$ and $y = x^2 + 3x$
19. $y = 12x - x^3$ and $y = 16$
20. Determine the value of

$$\int_{-2}^{2} \sqrt{4 - x^2} \, dx$$

from geometric considerations.

7-10. A Product Theorem

In the process of applying the definite integral to certain problems, we often encounter sums of the sort

$$\sum_{k=1}^{n} f(x_k^*)g(x_k^{**})(\Delta x)_k.$$

If f and g were both evaluated for the same values of x, say x_k^*, it would follow from Definition 7-2 and Theorem 7-1 that

$$\lim_{n \to \infty} \sum_{k=1}^{n} f(x_k^*)g(x_k^*)(\Delta x)_k = \int_{a}^{b} f(x)g(x) \, dx,$$

but in the sum given above it is indicated that f and g may be evaluated for different values of x. Therefore we cannot conclude so easily that the limit of this sum is the definite integral of $f \cdot g$.

In order to cover this situation we state the following theorem.

THEOREM 7-12. Let f and g be continuous on $a \leqslant x \leqslant b$ and let the sum

$$\sum_{k=1}^{n} f(x_k^*)g(x_k^{**})(\Delta x)_k,$$

for which $x_{k-1} \leqslant x_k^* \leqslant x_k$, $x_{k-1} \leqslant x_k^{**} \leqslant x_k$, be formed as in Sec. 7-4. Then

$$\lim_{n \to \infty} \sum_{k=1}^{n} f(x_k^*)g(x_k^{**})(\Delta x)_k = \int_a^b f(x)g(x)\, dx.$$

The proof of this theorem is beyond the scope of this book. However, we shall make use of it as occasion demands. We shall find our first need for it in the next section.

7-11. Work

When a constant force F is applied through a distance x we say that the work done is given by

$$W = F \cdot x.$$

If the force units are pounds and the distance units are feet, the work units are called foot-pounds. Similarly, we can have inch-pounds, foot-tons, and so forth.

When the force is not constant, we are confronted with an undefined situation. We shall make use of the definite integral to remove this deficiency. We consider only forces applied in a straight line and in only one direction on that line. Accordingly, there is no loss in taking the line as the x-axis and the direction as the positive one. This we shall usually do, and the force applied at the point x will be represented by $F(x)$.

Suppose an object is moved from the point $x = a$ to $x = b$ by the force $F(x)$. Let the interval be subdivided into n parts by

$$x_0 = a < x_1 < x_2 \cdots < x_n = b$$

and let x_k^* be any number satisfying

$$x_{k-1} \leqslant x_k^* \leqslant x_k, \qquad k = 1, 2, \ldots, n.$$

(See Fig. 7-12.) If F were a constant on each of the subintervals, the work done on the object in moving it from x_{k-1} to x_k could be represented by

$$W_k = F(x_k^*)(x_k - x_{k-1}) = F(x_k^*)(\Delta x)_k,$$

and the total work done from $x = a$ to $x = b$ would be given by

$$W = \sum_{k=1}^{n} W_k = \sum_{k=1}^{n} F(x_k^*)(\Delta x)_k .$$

If F is a continuous function, the shorter $(\Delta x)_k$ is, the less difference there is between values of $F(x)$ taken at various points of the subinterval. In other words, the shorter $(\Delta x)_k$ is, the closer $F(x)$ is to being a constant on that subinterval. This leads us to make the following definition.

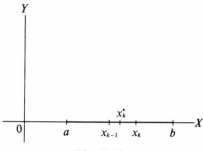

Fig. 7-12

DEFINITION 7-7. The work done on an object in moving it from $x = a$ to $x = b$, $a < b$, along the x-axis is given by

$$W = \lim_{n \to \infty} \sum_{k=1}^{n} F(x_k^*)(\Delta x)_k = \int_{a}^{b} F(x)\,dx,$$

where $F(x)$ is the force applied at the point x.

Example 7-10. The natural length of a spring is 2 ft. It stretches to a length of 3 ft when a 20-lb weight is suspended from it. Find the work done in stretching this spring from a length of 3 ft to 4 ft.

According to Hooke's law, the force required to stretch a spring x units from its natural length is

$$F(x) = kx,$$

where k is a constant depending on the physical characteristics of the spring and the units used. A force of 20 lb stretched this spring 1 ft. Therefore

$$F(1) = k = 20 \qquad \text{or} \qquad F(x) = 20x.$$

Then, from Definition 7-7,

$$W = \int_{1}^{2} 20x\,dx = 10x^2 \Big]_{1}^{2} = 30 \text{ ft-lb}.$$

Definition 7-7 is expressed in terms of the force function $F(x)$. However, in many problems a different point of view will simplify matters somewhat. Note that the terms in the sum used in this definition are *elements of work* in precisely the same sense as *elements of area* occurred in the definition of area. In certain situations it will be simpler to interpret the problem in terms of elements of work than to apply the definition directly. The next two examples illustrate this.

Example 7-11. A cylindrical tank 6 ft in diameter and 10 ft tall has its axis vertical and is full of water. Find the work done in delivering this water to a point 5 ft above the top of the tank. (Use 62.5 lb/cu ft as the weight of water.)

Let us choose the axis of the cylinder as the x-axis and the base as the origin (Fig. 7-13). If we intersect the cylinder with planes parallel to the base

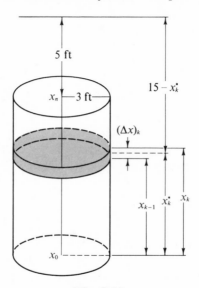

Fig. 7-13

and passing through the points of subdivision

$$x_0 = 0 < x_1 < x_2 < \cdots < x_n = 10,$$

we obtain n "disks of water" which are 6 ft in diameter and $(\Delta x)_k$ ft thick, $k = 1, 2, \ldots, n$. Let x_k^* be any number satisfying $x_{k-1} \leqslant x_k^* \leqslant x_k$. Now consider the work done in delivering a typical disk to the required point. Its weight is given by

$$3^2\pi(62.5)(\Delta x)_k \text{ lb,}$$

and the distance it must be lifted is approximately

$$(15 - x_k^*) \text{ ft}$$

so that the approximate work required for this disk is

$$W_k = 3^2\pi(62.5)(15 - x_k^*)(\Delta x)_k.$$

We wish to show that W_k is an *element of work*, that is, that the total work required is

$$W = \lim_{n \to \infty} \sum_{k=1}^{n} W_k = \int_0^{10} 3^2 \pi (62.5)(15 - x) \, dx.$$

Let

$$\underline{W}_k = 3^2 \pi (62.5)(15 - x_k)(\Delta x)_k,$$

and

$$\overline{W}_k = 3^2 \pi (62.5)(15 - x_{k-1})(\Delta x)_k.$$

Clearly,

$$\sum_{k=1}^{n} \underline{W}_k < W < \sum_{k=1}^{n} \overline{W}_k.$$

But, by Theorem 7-1,

$$\lim_{n \to \infty} \sum_{k=1}^{n} \underline{W}_k = \lim_{n \to \infty} \sum_{k=1}^{n} \overline{W}_k = \lim_{n \to \infty} \sum_{k=1}^{n} W_k$$

$$= \int_0^{10} 3^2 \pi (62.5)(15 - x) \, dx,$$

and we have the desired conclusion. Then

$$W = \int_0^{10} 3^2 \pi (62.5)(15 - x) \, dx$$

$$= 3^2 \pi (62.5) \left(15x - \frac{x^2}{2} \right) \Bigg]_0^{10} = 56{,}250\pi \text{ ft-lb.}$$

Example 7-12. A conical tank with its axis vertical is 6 ft in diameter at the top and 10 ft deep. If the tank is initially full of water, find the work done in delivering this water to a point 5 ft above the tank.

We choose the axis of the cone as the x-axis and the vertex as the origin (Fig. 7-14). We subdivide the interval from 0 to 10 into n parts, choosing

$$x_0 = 0 < x_1 < x_2 < \cdots < x_n = 10,$$

and pass planes through these points perpendicular to the axis. We obtain in this manner n frustums of the cone, which constitute its volume. Let r_k represent the radius of the section of the cone at a distance x_k from the vertex and let W_k represent the work required to deliver the section of water between x_{k-1} and x_k. We have

$$\pi (r_{k-1})^2 (62.5)(15 - x_k)(\Delta x)_k < W_k < \pi (r_k)^2 (62.5)(15 - x_{k-1})(\Delta x)_k,$$

that is, the total work W satisfies the inequality

$$\sum_{k=1}^{n} \pi (r_{k-1})^2 (62.5)(15 - x_k)(\Delta x)_k < W < \sum_{k=1}^{n} \pi (r_k)^2 (62.5)(15 - x_{k-1})(\Delta x)_k.$$

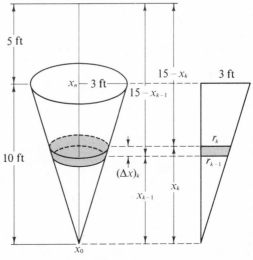

Fig. 7-14

From similar triangles,

$$\frac{r_k}{x_k} = \frac{3}{10},$$

so we may write

$$\sum_{k=1}^{n} \pi \left(\frac{3x_{k-1}}{10}\right)^2 (62.5)(15 - x_k)(\Delta x)_k < W < \sum_{k=1}^{n} \pi \left(\frac{3x_k}{10}\right)^2 (62.5)(15 - x_{k-1})(\Delta x)_k .$$

Now we take the limit as $n \to \infty$, making use of Theorem 7-12, and obtain

$$W = \int_0^{10} \pi \left(\frac{3x}{10}\right)^2 (62.5)(15 - x) \, dx.$$

Thus we can say that an element of work may be expressed in the form

$$\pi (r_k^*)^2 (62.5)(15 - x_k^*)(\Delta x)_k ,$$

where

$$r_{k-1} \leqslant r_k^* \leqslant r_k , \qquad x_{k-1} \leqslant x_k^* \leqslant x_k .$$

Finally

$$W = \frac{9\pi(62.5)}{100} \int_0^{10} (15x^2 - x^3) \, dx$$

$$= \frac{9\pi(62.5)}{100} \left[5x^3 - \frac{x^4}{4} \right]_0^{10}$$

$$= 14{,}062.5\pi \text{ ft-lb.}$$

The student should be able to adapt the method used on the last two examples to similar problems and write down the element of work, and consequently the integral for work, directly without a detailed analysis.

Exercises 7-6

1. Find the work done in stretching a spring from its natural length of 8 in. to a length of 12 in. if a force of 6 lb is needed to stretch the spring to a length of 10 in.

2. A spring of natural length 6 in. requires a force of 10 lb to stretch it 1 in. Find the work done in stretching the spring from a length of 7 in. to a length of 8 in.

3. If a force of 200 lb stretches a spring from its natural length of 5 ft to a length of 6 ft, find:
(a) The work done in stretching it to a length of 8 ft.
(b) The length of the spring when a 300-lb weight is suspended from it.
(c) The length of the spring when a total of 1600 ft-lb of work have been expended.

4. A chain 150 ft long weighing 3 lb/ft hangs in a mine shaft. Find the work done in winding this chain on a windlass until the end of the chain is 50 ft from the windlass.

5. Do the preceding exercise with a 350-lb bucket of ore attached to the end of the chain.

6. One end of a chain 60-ft long is attached to the top of a cage 20-ft high and the other end to a monkey weighing 25 lb. If the chain weighs 1/4 lb/ft, how much work does the monkey do in climbing the chain from the floor to the top of the cage?

7. A bag of sand weighing 250 lb initially is hoisted by means of a rope and pulley from a point 200 ft below the pulley at the rate of 4 fps. Sand leaks from a hole in the bag at the rate of 2 lb/sec and the rope weighs 1/4 lb/ft.
(a) Find the work done in hoisting the bag the first 100 ft.
(b) Find the work done in hoisting the bag the next 50 ft.

8. Two bodies of weights M_1 lb and M_2 lb which are r ft apart, attract each other with a force of KM_1M_2/r^2 lb. Find the work done in separating these bodies from a distance of d_1 ft to a distance of d_2 ft.

9. A vertical cylindrical tank 16 ft in diameter and 20 ft deep is half-full of water. Find the work done in pumping the water out over the top of the tank.

10. If the tank in the preceding exercise is empty, find the work done in filling it through a pipe in the bottom from a source 12 ft below the bottom.

11. A tank is in the form of a cone with its axis vertical. The top is 8 ft in diameter and it is 12 ft deep. If it contains water to a depth of 10 ft, how much work is done by pumping water into a pipe line 15 ft above the top of the tank until there remains only 4 ft of water in the tank?

12. If the tank in the preceding exercise is empty, find the work done in filling it through a pipe in the bottom from a source 6 ft below the bottom.

13. A tank has the shape of a paraboloid of revolution. (A paraboloid of revolution is the surface obtained by rotating a parabola about its axis of symmetry.) The radius of the circular top is 5 ft and it is 10 ft deep. If it is full of water, find the work done in emptying the tank at a point 4 ft above the top.

7-12. Applications of the Indefinite Integral

The applications in the preceding sections relied on the definite integral for their solution. In this section we shall see some applications of the indefinite integral and, it is hoped, learn more of the importance of the role of the arbitrary constant.

The relationships among variables in some problems are much more easily set up in terms of their derivatives and differentials than in terms of the variables themselves. Such relationships are called *differential equations*. Their solutions often hinge on indefinite integrals.

Example 7-13. Find the equations of curves which have the slope of the tangent at each point equal to twice the abscissa of the point.

We cannot immediately write down an equation $y = f(x)$ between the variables x and y, but we can write down at once the differential equation

$$\frac{dy}{dx} = f'(x) = 2x.$$

From this we can obtain the desired equation relating x and y. We have

$$y = f(x) = \int f'(x)\, dx = \int 2x\, dx$$

or

$$y = x^2 + c.$$

Thus we see that the curves defined by the conditions of the problem constitute a *family of parabolas*† (Fig. 7-15). This family of curves is sometimes called the set of *integral curves* of the differential equation.

If we add a *boundary condition* to this problem, that is, require further that the curves pass through a specified point, say $(1, 3)$, the problem reduces to determining a value, or values, of c such that the curves

$$y = x^2 + c$$

pass through the point $(1, 3)$. Substituting these coordinates in this equation, we obtain

$$3 = 1 + c \qquad \text{or} \qquad c = 2.$$

Hence there is a single member of the family, $y = x^2 + 2$, which satisfies the boundary condition.

---

† An equation containing a single arbitrary constant, often called a parameter, is said to define a one-parameter family of curves. All these curves have some common property. In this case, besides the defining property, the vertices of all the parabolas lie on the y-axis.

The arbitrary constant in the indefinite integral plays a vital role in this process. Without it the boundary condition could not be satisfied.

A particular problem which often leads to a differential equation is that of linear motion. If the distance s of a moving object from some fixed point on a line is a function of the time t, the velocity of the object is given by

$$v = \frac{ds}{dt}$$

and its acceleration a is given by

$$a = \frac{dv}{dt}.$$

These two equations enable us to solve linear motion problems when either the acceleration or velocity is known. For example, the problem of freely falling bodies near the surface of the earth (neglecting all forces except that of gravity) yield readily to these methods, since we know that the acceleration due to gravity is approximately 32 ft/sec².

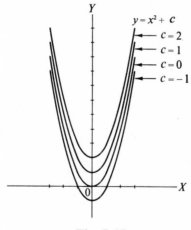

Fig. 7-15

Example 7-14. A ball is thrown upward with a velocity of 96 ft/sec from the top of a building 256 ft tall.
  (a) When will it reach its highest point?
  (b) When will it strike the ground?
  (c) With what velocity will it strike the ground?
  (d) What total distance will the ball travel?

An origin for the coordinate s must be chosen. In a problem like this it is usual to choose either the starting point or the ground. We elect to do the latter. Then at $t = 0$ we have the *initial conditions*

$$s = 256, \qquad v = 96.$$

A point to keep in mind is that $v = ds/dt$ will be positive in the direction of increasing s, and $a = dv/dt$ will be positive in the direction of increasing v.

Thus the solution of our problem is based on the equation

$$\frac{dv}{dt} = -32,$$

and the initial conditions given above. We have

$$v(t) = \int -32dt = -32t + c_1.$$

We apply the second initial condition and obtain

$$96 = -32(0) + c_1,$$

so that

$$v(t) = -32t + 96.$$

The ball will reach its highest point when $v = 0$, that is, when

$$-32t + 96 = 0.$$

Hence the answer to (a) is $t = 3$ sec after the ball was thrown.

The answer to question (b) will be obtained from $s = 0$. Continuing from above,

$$v(t) = \frac{ds}{dt} = -32t + 96,$$

or

$$s(t) = \int (-32t + 96)\, dt$$

$$= -16t^2 + 96t + c_2.$$

Since $s(0) = 256$, we have

$$256 = -16(0)^2 + 96(0) + c_2, \qquad \text{or} \qquad c_2 = 256.$$

Therefore the relationship expressing s as a function of t is

$$s(t) = -16t^2 + 96t + 256.$$

As noted above, the answer to question (b) is to be found in the roots of the equation

$$-16t^2 + 96t + 256 = 0$$

or

$$t^2 - 6t - 16 = 0.$$

The roots of this equation are $t = 8$ and $t = -2$. Since the ball was thrown at $t = 0$, the root $t = -2$ has to be discarded. Therefore the ball struck the ground 8 sec after it was thrown.

For part (c) we have only to substitute $t = 8$ into the formula for v and obtain

$$v(8) = -32(8) + 96 = -160 \text{ ft/sec.}$$

The ball was at its highest point when $t = 3$ sec. Thus its highest point was

$$s(3) = -16(3)^2 + 96(3) + 256 = 400 \text{ ft}$$

above the ground. Hence the total distance traveled is

$$400 + 144 = 544 \text{ ft.}$$

Notice again the importance of the constants entering from the indefinite integration. Without these constants it would be impossible to impress on the solution the special initial conditions which distinguish this problem from other similar ones.

Example 7-15. A firm estimates its marginal revenue at a sales level of x units per day to be

$$r = -\frac{2(x - 100)}{5}.$$

Find the revenue function.

Let R represent the total revenue from sales. Then

$$\frac{dR}{dx} = r = -\frac{2(x - 100)}{5}.$$

Hence

$$R = \int -\frac{2(x - 100)}{5} \, dx$$

$$= -\frac{2}{5}\left(\frac{x^2}{2} - 100x\right) + c,$$

where c must be determined to satisfy the conditions implied by the problem. Obviously, $R = 0$ when $x = 0$. Thus

$$0 = -\frac{2}{5}\left[\frac{0^2}{2} - 100(0)\right] + c, \quad \text{or} \quad c = 0.$$

Therefore the revenue function is expressed by the relation

$$R = -\frac{2}{5}\left(\frac{x^2}{2} - 100x\right)$$

or, completing the square in the right member,

$$R = \frac{1}{5}[10,000 - (x - 100)^2].$$

Example 7-16. A colony of animals increases from 400 to 900 individuals in 3 years. If it appears that the rate of increase is proportional to the 3/2 power of the number in the colony, find its expected population at the end of 5 years.

Let x represent the population at time t. Then the assumed growth rate may be expressed in the form

$$\frac{dx}{dt} = kx^{3/2}.$$

This differential equation is more amenable to solution if we use differentials to write it

$$x^{-3/2}\, dx = k\, dt.$$

This process is often referred to as *separation of variables*.

It is but a short step from Theorem 6-3 to the fact that *if the differentials of two functions are equal over an interval, the two functions differ at most by a constant.* We have

$$d(-2x^{-1/2}) = x^{-3/2}\, dx,$$

and

$$d(kt) = k\, dt.$$

Hence

$$-2x^{-1/2} = kt + c.$$

This is usually expressed by writing

$$\int x^{-3/2}\, dx = \int k\, dt + c,$$

which, of course, produces the same result.

$x = 400$ when $t = 0$, so

$$-2(400)^{-1/2} = k(0) + c,$$

or

$$c = -0.1.$$

Also $x = 900$ when $t = 3$ sc

$$-2(900)^{-1/2} = 3k - 0.1,$$

or

$$-\frac{2}{30} = 3k - 0.1,$$

from which

$$k = \frac{1}{90}.$$

Thus the fundamental relation between the population x and time t is

$$-2x^{-1/2} = \frac{t}{90} - 0.1.$$

At the end of 5 years we have

$$-2x^{-1/2} = \frac{5}{90} - 0.1 = -\frac{4}{90}$$

or

$$x^{-1/2} = \frac{2}{90} = \frac{1}{45}.$$

Hence

$$x^{1/2} = 45$$

or

$$x = (45)^2 = 2025,$$

the population to be expected at this time.

Exercises 7-7

1. The slope of a curve at any point is $2x + 1$ and it passes through the point $(-3, 2)$. Find its equation.

2. The slope of a curve at any point is $x^2 + 1$ and it passes through the point $(3, 8)$. Find its equation.

3. The slope of a curve at any point is $x - (1/x^2)$ and it passes through the point $(-2, 2)$. Find its equation.

4. Find the equation of the curve which has the slope -1 at the point $(2, -3)$ and for which $y'' = 2$ at all points.

5. Find the equation of the curve which is tangent to the line $2x - 3y + 5 = 0$ at the point $(-1, 1)$ and for which $y'' = -2$ at all points.

6. Find the equation of the curve which passes through the points $(-1, 2)$ and $(2, 5)$ and for which $y'' = 2 - 6x$ at all points.

7. A particle starts from rest at a point 20 ft from the origin and moves toward it on a straight line with a speed of $2t$ ft/sec. At what distance from the origin will the particle be at the end of 4 sec?

8. An object starts from rest and is accelerated 3 ft/sec^2. Find the formula which gives its distance from the starting point at the end of t sec if the motion is in a straight line.

9. An object moving in a straight line starts from rest with an acceleration which is proportional to the elapsed time. If $v = 6$ when $t = 2$, find the distance traveled in the first 5 sec.

10. An object slides down an inclined plane with an acceleration of 12 ft/sec^2. If it starts from rest and the plane is 54 ft long, what will be its velocity when it reaches the end of the plane?

11. A ball rolls down an inclined plane with an acceleration of 8 ft/sec². If it is given an initial velocity of 12 ft/sec and the plane is 160 ft long, how long will it take the ball to reach the end? What initial velocity must be imparted if the ball is to reach the end of the plane in 4 sec?

12. An object is thrown directly downward from the top of a building 256 ft tall with a speed of 96 ft/sec. When will the object strike the ground and with what speed?

13. An object is thrown vertically upward from a point 144 ft above the ground with a speed of 128 ft/sec. How high will it go and how fast will it be going when it hits the ground?

14. A ball is dropped from a balloon rising at the rate of 64 ft/sec at a height of 1872 ft. When and with what speed will it strike the ground?

15. A ball is thrown upward from the ground with an initial velocity of 48 ft/sec. At the same time a ball is dropped from a height of 144 ft. Prove that they hit the ground at the same time.

16. If a stone dropped from a balloon rising at the rate of 32 ft/sec reaches the ground in 8 sec, how high was the balloon when the stone was dropped?

17. The slope of the cost curve for a given operation is $2\sqrt{x}$. If the cost is 50 when $x = 9$, find the cost function.†

18. The rate of change of the slope of a revenue curve is $2 - 6x$. If the revenue is 208 and 300 when the sales x are 2 and 3, respectively, find the revenue function.

19. A firm estimates that the marginal cost of producing an item at the production level of x units per day is

$$m = 500 + 10x^{-1/2}.$$

If the fixed costs are \$1000 per day, how much will it cost to produce 100 units per day?

20. The marginal cost in units of \$500 for the production of an item is given by $x^2 - x + 1$, where x is the daily production in units of 1000 items. If the fixed daily costs are \$2000, what is the cost for a daily production of 3000 items?

21. A given population grows at a rate proportional to the square of the number of individuals x existing at time t, that is,

$$\frac{dx}{dt} = kx^2.$$

If there are initially 1000 individuals and this population doubles in one year, what will the population be in $1\frac{1}{2}$ years? What happens in 2 years?

---

† The units in this exercise and the next one are omitted intentionally. For example, x may be in units of 1000 items and the cost may be in units of \$10,000.

22. Suppose the population of a country increases from 1 million to $1\frac{1}{2}$ millions in the first 30 years of its existence and that the population growth follows the law stated in Exercise 21. How many people can be expected to celebrate its fiftieth anniversary? Does this country ever become overpopulated regardless of size? If so, when?

Chapter 8

THE TRIGONOMETRIC FUNCTIONS

8–1. Formulas and Definitions

The student who has not had trigonometry previously will not find the brief treatment here adequate for his needs. We shall merely restate some definitions for emphasis and enumerate a number of facts and formulas for reference. Standard books on trigonometry will supply additional details where needed.

We shall discuss angles first, since they play a key role in trigonometry. It is useful to think of an angle as being generated by a half-line rotating about its end point, the vertex of the angle. We refer to the initial position of the half-line as the *initial side* and its final position as the *terminal side*.

If an angle is generated by a clockwise rotation it is said to be a *negative angle*; a counterclockwise rotation produces, by definition, a *positive angle*. Some conventions such as these are needed to set up a correspondence between real numbers and angles.

An angle is said to be in *standard position* with respect to a coordinate system if the vertex is at the origin and the initial side lies along the positive *x*-axis.

The student is probably more familiar with the *degree-minute-second system* of angle measure. However, the system best suited for the operations of calculus is the *radian-measure system*.

DEFINITION 8–1. Place the vertex of the angle to be measured at the center of a circle of radius r. Let the length of the arc of the circle subtended by the two sides of the angle be denoted by s. Then the radian measure θ of the angle is given by $\theta = s/r$ (Fig. 8-1).

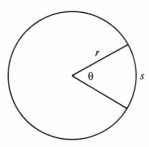

Fig. 8-1

Unless otherwise stated, it is understood from here on that the measure used for angles is radian measure. This does not mean, however, that we cannot use the familiar degree measure where convenient. It is very simple to convert from radians to degrees by using the factor $180/\pi$, or from degrees to radians by using the factor $\pi/180$. These factors arise from the fact that there are π radians in an angle whose degree measure is $180°$.

DEFINITION 8–2. Let t be any real number and place an angle whose radian measure is t in standard position. Let (x, y) be any point on the terminal side of this angle and let $r = \sqrt{x^2 + y^2}$ (Fig. 8-2). Then the six trigonometric functions of t are defined by

$$\sin t = \frac{y}{r}, \qquad \csc t = \frac{r}{y},$$

$$\cos t = \frac{x}{r}, \qquad \sec t = \frac{r}{x},$$

$$\tan t = \frac{y}{x}, \qquad \cot t = \frac{x}{y}.$$

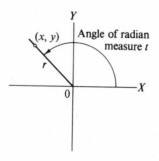

Fig. 8-2

If the angle A, whose radian measure is t, is an acute angle of a right triangle ABC (Fig. 8-3), the preceding definitions may be expressed as follows:

$$\sin t = \frac{\text{opp. side}}{\text{hypotenuse}}, \qquad \csc t = \frac{\text{hypotenuse}}{\text{opp. side}},$$

$$\cos t = \frac{\text{adj. side}}{\text{hypotenuse}}, \qquad \sec t = \frac{\text{hypotenuse}}{\text{adj. side}},$$

$$\tan t = \frac{\text{opp. side}}{\text{adj. side}}, \qquad \cot t = \frac{\text{adj. side}}{\text{opp. side}}.$$

These are useful and convenient where they apply, but the student must keep in mind their limitations.

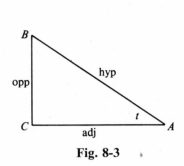

Fig. 8-3

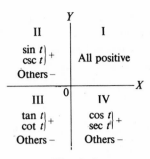

Fig. 8-4

It is clear from the definition that $\tan t$ and $\sec t$ are undefined for $t = \pm(2k + 1)\pi/2$, $k = 0, 1, 2, 3, \ldots$, for the reason that the related angle has its terminal side on the y-axis, with the result that any point on it has $x = 0$. For similar reasons $\cot t$ and $\csc t$ are undefined for $t = \pm k\pi$, $k = 0, 1, 2, 3, \ldots$.

An angle is said to lie in a given quadrant if, when the angle is placed in standard position, its terminal side lies in that quadrant.

The definitions readily yield the facts recorded in Fig. 8-4 regarding the signs of the functions of angles in the various quadrants.

A useful pairing of the trigonometric functions is the following:

$$\begin{cases} \sin t \\ \cos t \end{cases}, \quad \begin{cases} \tan t \\ \cot t \end{cases}, \quad \begin{cases} \sec t \\ \csc t \end{cases}.$$

These are the cofunctional pairs. The second of each pair is said to be the cofunction of the first, and vice versa. They are named accordingly. With this nomenclature in mind, denoting by $f(t)$ any one of the six functions, we can state the following important *reduction formulas*:

$$f\left(t \pm n \cdot \frac{\pi}{2}\right) = \begin{cases} \pm f(t), & n \text{ even} \\ \pm \text{co } f(t), & n \text{ odd} \end{cases}. \tag{8-1}$$

The $\pm$ sign in the second member must be chosen to correspond to the sign of $f(t \pm n(\pi/2))$.

Other important properties and facts regarding the trigonometric functions are listed below. They may be obtained readily from the definitions and known geometric properties of plane figures.

$$\begin{aligned} \sin(-t) &= -\sin t, & \csc(-t) &= -\csc t, \\ \cos(-t) &= \cos t, & \sec(-t) &= \sec t, \\ \tan(-t) &= -\tan t, & \cot(-t) &= -\cot t. \end{aligned} \tag{8-2}$$

$$\begin{aligned} \sin(t + 2\pi) &= \sin t, & \csc(t + 2\pi) &= \csc t, \\ \cos(t + 2\pi) &= \cos t, & \sec(t + 2\pi) &= \sec t, \\ \tan(t + \pi) &= \tan t, & \cot(t + \pi) &= \cot t \end{aligned} \tag{8-3}$$

Properties (8-3) state that the trigonometric functions are periodic, one of their most important properties. Many physical phenomona are periodic in character and are expressible in terms of these functions.

Numerical bounds on the trigonometric functions are as follows:

$$-1 \leqslant \sin t \leqslant 1, \qquad \csc t \leqslant -1 \quad \text{or} \geqslant 1,$$
$$-1 \leqslant \cos t \leqslant 1, \qquad \sec t \leqslant -1 \quad \text{or} \geqslant 1, \qquad (8\text{-}4)$$
$$-\infty < \tan t < \infty, \qquad -\infty < \cot t < \infty.$$

The following function values of special angles will be used repeatedly in the work to follow.

t	0	$\pi/6 \ (30°)$	$\pi/4 \ (45°)$	$\pi/3 \ (60°)$	$\pi/2 \ (90°)$
$\sin t$	0	$1/2$	$\sqrt{2}/2$	$\sqrt{3}/2$	1
$\cos t$	1	$\sqrt{3}/2$	$\sqrt{2}/2$	$1/2$	0
$\tan t$	0	$\sqrt{3}/3$	1	$\sqrt{3}$	undefined

Sets of identities, usually referred to as the *fundamental identities*, are

$$\sin t = \frac{1}{\csc t}, \qquad \cos t = \frac{1}{\sec t},$$
$$\tan t = \frac{1}{\cot t}, \qquad \tan t = \frac{\sin t}{\cos t}, \qquad (8\text{-}5)$$

and

$$\sin^2 t + \cos^2 t = 1,$$
$$1 + \tan^2 t = \sec^2 t, \qquad (8\text{-}6)$$
$$1 + \cot^2 t = \csc^2 t.$$

Other trigonometric formulas that are needed from time to time follow:

$$\sin(t_1 \pm t_2) = \sin t_1 \cos t_2 \pm \cos t_1 \sin t_2. \qquad (8\text{-}7)$$

$$\cos(t_1 \pm t_2) = \cos t_1 \cos t_2 \mp \sin t_1 \sin t_2. \qquad (8\text{-}8)$$

$$\sin 2t = 2 \sin t \cos t. \qquad (8\text{-}9)$$

$$\cos 2t = \cos^2 t - \sin^2 t$$
$$= 2 \cos^2 t - 1$$
$$= 1 - 2 \sin^2 t. \qquad (8\text{-}10)$$

$$\sin \frac{t}{2} = \pm \sqrt{\frac{1 - \cos t}{2}}. \qquad (8\text{-}11)$$

$$\cos \frac{t}{2} = \pm \sqrt{\frac{1 + \cos t}{2}}. \qquad (8\text{-}12)$$

$$\sin t_1 + \sin t_2 = 2 \sin \frac{t_1 + t_2}{2} \cos \frac{t_1 - t_2}{2}. \tag{8-13}$$

$$\sin t_1 - \sin t_2 = 2 \cos \frac{t_1 + t_2}{2} \sin \frac{t_1 - t_2}{2}. \tag{8-14}$$

$$\cos t_1 + \cos t_2 = 2 \cos \frac{t_1 + t_2}{2} \cos \frac{t_1 - t_2}{2}. \tag{8-15}$$

$$\cos t_1 - \cos t_2 = -2 \sin \frac{t_1 + t_2}{2} \sin \frac{t_1 - t_2}{2}. \tag{8-16}$$

Two useful trigonometric formulas related to the general triangle (Fig. 8-5)

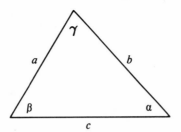

Fig. 8-5

are the following:

$$\text{Area} = \tfrac{1}{2}ab \sin \gamma. \tag{8-17}$$

$$c^2 = a^2 + b^2 - 2ab \cos \gamma. \tag{8-18}$$

Obviously each of these formulas has two other forms obtained by appropriate interchanges of sides and angles.

Finally, for reference, we give the graphs of $y = \sin x$, $y = \cos x$, $y = \tan x$ (Figs. 8-6, 8-7, 8-8).

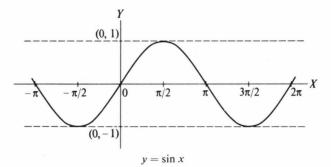

$y = \sin x$

Fig. 8-6

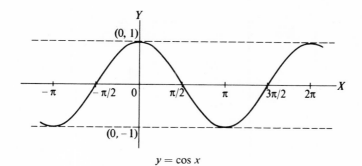

$y = \cos x$

Fig. 8-7

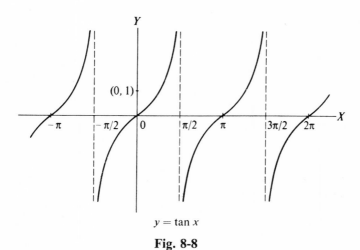

$y = \tan x$

Fig. 8-8

The following set of exercises will serve to review some of the operations the student will make use of in the remainder of this book.

Exercises 8-1

1. Convert the following to radians, expressing your result in terms of simple fractions of π, or as decimal fractions, using the fact that $\pi/180 = 0.01745$ and $\pi/60(180) = 0.00029$.
   - (a) $-90°$
   - (b) $330°$
   - (c) $-135°$
   - (d) $405°$
   - (e) $270°$
   - (f) $210°$
   - (g) $10°$
   - (h) $15°13'$
   - (i) $25°10'$

2. Convert the following angles, given in radian measure, to degree-minute measure.
   - (a) $3\pi/4$
   - (b) π
   - (c) $-4\pi/3$
   - (d) $5\pi/3$
   - (e) $-7\pi/4$
   - (f) 3π
   - (g) 0.14453
   - (h) 1.37253
   - (i) 0.51294

3. Make a table of values for $\sin t$, $\cos t$, and $\tan t$ for $t = 0$, $\pi/6$, $\pi/4$, $\pi/3$, $\pi/2$, $2\pi/3$, $3\pi/4$, $5\pi/6$, π, $7\pi/6$, $5\pi/4$, $4\pi/3$, $3\pi/2$, $5\pi/3$, $7\pi/4$, $11\pi/6$, 2π.

4. Without the use of tables, find two positive and two negative values of t satisfying:
   (a) $\sin t = \sqrt{2}/2$ (b) $\cos t = -\sqrt{3}/2$ (c) $\tan t = -1$
   (d) $\cos t = 1/2$ (e) $\tan t = -\sqrt{3}$ (f) $\sin t = -\sqrt{3}/2$

5. Derive the fundamental identities (8-5) and (8-6) directly from Definition 8-2.

6. Derive (8-9) and (8-10) from (8-7) and (8-8).

Make use of (8-5) to (8-16) to prove the statements in Exercises 7–10.

7. $4 \sin^4 t = 1 - 2 \cos 2t + \cos^2 2t$

8. $\dfrac{1}{1 + \cos t} - \dfrac{1}{1 - \cos t} = -2 \csc t \cot t$

9. $\cos^4 \dfrac{t}{2} + \sin^4 \dfrac{t}{2} = 1 - \dfrac{1}{2} \sin^2 t$

10. $\sin^2 \dfrac{t}{2} \cos^2 \dfrac{t}{2} = \dfrac{1}{8}(1 - \cos 2t)$

11. Without using tables, find the value of (a) $\sin 22\frac{1}{2}°$; (b) $\cos \pi/12$; (c) $\tan 5\pi/12$.

12. Write formulas in terms of $\sin t$ and $\cos t$ for
    (a) $\sin(3\pi/2 + t)$ (b) $\cos(5\pi/2 - t)$ (c) $\sin(3\pi/2 - t)$
    (d) $\sin(3\pi/4 + t)$ (e) $\sin(2\pi/3 - t)$ (f) $\cos(\pi/2 - t)$

Find all values of x, $|x| \leqslant 2\pi$, which satisfy the following equations:

13. $2 \cos^2 x = 1$ 14. $\cot x - 3 \tan x = 0$

15. $\cos 2x = \cos^2 x$ 16. $2 \sin^2 x - \cos 2x = 0$

17. $3 \sin x - 2 \cos^2 x = 0$ 18. $\sin 2x + 2 \cos^2 x = 2$

8-2. An Important Limit

We now seek to evaluate

$$\lim_{\alpha \to 0} \frac{\sin \alpha}{\alpha},$$

where α is measured in radians. This limit is fundamental to the calculus of trigonometric functions. It cannot be determined by Theorem 4-5 because the limit of the denominator is zero, but it can be calculated by a geometric device as follows.

Since $\alpha \to 0$ there is no loss in assuming $|\alpha| < \pi/2$; let us for the moment also assume $\alpha > 0$. In Fig. 8-9 let $A_1 = $ area $\triangle ABC$, $A_2 = $ area $\measuredangle ABD$, and

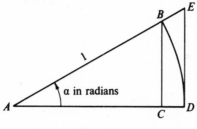

α in radians

Fig. 8-9

$A_3 = $ area $\triangle AED$. Then

$$A_1 < A_2 < A_3,$$

and since $AB = AD = 1$,

$$A_1 = \tfrac{1}{2}(AC)(CB) = \tfrac{1}{2} \cos \alpha \sin \alpha,$$

$$A_2 = \tfrac{1}{2}(AB)^2\alpha = \tfrac{1}{2}\alpha,\dagger$$

$$A_3 = \frac{1}{2}(AD)(ED) = \frac{1}{2}\tan \alpha = \frac{1}{2}\frac{\sin \alpha}{\cos \alpha}.$$

Therefore

$$\frac{1}{2}\cos \alpha \sin \alpha < \frac{1}{2}\alpha < \frac{1}{2}\frac{\sin \alpha}{\cos \alpha},$$

from which, dividing by the positive quantity $\tfrac{1}{2}\sin \alpha$, we obtain

$$\cos \alpha < \frac{\alpha}{\sin \alpha} < \frac{1}{\cos \alpha}.$$

But

$$\lim_{\alpha \to 0} \cos \alpha = \lim_{\alpha \to 0} \frac{1}{\cos \alpha} = 1,$$

so, by Theorem 4-9,

$$\lim_{\alpha \to 0} \frac{\alpha}{\sin \alpha} = 1, \qquad \alpha > 0.$$

$\dagger$ The area of the sector of a circle of radius r and central angle α (α measured in radians) is $\tfrac{1}{2}r^2\alpha$.

If we write

$$f(\alpha) = \frac{\alpha}{\sin \alpha},$$

we have

$$f(-\alpha) = \frac{-\alpha}{\sin(-\alpha)} = \frac{-\alpha}{-\sin \alpha} = \frac{\alpha}{\sin \alpha} = f(\alpha).$$

Hence f is an even function and a change in sign of α produces no change in the function value. Thus we are able to remove the restriction that α be positive.

Now

$$\lim_{\alpha \to 0} \frac{\sin \alpha}{\alpha} = \lim_{\alpha \to 0} \frac{1}{\dfrac{\alpha}{\sin \alpha}} = \frac{1}{\lim\limits_{\alpha \to 0} \dfrac{\alpha}{\sin \alpha}} = 1,$$

and we have the following theorem.

THEOREM 8–1. If α is measured in radians,

$$\lim_{\alpha \to 0} \frac{\sin \alpha}{\alpha} = 1.$$

Radian measure was essential to this proof because we used $\frac{1}{2}r^2\alpha$ for the area of a sector of a circle of central angle α and radius r. This formula is true only for angles measured in radians.

Let us calculate

$$\lim_{\alpha \to 0} \frac{\sin \alpha}{\alpha},$$

where α is in degrees. In order to apply Theorem 8-1 we must convert to radians. Corresponding to α degrees we have $\alpha\pi/180$ radians, both measures representing the same angle. Thus

$$\sin(\alpha)^\circ = \sin\left(\frac{\alpha\pi}{180}\right) \text{ radians.}$$

Hence, when we convert from degrees to radians, we have

$$\lim_{\alpha \to 0} \frac{\sin \alpha}{\alpha} = \lim_{\alpha \to 0} \frac{\pi}{180} \frac{\sin(\alpha\pi/180)}{\alpha\pi/180} = \frac{\pi}{180} \lim_{\alpha\pi/180 \to 0} \frac{\sin(\alpha\pi/180)}{\alpha\pi/180} = \frac{\pi}{180}.$$

This illustrates why we prefer radian measure if this limit is involved.
COROLLARY

$$\lim_{\alpha \to 0} \frac{1 - \cos \alpha}{\alpha} = 0.$$

To prove this we write

$$\lim_{\alpha \to 0} \frac{1 - \cos \alpha}{\alpha} = \lim_{\alpha \to 0} \frac{1 - \cos \alpha}{\alpha} \cdot \frac{1 + \cos \alpha}{1 + \cos \alpha}$$

$$= \lim_{\alpha \to 0} \frac{\sin^2 \alpha}{\alpha(1 + \cos \alpha)}$$

$$= \lim_{\alpha \to 0} \frac{\sin \alpha}{\alpha} \lim_{\alpha \to 0} \frac{\sin \alpha}{1 + \cos \alpha}$$

$$= 1 \cdot \frac{0}{2} = 0.$$

8-3. Derivatives of Trigonometric Functions

In this section we shall obtain formulas for calculating derivatives of the trigonometric functions. We shall make use of the two limits found in Sec. 8-2 to determine the derivative of the sine function and then the derivatives of the other functions will follow easily from the differentiation formulas of Chapter 5 and known trigonometric relations.

Let

$$y = \sin x.$$

Then

$$\frac{\Delta y}{\Delta x} = \frac{\sin(x + \Delta x) - \sin x}{\Delta x}$$

$$= \frac{\sin x \cos \Delta x + \cos x \sin \Delta x - \sin x}{\Delta x},$$

where we have made use of (8-7). Thus we may write

$$\frac{\Delta y}{\Delta x} = (\sin x)\left(\frac{\cos \Delta x - 1}{\Delta x}\right) + (\cos x)\left(\frac{\sin \Delta x}{\Delta x}\right),$$

and

$$\frac{dy}{dx} = \lim_{\Delta x \to 0} \frac{\Delta y}{\Delta x} = \lim_{\Delta x \to 0} \left\{ (\sin x)\left(\frac{\cos \Delta x - 1}{\Delta x}\right) + (\cos x)\left(\frac{\sin \Delta x}{\Delta x}\right) \right\}$$

$$= (\sin x) \lim_{\Delta x \to 0} \left(\frac{\cos \Delta x - 1}{\Delta x}\right) + (\cos x) \lim_{\Delta x \to 0} \frac{\sin \Delta x}{\Delta x}.$$

Hence, applying Theorem 8-1 and its corollary,

$$\frac{dy}{dx} = (\sin x)(0) + (\cos x)(1)$$

$$= \cos x,$$

and we have the next theorem.

THEOREM 8–2

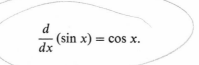

$$\frac{d}{dx}(\sin x) = \cos x.$$

Application of Theorem 8-2 and the chain rule gives a more general form of this theorem.

THEOREM 8–3. If u is a differentiable function of x,

$$\frac{d}{dx}(\sin u) = \cos u \, \frac{du}{dx}.$$

Now let

$$y = \cos u.$$

Then, since

$$\cos u = \sin\left(\frac{\pi}{2} - u\right),$$

we have

$$y = \sin\left(\frac{\pi}{2} - u\right),$$

and applying Theorem 8-3,

$$\frac{dy}{dx} = \cos\left(\frac{\pi}{2} - u\right)\frac{d}{dx}\left(\frac{\pi}{2} - u\right)$$

$$= -\sin u \, \frac{du}{dx}.$$

THEOREM 8–4. If u is a differentiable function of x,

$$\frac{d}{dx}(\cos u) = -\sin u \, \frac{du}{dx}.$$

To obtain the formula for differentiating the tangent function we write

$$y = \tan u = \frac{\sin u}{\cos u},$$

and, from Theorems 5-6, 8-3, 8-4,

$$\frac{dy}{dx} = \frac{\cos u \dfrac{d(\sin u)}{dx} - \sin u \dfrac{d(\cos u)}{dx}}{\cos^2 u}$$

$$= \frac{\cos^2 u \dfrac{du}{dx} + \sin^2 u \dfrac{du}{dx}}{\cos^2 u}$$

$$= \frac{\dfrac{du}{dx}}{\cos^2 u} = \sec^2 u \frac{du}{dx}.$$

THEOREM 8-5. If u is a differentiable function of x,

$$\frac{d}{dx}(\tan u) = \sec^2 u \frac{du}{dx}.$$

The same general procedures applied to the remaining three functions will result in the following theorem.

THEOREM 8-6. If u is a differentiable function of x,

(a)
$$\frac{d}{dx}(\cot u) = -\csc^2 u \frac{du}{dx};$$

(b)
$$\frac{d}{dx}(\sec u) = \sec u \tan u \frac{du}{dx};$$

(c)
$$\frac{d}{dx}(\csc u) = -\csc u \cot u \frac{du}{dx}.$$

The proof of this theorem is left as an exercise for the student.

The differentiation of expressions involving trigonometric functions will require the use not only of Theorems 8-2 through 8-6 but also of the formulas of Chapter 5. The following examples will illustrate this.

Example 8-1. Find dy/dx if $y = \sin x^2$.

This may be interpreted as $y = \sin u$, $u = x^2$. Therefore

$$\frac{dy}{dx} = \cos x^2 \frac{d}{dx}(x^2) = 2x \cos x^2,$$

by Theorems 8-3 and 5-4.

Example 8–2. Find dy/dx if $y = \sin^2 x$.

First we consider this as $y = u^2$, $u = \sin x$. Then, by Theorem 5-8,

$$\frac{dy}{dx} = 2 \sin x \frac{d}{dx} (\sin x),$$

which, by Theorem 8-2, gives

$$\frac{dy}{dx} = 2 \sin x \cos x = \sin 2x.$$

Example 8–3. Find dy/dx, where

$$y = \cos x \sin^2 2x.$$

First we apply the rule for a product and obtain

$$\frac{dy}{dx} = \cos x \frac{d}{dx} (\sin^2 2x) + \sin^2 2x \frac{d}{dx} (\cos x).$$

Then we apply the rule for a power to the first term in the left member and get

$$\frac{dy}{dx} = (\cos x)(2 \sin 2x) \frac{d}{dx} (\sin 2x) + \sin^2 2x \frac{d}{dx} (\cos x).$$

Finally, Theorems 8-3 and 8-4 give

$$\frac{dy}{dx} = (\cos x)(2 \sin 2x)(2 \cos 2x) - (\sin^2 2x)(\sin x)$$

$$= 2 \cos x \sin 4x - \sin x \sin^2 2x.$$

The standard trigonometric identities can be used to express this result in a variety of forms.

Many applications are most readily solved by the introduction of trigonometric functions. The following is an example of this.

Example 8–4. A gutter in the form of an isosceles trapezoid is to be made from a strip of sheet metal by bending up one third of the strip on either side. If the strip is 15 in. wide, how wide should the gutter be across the top if its capacity is to be a maximum?

Clearly the capacity of the gutter will be a maximum if its cross-sectional area is a maximum. From Fig. 8-10, the area of a cross section is given by

$$A = \left(\frac{5 + w}{2}\right) h$$

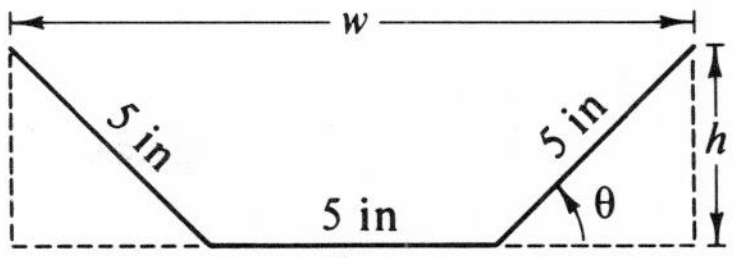

Fig. 8-10

where w is the upper base of the trapezoid and h is its altitude. If we denote the exterior base angle of the trapezoid by θ, we have

$$h = 5 \sin \theta,$$

and

$$w = 5 + 2(5 \cos \theta).$$

Thus we may express A in terms of trigonometric functions:

$$A = 25 \sin \theta (1 + \cos \theta),$$

so our problem reduces to finding the values of θ that make A a maximum. Following the procedures developed in Chapter 6, we obtain

$$\frac{dA}{d\theta} = 25[\cos \theta (1 + \cos \theta) + \sin \theta (-\sin \theta)]$$

$$= 25(\cos \theta + \cos^2 \theta - \sin^2 \theta).$$

The critical values of θ are the roots of the equation

$$25(\cos \theta + \cos^2 \theta - \sin^2 \theta) = 0.$$

If we divide by 25 and set $\sin^2 \theta = 1 - \cos^2 \theta$, we get

$$2 \cos^2 \theta + \cos \theta - 1 = 0,$$

or

$$(2 \cos \theta - 1)(\cos \theta + 1) = 0,$$

or

$$\cos \theta = \frac{1}{2}, \qquad \cos \theta = -1.$$

Obviously, from the geometry of the problem, the roots

$$\theta = \pm \pi, \pm 3\pi, \dots ,$$

arising from $\cos \theta = -1$, do not provide the solution to our problem. But the root $\theta = \pi/3$ resulting from $\cos \theta = 1/2$ does give a maximum A. This may be verified by geometrical considerations or the methods of Chapter 6. For example,

$$\frac{d^2 A}{d\theta^2} = -25(\sin \theta + 4 \sin \theta \cos \theta).$$

Therefore

$$\frac{d^2 A}{d\theta^2}\bigg]_{\theta=\pi/3} = -25\left[\frac{\sqrt{3}}{2} + 4\left(\frac{\sqrt{3}}{2}\right)\left(\frac{1}{2}\right)\right] < 0.$$

Thus A is a maximum when $\theta = \pi/3$, by Theorem 6-7, and

$$w = 5 + 2(5 \cos (\pi/3)) = 10 \text{ in.}$$

Exercises 8-2

Evaluate the limits in Exercises 1–6.

1. $\displaystyle\lim_{\theta \to 0} \frac{\sin 2\theta}{\theta}$ 　　　　　　　　 **2.** $\displaystyle\lim_{t \to 0} \frac{\sin t}{2t}$

3. $\displaystyle\lim_{\alpha \to 0} \frac{\tan \alpha}{\alpha}$ 　　　　　　　　 **4.** $\displaystyle\lim_{x \to 0} \frac{\sin^2 x}{x}$

5. $\displaystyle\lim_{x \to 0} \frac{\sin x^2}{x}$ 　　　　　　　　 **6.** $\displaystyle\lim_{y \to 0} \frac{1 - \cos y}{y^2}$

7. Prove Theorem 8-6.

Calculate dy/dx in each of Exercises 8–25.

8. $y = \sin(2x - 1)$ 　　　　　　　　 **9.** $y = \cos(3x + 2)$

10. $y = \cos^3 2x$ 　　　　　　　　 **11.** $y = \tan^2 3x$

12. $y = \sin x \tan x$ 　　　　　　　　 **13.** $y = x \cot x$

14. $y = \sec^2 x$ 　　　　　　　　 **15.** $y = \csc^3 2x$

16. $y = 2 \sin 3x + \tan x$ 　　　　　 **17.** $y = \sin x \cos x + \cot^2 2x$

18. $y = \sqrt{1 + \cos x}$ 　　　　　　 **19.** $y = \sin\sqrt{1 + x^2}$

20. $y = \dfrac{\tan 2x}{\cos 2x}$ 　　　　　　　 **21.** $y = \dfrac{x}{\sin x}$

22. $y = \sin^2\left(\dfrac{1}{x}\right)$ 　　　　　　 **23.** $y = \dfrac{1 - \tan^2 2x}{\tan 2x}$

24. $y = x(1 + \sec x)^3$ 　　　　　　 **25.** $y = \cos(x\sqrt{1 - x})$

Find dy/dx by implicit differentiation in each of Exercises 26–29.

26. $x \sin y + y \sin x = 0$ 　　　　 **27.** $x^2 \tan y + x^2 - y^2 = 0$

28. $\cos(2x + y) - \sin(2x + y) = 0$ 　　 **29.** $\cot x \sec y + \sin xy + 5 = 0$

30. Find the maximum and minimum values of $y = \sin x + \cos x$.

31. Determine the relative maximum and minimum points on the graph of $y = x + \cos x$. Sketch the curve.

32. A free balloon is released and rises at the constant rate of 10 ft/sec in a vertical line. An observer watches the balloon from a point 100 ft from the point of release. At what rate is the angle of elevation of the observer's line of sight increasing 3 sec after the release. (Neglect the height of the observer.)

33. A radar antenna located on a ship 3 mi directly off a point P on a straight shoreline rotates 30 times per minute. How fast is the radar beam moving along the shoreline at the instant it is 4 mi from P?

34. A gutter in the form of an isosceles trapezoid is to be made from a strip of sheet metal 13 in. wide by bending up 3 in. on either side. How wide should the gutter be at the top if its capacity is to be a maximum?

35. Find the base radius and altitude of the right-circular cylinder of maximum volume which may be inscribed in a sphere of radius a.

36. Find the altitude of the right-circular cone of minimum volume which may be circumscribed about a sphere of radius a.

37. Find the altitude of the right-circular cone of maximum volume which may be inscribed in a sphere of radius a.

38. A fence 8 ft high is 27 ft from a building. If the ground is level, what is the length of the shortest ladder that will reach the wall of the building from outside the fence?

39. Two hallways $4\frac{1}{2}$ ft and $10\frac{2}{3}$ ft wide, respectively, meet at right angles. Find the length of the longest thin rod that can be carried from one hallway to the other in a horizontal position.

40. A man in a rowboat is 3 mi from the closest point A on a straight shoreline. Point B is 10 mi from A on this shoreline. If he can row 3 mph and walk 5 mph, where should he land in order to reach B in the minimum time? Use trigonometric functions to solve this problem.

8-4. Inverse Trigonometric Functions

The need for *inverse functions* occurs frequently in mathematics. If we have a function f and

$$y = f(x),$$

we ask if this relation defines x as a function of y, that is, is there a function g such that

$$x = g(y)?$$

If so, we say that g is the inverse of f, and we often indicate this relationship by writing

$$g = f^{-1}.$$

This symbolism is not to be confused with the exponent -1.

It is not our purpose to go into the question of when a relation $y = f(x)$ defines x as a function of y, that is, when f has an inverse. We shall content ourselves with stating that many functions have inverses, and we shall see some of the important ones as we progress through the remaining pages of this book.

Let us look at this question as it relates to the sine function. We reverse the roles of x and y and write

$$x = \sin y \tag{8-19}$$

so that x will appear as the independent variable in the inverse function. The graph of this equation appears in Fig. 8-11. The domain of $\sin y$ is $-\infty < y < \infty$ and its range (the range of x) is $-1 \leqslant x \leqslant 1$. It is clear from the figure that any x, $-1 \leqslant x \leqslant 1$, has infinitely many values of y related to it by (8-19), since a vertical line $x = x_0$, $-1 \leqslant x_0 \leqslant 1$, intersects this curve in infinitely many points. Therefore, unless some restrictions are imposed, (8-19) does not define y as a function of x. However, it is equally clear from the graph that when y is restricted to the values $-\pi/2 \leqslant y \leqslant \pi/2$, equation (8-19) relates a single value of y to each value of x in its range, and consequently, under this restriction, y is defined as a function of x.

We write

$$y = \text{Arcsin } x$$

and read it *y equals Arcsine of x, or y is the number whose sine is x.*

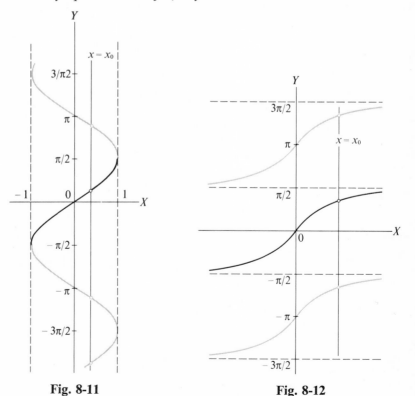

Fig. 8-11   Fig. 8-12

DEFINITION 8–3. Arcsin x is defined to be the number y, $-(\pi/2) \leqslant y \leqslant \pi/2$, such that $x = \sin y$, $-1 \leqslant x \leqslant 1$.†

Next let us consider the tangent function and the equation

$$x = \tan y. \qquad (8\text{-}20)$$

As before, we note from the graph (Fig. 8-12) that this relation does not define y as a function of x since any vertical line will intersect the curve in infinitely many points. But if we restrict y to the range $-(\pi/2) < y < \pi/2$, there is but one value of y associated with each value of x by (8-20). Therefore, under this condition, the tangent function has an inverse and we make the following definition.

DEFINITION 8–4. Arctan x is defined to be the number y, $-(\pi/2) < y < \pi/2$, such that $x = \tan y$.

Similar reasoning coupled with the graph of $x = \sec y$ (Fig. 8-13) will lead to a definition of an inverse of the secant function.

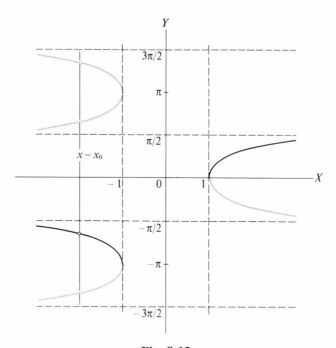

Fig. 8-13

---

† Many people use the notation $\text{Sin}^{-1} x$ for Arcsin x in accordance with the convention regarding inverse functions already mentioned in this section. However, we prefer the notation Arcsin x and shall use it consistently.

DEFINITION 8-5. Arcsec x is defined to be the number y, $-\pi \leqslant y < -(\pi/2)$ for $x \leqslant -1$, or $0 \leqslant y < \pi/2$ for $x \geqslant 1$, such that $x = \sec y$.

Other inverse trigonometric functions may be defined similarly but little is gained by doing so. Therefore we shall restrict ourselves to the three above.

Example 8-5. Determine (a) Arcsin $1/2$; (b) Arctan(-1); (c) Arcsec(-2).

(a) Arcsin $1/2 = \pi/6$ since $-(\pi/2) \leqslant \pi/6 \leqslant \pi/2$ and $\sin \pi/6 = 1/2$.
(b) Arctan$(-1) = -(\pi/4)$ because $\tan(-(\pi/4)) = -1$ and $-(\pi/2) < -(\pi/4) < \pi/2$.
(c) Arcsec$(-2) = -(2\pi/3)$ since $\sec(-(2\pi/3)) = -2$ and $-\pi \leqslant -(2\pi/3) < -(\pi/2)$.

8-5. Differentiation of the Inverse Trigonometric Functions

In order to compute the derivatives of the inverse trigonometric functions we employ the method of implicit differentiation coupled with the formulas developed for the derivatives of the trigonometric functions.

If

$$y = \text{Arcsin } x,$$

we write

$$x = \sin y$$

and differentiate implicitly with respect to x. We obtain

$$1 = \cos y \frac{dy}{dx},$$

from which

$$\frac{dy}{dx} = \frac{1}{\cos y}.$$

From (8-6), we have

$$\cos y = \pm\sqrt{1 - \sin^2 y}.$$

Moreover, from Definition 8-3, $-(\pi/2) \leqslant y \leqslant \pi/2$. Therefore $\cos y \geqslant 0$, so that we may eliminate the double sign and write

$$\frac{dy}{dx} = \frac{1}{\sqrt{1 - \sin^2 y}} = \frac{1}{\sqrt{1 - x^2}}, \qquad |x| < 1.$$

Thus, applying the chain rule, we may make the following statement.

THEOREM 8-7. If u is a differentiable function of x,

$$\frac{d}{dx}(\text{Arcsin } u) = \frac{du/dx}{\sqrt{1 - u^2}}, \qquad |u| < 1.$$

Similarly, if

$$y = \text{Arctan } x,$$

we have

$$x = \tan y$$

and, differentiating implicitly,

$$1 = \sec^2 y \frac{dy}{dx}.$$

Then, using (8-6),

$$\frac{dy}{dx} = \frac{1}{\sec^2 y} = \frac{1}{1 + \tan^2 y} = \frac{1}{1 + x^2}.$$

More generally, we again apply the chain rule and obtain the following theorem.

THEOREM 8-8. If u is a differentiable function of x,

$$\frac{d}{dx} (\text{Arctan } u) = \frac{du/dx}{1 + u^2}.$$

Let

$$y = \text{Arcsec } x.$$

Then

$$x = \sec y,$$

from which, differentiating implicitly, we obtain

$$1 = \sec y \tan y \frac{dy}{dx}$$

or

$$\frac{dy}{dx} = \frac{1}{\sec y \tan y}.$$

According to Definition 8-5, $-\pi \leqslant y < -(\pi/2)$ or $0 \leqslant y < \pi/2$, and for these values of y, $\tan y \geqslant 0$. Thus we may write

$$\frac{dy}{dx} = \frac{1}{\sec y \sqrt{\sec^2 y - 1}} = \frac{1}{x\sqrt{x^2 - 1}}, \qquad |x| > 1.$$

Proceeding as before we have the following theorem.

THEOREM 8-9. If u is a differentiable function of x,

$$\frac{d}{dx} (\text{Arcsec } u) = \frac{du/dx}{u\sqrt{u^2 - 1}}, \qquad |u| > 1.$$

Example 8–6. Find dy/dx if $y = \text{Arcsin} \sqrt{x}$.

From Theorem 8-7,

$$\frac{dy}{dx} = \frac{\dfrac{d}{dx}(\sqrt{x})}{\sqrt{1 - (\sqrt{x})^2}}$$

$$= \frac{(1/2)x^{-1/2}}{\sqrt{1 - x}} = \frac{1}{2\sqrt{x(1 - x)}}.$$

Example 8–7. Find dy/dx if $y = \text{Arctan}(1/x)$.

From Theorem 8-8,

$$\frac{dy}{dx} = \frac{\dfrac{d}{dx}\left(\dfrac{1}{x}\right)}{1 + (1/x)^2}$$

$$= \frac{-(1/x^2)}{(x^2 + 1)/x^2} = -\frac{1}{1 + x^2}.$$

Example 8–8. A rotating light is 120 ft from a straight fence upon which it casts a moving spot of light. If the light is rotating at a constant speed of 3 rpm, how fast is the spot moving along the fence at the point nearest the light? How fast is it moving when it has moved 30 ft beyond this point?

In order to answer the questions in this example we need to calculate ds/dt for $s = 0$ and $s = 30$ (Fig. 8-14). Although not essential to the solution, the inverse tangent function provides a convenient means of writing a relationship among the variables in this problem. We have

$$\theta = \text{Arctan} \frac{s}{120},$$

and thus

$$\frac{d\theta}{dt} = \frac{\dfrac{d}{dt}\left(\dfrac{s}{120}\right)}{1 + (s/120)^2} = \frac{120(ds/dt)}{(120)^2 + s^2}.$$

We may write

$$\frac{d\theta}{dt} = 6\pi \quad \text{rad/min},$$

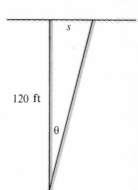

120 ft

Fig. 8-14

since the light makes 3 rpm. Then the general formula for ds/dt is

$$\frac{ds}{dt} = \frac{6\pi[(120)^2 + s^2]}{120},$$

and in particular

$$\frac{ds}{dt}\bigg]_{s=0} = 6\pi(120) = 720\pi \quad \text{ft/min};$$

$$\frac{ds}{dt}\bigg]_{s=30} = \frac{6\pi[(120)^2 + (30)^2]}{120} = 765\pi \quad \text{ft/min}.$$

Exercises 8-3

1. Evaluate the following:

(a) $\text{Arctan}(1)$ (b) $\text{Arcsin}\dfrac{\sqrt{3}}{2}$

(c) $\text{Arcsec}\sqrt{2}$ (d) $\text{Arcsin}\left(\dfrac{-\sqrt{2}}{2}\right)$

2. Evaluate the following:

(a) $\cos(\text{Arcsin}(-1))$ (b) $\cot\left(\text{Arctan}\left(\dfrac{-\sqrt{3}}{3}\right)\right)$

(c) $\tan(\text{Arcsec }2)$ (d) $\csc(\text{Arctan}(-\sqrt{3}))$

Calculate dy/dx in each of Exercises 3–16.

3. $y = \text{Arcsin }3x$ **4.** $y = \text{Arctan }x^2$

5. $y = \text{Arcsec }\sqrt{x}$ **6.** $y = \text{Arcsin }\dfrac{1}{x+1}$

7. $y = \text{Arctan }\sqrt{x^2 - 1}$ **8.** $y = \text{Arcsec}\left(\dfrac{1}{x}\right)$

9. $y = \text{Arctan}\left(x + \dfrac{1}{x}\right)$ **10.** $y = \text{Arcsin }\dfrac{x}{x-1}$

11. $y = \text{Arcsin}\left(x - \dfrac{1}{x}\right)$ **12.** $y = \text{Arcsec }\dfrac{1}{\sqrt{1-x^2}}$

13. $y = x^2 \text{ Arcsin }3x$ **14.** $y = \sqrt{x^2 - 1} + \text{Arcsin }\dfrac{1}{x}$

15. $y = 2\sqrt{x} - 2 \text{ Arctan}\sqrt{x}$ **16.** $y = x\sqrt{1 - x^2} + \text{Arcsin }x$

17. A balloon leaving the ground 750 ft from an observer rises vertically at the rate of 80 ft/min. At what rate is the angle of elevation of the observer's line of sight increasing after 10 min?

18. A flag pole 45 ft tall stands on the roof of a building. If the base of the pole is 80 ft above the observer's eye, how far from the building should he stand in order that the pole subtend the maximum angle between his lines of sight to its base and top?

19. The legs of a right triangle are 5 and 6 ft, respectively. Find approximately the change in the larger acute angle when the shorter side is changed from 5 ft to 5.05 ft.

20. A man standing on a pier is retrieving a rope attached to a skiff at the rate of 3 ft/min. If the vertical distance between the man's hands and the point of attachment to the skiff is 12 ft, how fast is the angle between the rope and a horizontal line changing at the instant there is 20 ft of rope still out?

21. Do Problem 32, Exercises 8-2 by using inverse trigonometric functions.

22. Do Problem 33, Exercises 8-2 by using inverse trigonometric functions.

23. A painting 4-ft high is hung with its base 3 ft above the observer's eye. At what rate is the angle subtended at his eye by the painting changing as he walks toward it at the rate of 5 ft/sec at the instant he is 10 ft from the wall on which it hangs?

8–6. Some Integrals

In Chapter 7 we noted that each differentiation formula may be restated as an integration formula. Therefore the differentiation formulas for the trigonometric and inverse trigonometric functions obtained in the preceding sections may be reformulated in the following theorem.

THEOREM 8–10. If u is a differentiable function of x,

(a) $\int \sin u \, du = -\cos u + C$

(b) $\int \cos u \, du = \sin u + C$

(c) $\int \sec^2 u \, du = \tan u + C$

(d) $\int \csc^2 u \, du = -\cot u + C$

(e) $\int \sec u \tan u \, du = \sec u + C$

(f) $\int \csc u \cot u \, du = -\csc u + C$

(g) $\displaystyle\int \frac{du}{\sqrt{a^2 - u^2}} = \text{Arcsin} \, \frac{u}{a} + C, \quad |u| < |a|$

(h) $\displaystyle\int \frac{du}{a^2 + u^2} = \frac{1}{a} \, \text{Arctan} \, \frac{u}{a} + C$

(i) $\displaystyle\int \frac{du}{u\sqrt{u^2 - a^2}} = \frac{1}{a} \operatorname{Arcsec} \frac{u}{a} + C, \quad |u| > |a|$

Parts (g), (h), and (i) are slight modifications of the corresponding differentiation formulas. They may be verified readily by differentiation. For example,

$$d\left(\frac{1}{a} \operatorname{Arctan} \frac{u}{a}\right) = \frac{1}{a} \frac{\frac{d}{dx}\left(\frac{u}{a}\right)}{1 + (u^2/a^2)} \, dx$$

$$= \frac{(1/a^2)\, du}{(a^2 + u^2)/a^2} = \frac{du}{a^2 + u^2},$$

which verifies (h).

Some of the ways in which these formulas, as well as those of Chapter 7, may be used to evaluate certain types of integrals are indicated in the following examples.

Example 8–9. Evaluate

$$\int \sin 3x \, dx.$$

If we let $u = 3x$, we have

$$du = 3dx$$

and we may write, applying Theorem 8-10(a),

$$\int \sin 3x \, dx = \int \sin u \, \frac{du}{3} = \frac{1}{3} \int \sin u \, du$$

$$= \tfrac{1}{3}(-\cos u) + C = -\tfrac{1}{3} \cos 3x + C.$$

Example 8–10. Evaluate

$$\int \sec 2x \tan 2x \, dx.$$

We let $u = 2x$. Then

$$du = 2dx$$

and we may write, appealing to Theorem 8-10(e),

$$\int \sec 2x \tan 2x \, dx = \int \sec u \tan u \, \frac{du}{2}$$

$$= \tfrac{1}{2} \sec u + C = \tfrac{1}{2} \sec 2x + C.$$

Example 8-11. Evaluate

$$\int x \cos(1 - x^2) \, dx.$$

We let $u = 1 - x^2$. Then

$$du = -2x \, dx \qquad \text{or} \qquad x \, dx = -\frac{du}{2}.$$

We may now write

$$\int x \cos(1 - x^2) \, dx = -\frac{1}{2} \int \cos u \, du$$

$$= -\tfrac{1}{2} \sin u + C = -\tfrac{1}{2} \sin(1 - x^2) + C.\dagger$$

Example 8-12. Evaluate

$$\int \sec^3 2x \tan 2x \, dx.$$

On the surface this may appear to be similar to Example 8-10. However, this integral is not reducible to any of the forms of Theorem 8-10. This problem may be solved by reducing it to the form of Theorem 7-10.

First we note that

$$d(\sec 2x) = \sec 2x \tan 2x \cdot 2 \, dx,$$

so our integral may be written in the form

$$\frac{1}{2} \int \sec^2 2x \, d(\sec 2x).$$

Now, if we let $u = \sec 2x$, we have

$$\int \sec^3 2x \tan 2x \, dx = \frac{1}{2} \int u^2 \, du$$

$$= \frac{1}{2} \frac{u^3}{3} + C = \frac{1}{6} \sec^3 2x + C.$$

$\dagger$ Note that $\int \cos(1 - x^2) \, dx$ cannot be evaluated by this substitution. When we set $u = 1 - x^2$ in this integral we have

$$\int \cos u \left(-\frac{du}{2x} \right) = -\frac{1}{2} \int \cos u \, \frac{du}{\pm\sqrt{1 - u}},$$

which is not one of the forms of Theorem 8-10. However, this does not preclude its evaluation by more powerful methods which might be developed later.

Example 8-13. Evaluate

$$\int \frac{dx}{\sqrt{1 - 9x^2}}.$$

This integral reduces to the form of Theorem 8-10(g) if we set $u = 3x$. We have

$$du = 3dx$$

and

$$\int \frac{dx}{\sqrt{1 - 9x^2}} = \frac{1}{3} \int \frac{du}{\sqrt{1 - u^2}} = \frac{1}{3} \text{Arcsin } u + C$$

$$= \tfrac{1}{3} \text{Arcsin } 3x + C.$$

Example 8-14. Evaluate

$$\int_0^{3\sqrt{3}/4} \frac{dx}{9 + 16x^2}.$$

This is of the form of Theorem 8-10(h) with $a = 3$ and $u = 4x$. Following the same procedure used in the preceding examples, we obtain

$$\int \frac{dx}{9 + 16x^2} = \int \frac{du/4}{3^2 + u^2} = \frac{1}{4} \left(\frac{1}{3} \text{Arctan } \frac{u}{3} \right) + C$$

$$= \frac{1}{12} \text{Arctan } \frac{4x}{3} + C.$$

Thus

$$\int_0^{3\sqrt{3}/4} \frac{dx}{9 + 16x^2} = \frac{1}{12} \text{Arctan } \frac{4x}{3} \Bigg]_0^{3\sqrt{3}/4}$$

$$= \frac{1}{12} [\text{Arctan } \sqrt{3} - \text{Arctan } 0]$$

$$= \frac{1}{12} \left(\frac{\pi}{3} - 0 \right) = \frac{\pi}{36}.$$

Since $u = 0$ when $x = 0$, and $u = 3\sqrt{3}$ when $x = 3\sqrt{3}/4$, one step in this process could have been omitted by writing

$$\int_0^{3\sqrt{3}/4} \frac{dx}{9 + 16x^2} = \frac{1}{4} \int_0^{3\sqrt{3}} \frac{du}{3^2 + u^2} = \frac{1}{12} \text{Arctan } \frac{u}{3} \Bigg]_0^{3\sqrt{3}}.$$

Exercises 8-4

Evaluate the indefinite integrals in Exercises 1–30.

1. $\displaystyle\int \cos 3x \, dx$

2. $\displaystyle\int \sin \frac{x}{2} \, dx$

3. $\displaystyle\int \sec^2 4x \, dx$

4. $\displaystyle\int \csc^2 2x \, dx$

5. $\displaystyle\int \sec 2x \tan 2x \, dx$

6. $\displaystyle\int \csc \frac{x}{3} \cot \frac{x}{3} \, dx$

7. $\displaystyle\int \sin 4x \, dx$

8. $\displaystyle\int \cos \frac{x}{5} \, dx$

9. $\displaystyle\int \cos(1 - x) \, dx$

10. $\displaystyle\int \sec^2(x + 4) \, dx$

11. $\displaystyle\int \csc(2x - 1)\cot(2x - 1) \, dx$

12. $\displaystyle\int \sin(9 - 4x) \, dx$

13. $\displaystyle\int x \cos x^2 \, dx$

14. $\displaystyle\int x \sin(3x^2 - 2) \, dx$

15. $\displaystyle\int x^2 \sec^2 x^3 \, dx$

16. $\displaystyle\int x \csc^2(9 - 4x^2) \, dx$

17. $\displaystyle\int \sin x \cos x \, dx$

18. $\displaystyle\int \tan^2 2x \sec^2 2x \, dx$

19. $\displaystyle\int \frac{\sin x \, dx}{\sqrt{1 - \cos x}}$

20. $\displaystyle\int \sec^4 2x \tan 2x \, dx$

21. $\displaystyle\int \frac{\sin 3x}{\cos^3 3x} \, dx$

22. $\displaystyle\int \frac{\sec^2 x}{\sqrt{\tan x + 1}} \, dx$

23. $\displaystyle\int \frac{dx}{\sqrt{4 - x^2}}$

24. $\displaystyle\int \frac{x \, dx}{\sqrt{4 - x^2}}$

25. $\displaystyle\int \frac{dx}{4 + x^2}$

26. $\displaystyle\int \frac{x \, dx}{(4 + x^2)^2}$

27. $\displaystyle\int \frac{dx}{9 + 4x^2}$

28. $\displaystyle\int \frac{dx}{\sqrt{9 - 4x^2}}$

29. $\displaystyle\int \frac{dx}{x\sqrt{4x^2 - 1}}$

30. $\displaystyle\int \frac{dx}{x\sqrt{9x^2 - 25}}$

Evaluate the definite integrals in Exercises 31–38.

31. $\displaystyle\int_{-\pi/2}^{\pi/2} \sec^2 \frac{x}{2} \tan \frac{x}{2}\, dx$

32. $\displaystyle\int_0^{\sqrt{3}} \frac{dx}{1+x^2}$

33. $\displaystyle\int_0^{\sqrt{3}} \frac{x\, dx}{(1+x^2)^2}$

34. $\displaystyle\int_0^2 \frac{dx}{\sqrt{16-x^2}}$

35. $\displaystyle\int_{\pi/4}^{\pi/2} \frac{\csc^2 x\, dx}{(\cot x + 1)^3}$

36. $\displaystyle\int_{-3}^3 \frac{dx}{9+x^2}$

37. $\displaystyle\int_{\pi/12}^{\pi/6} (1 + \cos 2x)\sin 2x\, dx$

38. $\displaystyle\int_0^{2\pi/3} \sin \frac{x}{2} \cos^2 \frac{x}{2}\, dx$

39. Find the area under one arch of the curve $y = \sin 2x$.

40. Find the first quadrant area bounded by $y = \cos x$, $y = \sin x$, $x = 0$.

41. Find the other area bounded by the curves of Exercise 40.

42. Find the area bounded by $y = \sec^2 x$, $y = \cos x$, $x = \pi/3$.

43. Find the area under the curve $y = 12/(9 + x^2)$ from $x = 0$ to $x = 3$.

44. Find the area under the curve $y = 12/\sqrt{9 - x^2}$ from $x = 0$ to $x = 1.5$.

Chapter 9

THE EXPONENTIAL AND LOGARITHM FUNCTIONS

9–1. The Exponential Function

In algebra we defined the quantity a^x, $a > 0$, uniquely for x any rational number. We now have need to define a^x, $a > 0$, where x is any real number. In this connection we shall make some assumptions, none of which will be proved but all of which are proved in more advanced and complete treatments of the subject.

The first of these assumptions is that *any real number is the limit of a sequence* $\{x_n\}$ *of rational numbers*. One way of defining real numbers is in terms of such a sequence. For example, the first few terms in a sequence defining the irrational number $\sqrt{2}$ are

$$1, \; 1.4, \; 1.41, \; 1.414, \; 1.4142, \; 1.41421, \; \ldots .$$

The law for obtaining the terms in this sequence is the usual arithmetic process of computing the square root of a number.

DEFINITION 9–I

$$a^x = \lim_{n \to \infty} a^{x_n}, \qquad a > 0.$$

where $\lim_{n \to \infty} x_n = x$.

This involves the assumption that this limit exists. A few of the terms in the sequence defining $3^{\sqrt{2}}$ are

$$3, \; 3^{1.4}, \; 3^{1.41}, \; 3^{1.414}, \; 3^{1.4142}, \; \ldots .$$

Each one of the terms in this sequence is uniquely defined because the exponent is rational. This limit does indeed exist and a^x, $a > 0$, exists uniquely in the sense just indicated for all real numbers x.

The function f defined by

$$f(x) = a^x, \qquad a > 0,$$

is called the *exponential function*. It has a number of important properties:

(a) The exponential function obeys the laws of exponents already established for rational numbers, for example, $a^{x_1+x_2} = a^{x_1} \cdot a^{x_2}$.

(b) $a^0 = 1$.

(c) $a^x > 0$ for all x.

(d) $\lim_{x \to -\infty} a^x = 0$, $a > 1$.

(e) $\lim_{x \to \infty} a^x = \infty$, $a > 1$.

(f) a^x is continuous for all x.

To avoid endless repetition we shall henceforth assume $a > 0$. The graph of

$$y = a^x, \quad a > 1,$$

is shown in Fig. 9-1. From property (b) it is clear that this graph will always pass through the point $(0, 1)$ regardless of the value of a.

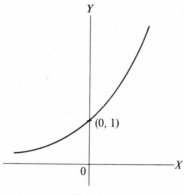

Fig. 9-1

9-2. The Logarithm Function

The *logarithm function* is defined to be the inverse of the exponential function.

DEFINITION 9-2

$$y = \log_a x \quad \text{if } a^y = x, \quad a > 1.\dagger$$

The statement $y = \log_a x$ is read "*y is the logarithm of x to the base a.*

A fundamental theorem related to the definition of the logarithm function is the following, which we state without proof.

THEOREM 9-1. For each $x > 0$ the equation $a^y = x$, $a > 0$, $a \neq 1$, has exactly one real solution.

---

† The condition $a > 1$ is introduced to eliminate ambiguity in certain properties of the logarithm function. Actually $a > 0$, $a \neq 1$, are the only restrictions necessary.

Application of this theorem, combined with Definition 9-2 and the properties of the exponential function, yields the following facts regarding the logarithm function:

(a) $\log_a x$ is continuous, $x > 0$.
(b) $\log_a x < 0, 0 < x < 1$.
(c) $\log_a x > 0, x > 1$.
(d) $\log_a 1 = 0$.
(e) $\lim\limits_{x \to 0^+} \log_a x = -\infty$.
(f) $\lim\limits_{x \to \infty} \log_a x = \infty$.

The familiar *laws of logarithms* follow immediately from the definition and the laws of exponents. We list them here for reference.

$$\log_a x_1 x_2 = \log_a x_1 + \log_a x_2. \tag{9-1}$$

$$\log_a \frac{x_1}{x_2} = \log_a x_1 - \log_a x_2. \tag{9-2}$$

$$\log_a x^n = n \log_a x. \tag{9-3}$$

The following properties are also useful when working with these functions.

$$\log_a a^x = x. \tag{9-4}$$

$$a^{\log_a x} = x. \tag{9-5}$$

$$\log_a x = \log_a b \log_b x. \tag{9-6}$$

Formula (9-6) expresses the relationship between the logarithms of x to two different bases. If we have a table of logarithms to the base a we can then, by means of (9-6), determine the logarithm of x to any other base b. A proof of this formula will serve as an example of the general method by means of which proofs of statements about logarithms are obtained. The statements

$$y_1 = \log_a x, \qquad y_2 = \log_b x,$$

may be written in equivalent form

$$a^{y_1} = x, \qquad b^{y_2} = x,$$

so

$$a^{y_1} = b^{y_2}.$$

If we take the logarithm of both sides of this equation to the base a, we have, making use of (9-3) and (9-4),

$$y_1 \log_a a = y_2 \log_a b,$$

or

$$y_1 = y_2 \log_a b, \qquad \text{or} \qquad \log_a x = \log_b x \log_a b.$$

A special case of this formula is convenient at times. We take $x = a$ and obtain

$$\log_a a = \log_b a \log_a b, \quad \text{or} \quad 1 = \log_b a \log_a b.$$

Thus

$$\log_a b = \frac{1}{\log_b a}. \tag{9-7}$$

The graph of $y = \log_a x$ is shown in Fig. 9-2. Regardless of the base, the curve passes through $(1, 0)$ and, of course, lies completely to the right of the y-axis.

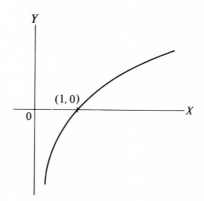

Fig. 9-2

9–3. The Number *e*

In the next section we shall need to evaluate a limit of the form $\lim_{t \to 0} (1 + t)^{1/t}$. This involves, first of all, the question of the existence of the limit, and secondly, its value. It is not within the scope of this book to do either. The student will find complete treatments of both questions in more advanced books (for example, *Differential and Integral Calculus*, Vol. I, by R. Courant).

This limit does exist and is usually represented by the letter *e*. We shall make use of these results, to be found elsewhere, and write

$$\lim_{t \to 0} (1 + t)^{1/t} = e = 2.71828 \cdots. \tag{9-8}$$

This number *e* is used as the base for the system of *natural logarithms*. This system of logarithms, as we shall see, proves to be the most convenient one for the operations of calculus. *From here on, if no base is indicated, it is to be understood that it is e.*

The number *e* appears in many contexts. Consider that of compound interest. Let a sum of money *P* be invested at an annual rate of $100i$ percent compounded n times a year. Then the rate per conversion period is $100i/n$ percent and the amount of money accumulated at the end of one period is

$$S = P + \frac{Pi}{n} = P\left(1 + \frac{i}{n}\right).$$

At the end of two conversion periods, we have

$$S = P\left(1 + \frac{i}{n}\right) + P\left(1 + \frac{i}{n}\right)\frac{i}{n}$$

$$= P\left(1 + \frac{i}{n}\right)^2.$$

Following this procedure, we see that

$$S = P\left(1 + \frac{i}{n}\right)^{nt}$$

at the end of t years (nt conversion periods).

Now let us consider what happens when we allow the length of the conversion period to approach zero, that is, when $n \to \infty$. This amounts to *continuously converted compound interest*. We have

$$S = P \lim_{n \to \infty}\left(1 + \frac{i}{n}\right)^{nt}$$

$$= P \lim_{n \to \infty}\left[\left(1 + \frac{i}{n}\right)^{n/i}\right]^{it}.$$

Let $r = i/n$. Then, as $n \to \infty$, $r \to 0$ and we may write

$$S = P\left[\lim_{r \to 0}(1 + r)^{1/r}\right]^{it}.$$

Hence, from (9-8),

$$S = Pe^{it}.$$

Thus, the accumulated sum under continuous compound interest involves not only the amount invested, the interest rate, and the time, *but also the constant e*.

Exercises 9-1

Without the use of tables, find x in Exercises 1–5.

1. (a) $\log_{10} x = 2$; (b) $\log_{10} x = -2$; (c) $\log_{10} x = 1/2$

2. (a) $\log x = 0$; (b) $\log_a x = 1$; (c) $\log_3 x = -1$

3. $\log x = \log 4 + \log 3 - \log 2$

4. $\log_{10} x = \frac{1}{2}\log_{10} 9 + 3 \log_{10} 4 - \log_{10} 8$

5. $\log_2 x = \dfrac{1}{2} - 3 \log_2 5 + \log_2 50$

6. Simplify: (a) $e^{\log 5}$; (b) $10^{\log_{10} x}$; (c) $\log_2 2$; (d) $\log_{10} 1000$; (e) $e^{x + \log x}$

In Exercises 7–10 reduce the logarithms given to sums and differences of simpler logarithms.

7. $\log \dfrac{(x^2 + 2)^2}{x - 1}$

8. $\log \dfrac{\sqrt{x^2 + 2}}{x^2 - 1}$

9. $\log \sqrt{\dfrac{(x + 1)(x - 1)}{x^2 + 1}}$

10. $\log \left[\dfrac{\sqrt{(2x^2 + 3)}(x^2 + 4)}{x^3} \right]^2$

In Exercises 11–18, solve for x.

11. $2^x = 8^{2x+1}$

12. $3^x = 9^{x+3}$

13. $5^x = 4$

14. $2^x = 10$

15. $\log(x + 1) + \log(x - 1) = 0$

16. $\log_{10} x - \log_{10}(x + 1) = 1$

17. $3^x + 3^{2x} = 3^{2x+1}$

18. $2^{2x+1} + 2^{x+1} = 4$

In Exercises 19–24, solve for x in terms of y.

19. $y = 3^x$

20. $y = 10^{x+1}$

21. $y = 10^{\log_{10} x}$

22. $y = e^{x^2}$

23. $y = \dfrac{e^x - e^{-x}}{2}$

24. $y = \dfrac{2^x + 2^{-x}}{2}$

In Exercises 25–28, sketch the graph of the given equation.

25. $y = 2^x$

26. $y = 3^{-x}$

27. $y = \log x^2$

28. $y = \dfrac{e^x + e^{-x}}{2}$

9-4. Derivative of the Logarithm Function

Let

$$y = \log_a x.$$

Then

$$\frac{\Delta y}{\Delta x} = \frac{\log_a(x + \Delta x) - \log_a x}{\Delta x}$$

$$= \frac{1}{\Delta x} \log_a \left(\frac{x + \Delta x}{x} \right),$$

by (9-2). If we multiply and divide by x, we may write

$$\frac{\Delta y}{\Delta x} = \frac{1}{x}\frac{x}{\Delta x}\log_a\left(\frac{x + \Delta x}{x}\right)$$

$$= \frac{1}{x}\log_a\left(1 + \frac{\Delta x}{x}\right)^{x/\Delta x},$$

where the last operation is justified by (9-3). Now let us set

$$t = \frac{\Delta x}{x}$$

and note that $t \to 0$ as $\Delta x \to 0$. Thus

$$\frac{dy}{dx} = \lim_{\Delta x \to 0}\frac{\Delta y}{\Delta x} = \lim_{t \to 0}\frac{1}{x}\log_a(1 + t)^{1/t}$$

$$= \frac{1}{x}\lim_{t \to 0}\log_a(1 + t)^{1/t}.$$

Since the logarithm function is continuous, we have

$$\lim_{t \to 0}\log_a(1 + t)^{1/t} = \log_a\lim_{t \to 0}(1 + t)^{1/t} = \log_a e.$$

Therefore

$$\frac{dy}{dx} = \frac{1}{x}\log_a e,$$

and application of the chain rule gives the following theorem.

THEOREM 9–2. If u is a differentiable function of x, $u > 0$,

$$\frac{d}{dx}(\log_a u) = \frac{1}{u}\frac{du}{dx}\log_a e.$$

Since $\log e = 1$, we have immediately the next theorem.

THEOREM 9–3. If u is a differentiable function of x, $u > 0$,

$$\frac{d}{dx}(\log u) = \frac{1}{u}\frac{du}{dx}.$$

The reason for preferring *natural* logarithms over any other system for the operations of calculus is apparent in the simplicity of the result in Theorem 9-3 as compared with that of Theorem 9-2. There will be additional reasons apparent as we proceed.

Example 9-1. Find dy/dx if $y = \log \sqrt{x^2 + 1}$.

In order to apply Theorem 9-3 we set

$$u = \sqrt{x^2 + 1}.$$

Then

$$\frac{du}{dx} = \frac{1}{2}(x + 1)^{-1/2}(2x) = \frac{x}{\sqrt{x^2 + 1}}.$$

Therefore

$$\frac{dy}{dx} = \frac{1}{u}\frac{du}{dx} = \frac{1}{\sqrt{x^2 + 1}} \cdot \frac{x}{\sqrt{x^2 + 1}}$$

$$= \frac{x}{x^2 + 1}.$$

An alternate solution results from writing

$$\log \sqrt{x^2 + 1} = \frac{1}{2}\log(x^2 + 1).$$

Then

$$\frac{dy}{dx} = \frac{1}{2} \cdot \frac{1}{x^2 + 1}\frac{d}{dx}(x^2 + 1)$$

$$= \frac{x}{x^2 + 1}$$

as before.

Example 9-2. Find dy/dx if $y = \log_{10} \sin 2x$.

We set

$$u = \sin 2x$$

and obtain

$$\frac{du}{dx} = 2\cos 2x.$$

Theorem 9-2 gives

$$\frac{dy}{dx} = \frac{1}{u}\frac{du}{dx}\log_{10} e = \frac{1}{\sin 2x} \cdot 2\cos 2x \log_{10} e$$

$$= 2\log_{10} e \cot 2x.$$

Example 9-3. Find dy/dx if

$$y = \log \frac{(x^2 - 1)(2x + 1)}{\sqrt{x^2 + 1}}.$$

If we apply the laws of logarithms to this expression, we obtain

$$y = \log(x^2 - 1) + \log(2x + 1) - \frac{1}{2}\log(x^2 + 1),$$

and

$$\frac{dy}{dx} = \frac{1}{x^2 - 1}\frac{d}{dx}(x^2 - 1) + \frac{1}{2x + 1}\frac{d}{dx}(2x + 1) - \frac{1}{2}\cdot\frac{1}{x^2 + 1}\frac{d}{dx}(x^2 + 1)$$

$$= \frac{2x}{x^2 - 1} + \frac{2}{2x + 1} - \frac{x}{x^2 + 1}.$$

This form of the solution is usually satisfactory. However, if necessary, we can combine these into a single fraction and get

$$\frac{dy}{dx} = \frac{4x^4 + x^3 + 6x^2 + 3x - 2}{(x^2 - 1)(x^2 + 1)(2x + 1)}.$$

Applying the laws of logarithms before differentiation brings about a drastic simplification in the algebra involved.

Exercises 9-2

Find dy/dx in Exercises 1–24.

1. $y = \log 2x$

2. $y = \log x^2$

3. $y = x \log 2x$

4. $y = \log_{10}(2x^2 - 1)$

5. $y = \log \cos 3x$

6. $y = \log \tan \dfrac{x}{2}$

7. $y = \log \dfrac{x + 1}{x - 1}$

8. $y = \log_{10}\sqrt{9 - x^2}$

9. $y = \log_{10}\sqrt[3]{(x + 1)^2}$

10. $y = \log(\tan x + \sec x)$

11. $y = \log(\csc 2x - \cot 2x)$

12. $y = (\log x)^2$

13. $y = \log \dfrac{\sqrt{x^2 + 16}}{x + 1}$

14. $y = \log\sqrt[3]{\dfrac{2x + 1}{x - 1}}$

15. $y = \dfrac{(x + 2)\sqrt{3x + 2}}{\sqrt[3]{x^2 + 1}}$

16. $y = \log \log x$

17. $y = \log_{10}\cos^2 2x$

18. $y = \log\sqrt{\dfrac{1 + \sin x}{1 - \sin x}}$

19. $y = \dfrac{\log x^2}{x^2}$

20. $y = 4\log\sqrt{\dfrac{1 + x^4}{1 - x^4}}$

21. $\log xy - x - y = 0$ **22.** $x \log y + y \log x = 0$

23. $\log(\csc y - \cot y) + 2x + \cos 2x = 5$

24. $x^2 + y^2 + x \log \dfrac{y}{x} = 1$

Sketch each of the curves in Exercises 25–28. In the process, determine any relative maxima, relative minima, and points of inflection that may exist.

25. $y = \log(x - 2)$ **26.** $y = \log(x + 3)$

27. $y = x - \log x$ **28.** $y = \dfrac{\log x}{x}$

29. If $\log x = kt$, k a constant, show that the rate of change of x with respect to t is proportional to x.

30. Show that $\log(x + 1)$ is approximately equal to $\log x + (1/x)$.

31. Find the equation of the tangent to $y = \log x$ which is parallel to the line $3x - y = 5$.

32. Find the equation of the tangent to $y = \log x$ which is perpendicular to the line $x - 2y = 1$.

9-5. Derivative of the Exponential Function

Let

$$y = a^u,$$

where u is a differentiable function of x. Then

$$u = \log_a y$$

and, differentiating implicitly with respect to x, we obtain

$$\frac{du}{dx} = \frac{1}{y}\frac{dy}{dx} \log_a e,$$

or

$$\frac{dy}{dx} = \frac{y}{\log_a e}\frac{du}{dx}.$$

Appealing to the original definition of y and (9-7), this may be written as

$$\frac{dy}{dx} = a^u \log a \frac{du}{dx}.$$

THEOREM 9–4. If u is a differentiable function of x,

$$\frac{d}{dx}(a^u) = a^u \log a \frac{du}{dx}, \qquad a > 0.$$

Since $\log e = 1$, we have

THEOREM 9–5. If u is a differentiable function of x,

$$\frac{d}{dx}(e^u) = e^u \frac{du}{dx}.$$

Again we note that a simplification is achieved by the use of e rather than a general a.

Example 9–4. Find dy/dx if $y = 10^{2x^2}$.

We let $u = 2x^2$ and refer to Theorem 9-4. We have

$$\frac{dy}{dx} = 10^{2x^2} \log 10 \frac{d}{dx}(2x^2)$$

$$= 10^{2x^2} 4x \log 10.$$

Example 9–5. Find dy/dx if $y = e^{2 \cos^2 x}$.

Here we let $u = 2 \cos^2 x$ and make use of Theorem 9-5. We get

$$\frac{du}{dx} = -4 \cos x \sin x = -2 \sin 2x,$$

and

$$\frac{dy}{dx} = -2e^{2 \cos^2 x} \sin 2x.$$

Example 9–6. Sketch the graph of $y = e^{-x} \cos x$, $x \geq 0$.

We note first that $|e^{-x} \cos x| \leq e^{-x}$, since $|\cos x| \leq 1$. Therefore the required curve lies completely between the two curves $y = e^{-x}$ and $y = -e^{-x}$. Moreover, it touches these two curves at the points at which $\cos x = \pm 1$, that is, at $x = 0, \pi, 2\pi, \ldots$.

Since $e^{-x} \neq 0$ for any x, $y = 0$ when $\cos x = 0$, that is, at odd multiples of $\pi/2$.

Now let us look for relative maxima and minima. We have

$$\frac{dy}{dx} = -e^{-x} \cos x - e^{-x} \sin x$$

$$= -e^{-x}(\cos x + \sin x).$$

Thus the critical values of x are the roots of the equation

$$\cos x + \sin x = 0,$$

again noting $e^{-x} \neq 0$ for any x. This equation may be written $\tan x = -1$, so the critical values of x are those multiples of $\pi/4$ which lie in the second and fourth quadrants, namely,

$$x = \frac{3\pi}{4}, \frac{7\pi}{4}, \frac{11\pi}{4}, \ldots .$$

To identify these critical values we calculate the second derivative. We have

$$\frac{d^2 y}{dx^2} = 2e^{-x} \sin x.$$

This will be positive for x in the second quadrant and negative for x in the fourth quadrant. Hence, we have a relative minimum at $x = 3\pi/4$, and relative maximum at $x = 7\pi/4$, and so on.

Putting all this together, we obtain the graph shown in Fig. 9-3.

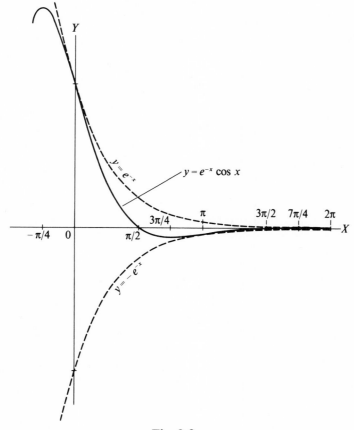

Fig. 9-3

This is known as a damped wave, and such curves appear in many applications.

Exercises 9-3

Find dy/dx in Exercises 1–16.

1. $y = e^{1/x}$

2. $y = 10^{-x^3}$

3. $y = e^{\sin^2 2x}$

4. $y = e^{\tan x}$

5. $y = e^{\text{Arctan } x}$

6. $y = x^2 e^{2x}$

7. $y = e^{2x} \sin 2x$

8. $y = 3^{\sqrt{x^2-4}}$

9. $y = 2(e^{x/2} + e^{-x/2})$

10. $y = e^{e^x}$

11. $y = (e^e)^x$

12. $y = \log \dfrac{1 + e^x}{1 - e^x}$

13. $y = \log_a(e^x + e^{-x})$

14. $y = 2 \sin e^{x/2}$

15. $x^2 e^y + y^2 e^x = 1$

16. $xe^{xy} + e^x = 3$

17. $1000 is invested at 5 percent interest. At what rate is this investment growing at the end of five years?

18. A firm finds that the extra sales S produced by a special sales campaign follow the law

$$S = 1500(3^{-x}),$$

where x is the number of days elapsed since the end of the campaign. At what rate are sales decreasing after four days have elapsed?

19. The population N of the United States has been observed to follow very closely the law

$$N = \frac{197{,}273{,}000}{1 + e^{-0.03134t}}$$

for the period of years 1790 to 1910 where t is measured in years from April 1, 1914 (t negative for years prior to this date). Find the general formula for the population growth rate during this period.

20. It has been observed that the amount of food consumed by chickens during a specified period of time appears to be represented by

$$f = C(1 - e^{-mt})$$

where f is the amount of food consumed, t is time, C is a constant determined by physiological capacity for food, and m is a voracity constant. Find the rate of consumption of food in terms of m, C, and f.

21. Show that $y = e^{2x} \sin x$ satisfies the differential equation

$$\frac{d^2 y}{dx^2} - 4 \frac{dy}{dx} + 5y = 0.$$

22. Show that $y = e^{-2x} \cos 2x$ satisfies the differential equation

$$\frac{d^2 y}{dx^2} + 4 \frac{dy}{dx} + 8y = 0.$$

23. Show that $y = (2x + (1/2)\tan x)e^x$ satisfies the differential equation

$$\frac{d^2y}{dx^2} - 2\frac{dy}{dx} + y = e^x \sec^2 x \tan x.$$

24. Find the maximum value of the function defined by $f(x) = (\log x)/x$.

25. Find the maximum and minimum values of the function defined by $f(x) = x^2 e^{-2x}$.

26. Find the maximum and minimum values of the function defined by $f(x) = x^3 e^{-3x^2/2}$.

Draw graphs of the equations in Exercises 27–32, making use of maximum and minimum points, direction of concavity, and so on.

27. $y = e^{-x}$ **28.** $y = e^{-x^2}$

29. $y = xe^{-x}$ **30.** $y = \dfrac{e^x + e^{-x}}{2}$

31. $y = \dfrac{e^x - e^{-x}}{2}$ **32.** $y = e^{-x}\sin x$

9–6. More Integrals

Theorems 9-4 and 9-5 give us immediately the following integration theorem.

THEOREM 9–6. If u is a differentiable function

(a)
$$\int a^u \, du = \frac{a^u}{\log a} + C;$$

(b)
$$\int e^u \, du = e^u + C.$$

In the process of converting Theorem 9-3 into a theorem on integration, let us note something about the differentiation of $\log |u|$. If $u < 0$, $|u| = -u$, and

$$d(\log |u|) = \frac{1}{|u|}\,d|u| = \frac{1}{-u}\,d(-u) = \frac{-du}{-u} = \frac{du}{u}.$$

If $u > 0$, $|u| = u$ and

$$d(\log |u|) = \frac{1}{|u|}\,d|u| = \frac{du}{u}.$$

Therefore we may state the next theorem.

THEOREM 9–7. If u is a differentiable function,

$$\int \frac{du}{u} = \log |u| + C.$$

The student should note that Theorem 9-7 fills the gap existing in Theorems 7-9 and 7-10. In those two theorems it was necessary to include the restriction $n \neq -1$, which is precisely the case dealt with in Theorem 9-7.

Example 9–7. Evaluate $\int e^{x^2}x \, dx$.

Theorem 9-6(b) covers this problem. We let $u = x^2$ and then $du = 2x \, dx$, so we may write

$$\int e^{x^2}x \, dx = \int e^u \frac{du}{2} = \frac{1}{2}\int e^u \, du$$

$$= \frac{1}{2}e^u + C = \frac{1}{2}e^{x^2} + C.$$

Example 9–8. Evaluate $\int \frac{e^{2x}\,dx}{e^{2x}+3}$.

We let $u = e^{2x} + 3$. Then $du = 2e^{2x}\,dx$ and, applying Theorem 9-7,

$$\int \frac{e^{2x}\,dx}{e^{2x}+3} = \int \frac{du/2}{u} = \frac{1}{2}\int \frac{du}{u}$$

$$= \frac{1}{2}\log |u| + C = \frac{1}{2}\log |e^{2x}+3| + C$$

$$= \frac{1}{2}\log(e^{2x}+3) + C.$$

The last statement is true because $e^{2x} + 3$ is positive for all values of x.

Example 9–9. Evaluate $\int \frac{e^x\,dx}{(e^x+1)^3}$.

This example illustrates the necessity of keeping in mind the general "power" formula (Theorem 7-10). If we let $u = e^x + 1$, $du = e^x \, dx$, and the integral may be written

$$\int \frac{e^x\,dx}{(e^x+1)^3} = \int \frac{du}{u^3} = \int u^{-3}\,du$$

$$= \frac{u^{-2}}{-2} + C = -\frac{1}{2(e^x+1)^2} + C.$$

Exercises 9-4

Evaluate the indefinite integrals in Exercises 1–30.

1. $\int e^{3x}\,dx$ 　　　　　　　　　　　**2.** $\int 2^x\,dx$

3. $\int e^{-2x}\,dx$ 　　　　　　　　　　**4.** $\int 3^{3x}\,dx$

5. $\displaystyle\int xe^{-x^2}\,dx$

6. $\displaystyle\int e^{2-3x}\,dx$

7. $\displaystyle\int 2^x e^x\,dx$

8. $\displaystyle\int 3^{-2x}e^{-2x}\,dx$

9. $\displaystyle\int \frac{e^{\log 3x}}{x}\,dx$

10. $\displaystyle\int \frac{dx}{e^{2x-1}}$

11. $\displaystyle\int \frac{e^{2x}\,dx}{e^{2x}-1}$

12. $\displaystyle\int e^{x+e^x}\,dx$

13. $\displaystyle\int \frac{x\,dx}{x^2+1}$

14. $\displaystyle\int \frac{x\,dx}{\sqrt{x^2+1}}$

15. $\displaystyle\int \frac{dx}{1-x}$

16. $\displaystyle\int \frac{dx}{(1-x)^2}$

17. $\displaystyle\int \frac{\sin x\,dx}{\cos x}$

18. $\displaystyle\int \frac{x^2\,dx}{x^3+27}$

19. $\displaystyle\int \frac{(1+2x)\,dx}{x^2+x}$

20. $\displaystyle\int \frac{x\,dx}{x^2-9}$

21. $\displaystyle\int \frac{x\,dx}{\sqrt{x^2-9}}$

22. $\displaystyle\int \frac{(x+3)\,dx}{x^2+6x+4}$

23. $\displaystyle\int \frac{(x^2+3x+1)\,dx}{x^2}$

24. $\displaystyle\int \frac{\sec^2 x\,dx}{\tan x+1}$

25. $\displaystyle\int (e^x+1)^2\,dx$

26. $\displaystyle\int \frac{(e^x-e^{-x})\,dx}{e^x+e^{-x}}$

27. $\displaystyle\int \frac{(\sec^2 x+\sec x\tan x)\,dx}{\sec x+\tan x}$

28. $\displaystyle\int \frac{(x^2+4)\,dx}{x-2}$

29. $\displaystyle\int e^{\cos x}\sin x\,dx$

30. $\displaystyle\int \frac{(e^{2x}+1)\,dx}{e^{2x}-1}$

Evaluate the definite integrals in Exercises 31–40.

31. $\displaystyle\int_0^1 \frac{dx}{2-x}$

32. $\displaystyle\int_{5/2}^3 \frac{dx}{2x-4}$

33. $\displaystyle\int_0^1 x^2 e^{x^3}\,dx$

34. $\displaystyle\int_1^2 \frac{e^x\,dx}{e^x+1}$

35. $\displaystyle\int_{-1}^1 2^{-x}\,dx$

36. $\displaystyle\int_0^2 \frac{(e^x-e^{-x})\,dx}{2}$

37. $\int_0^3 (3^x + x)\, dx$ **38.** $\int_0^1 e^{2x-1}\, dx$

39. $\int_0^1 \dfrac{x^3\, dx}{x^2 - 4}$ **40.** $\int_3^4 \dfrac{(2x + 4)\, dx}{x^2 + 4x - 20}$

Find the area bounded by the x-axis and the given curves in Exercises 41–44.

41. $y = xe^{x^2},\ x = 0,\ x = 1$ **42.** $y = \dfrac{x}{x^2 + 1},\ x = 1,\ x = 3$

43. $y = \dfrac{e^x + e^{-x}}{2},\ x = 0,\ x = 2$ **44.** $y = \tan x,\ x = 0,\ x = \pi/4$

45. One hypothesis in the study of reaction to stimuli assumes that the rate of change of sensation γ with respect to stimulus β is inversely proportional to the stimulus, that is,

$$\frac{d\gamma}{d\beta} = \frac{k}{\beta}.$$

If β_0 is the level of stimulus at which sensation begins and ceases, find the relationship between γ and β (Fechner's Law).

46. An animal performs a series of acts in an attempt to achieve a given goal. If the number of those acts that can be considered as leading to success is denoted by s, those leading to failure by f, then one analysis of this problem leads to the relation

$$\frac{ds}{df} = -\frac{s}{f}.$$

Find the fundamental relation between s and f.

47. One mathematical model of the predator-prey relationship is given by

$$\frac{dA}{dP} = -\frac{bA}{P}$$

where P is the number of predators looking for prey, A is the coefficient of attack, that is, the number of attacks per P, and b is a positive constant. Find the relationship between A and P.

9–7. Logarithmic Differentiation

We shall discuss at this point a device known as *logarithmic differentiation*, which enables us to fill in certain voids left by our differentiation theorems, and which may also be used to simplify certain types of problems. We shall discuss this method and illustrate its application by means of examples.

As a first example let us consider the function u^n, where u is a differentiable function of x, $u > 0$, and n is any real number.† We write

$$y = u^n,$$

from which we obtain

$$\log y = n \log u$$

by taking the natural logarithm of each side. Now we differentiate implicitly with respect to x and get

$$\frac{1}{y} \frac{dy}{dx} = \frac{n}{u} \frac{du}{dx}, \qquad \text{or} \qquad \frac{dy}{dx} = \frac{ny}{u} \frac{du}{dx}.$$

Next we substitute the original value of y in the right member. We have, then,

$$\frac{dy}{dx} = \frac{nu^n}{u} \frac{du}{dx} = nu^{n-1} \frac{du}{dx}.$$

We may now state a variation of Theorem 5-9.

THEOREM 9–8. If u is a differentiable function of x, $u > 0$,

$$\frac{d}{dx}(u^n) = nu^{n-1} \frac{du}{dx}$$

for all real n.

There remains one type of exponential function for which we have not developed a method for calculating its derivative. This is a function of the form u^v, where both u and v are functions of x. Although it is not difficult, we shall neither prove nor state a theorem for this case. We prefer to differentiate such functions by "method" rather than by formula. This method is that of logarithmic differentiation.

Example 9–10. Calculate dy/dx, where $y = x^{\sin x}$, $x > 0$.

We take the natural logarithm of both members and get

$$\log y = \sin x \log x.$$

Then, differentiating implicitly with respect to x, we obtain

$$\frac{1}{y} \frac{dy}{dx} = \cos x \log x + \frac{1}{x} \sin x$$

or

$$\frac{dy}{dx} = y \left[\cos x \log x + \frac{\sin x}{x} \right]$$

$$= x^{\sin x} \left[\cos x \log x + \frac{\sin x}{x} \right].$$

---

† There are cases where $u < 0$ for which u^n is defined and for which the method under discussion may be modified. However, for simplicity, we retain the condition $u > 0$.

The differentiation of a function defined by combinations of products, quotients, powers, and roots may frequently be simplified by the device of logarithmic differentiation.

Example 9–11. Calculate dy/dx, where

$$y = \frac{\sqrt[3]{(x^2 + 1)}(x^3 - 1)}{2x + 3}.$$

The algebra involved in calculating this derivative is greatly reduced if we first take the natural logarithm of both members. We have

$$\log y = \tfrac{1}{3} \log (x^2 + 1) + \log (x^3 - 1) - \log (2x + 3).$$

Then, differentiating implicitly,

$$\frac{1}{y}\frac{dy}{dx} = \frac{1}{3}\frac{2x}{x^2 + 1} + \frac{3x^2}{x^3 - 1} - \frac{2}{2x + 3},$$

or

$$\frac{dy}{dx} = \frac{\sqrt[3]{(x^2 + 1)}(x^3 - 1)}{2x + 3}\left[\frac{2x}{3(x^2 + 1)} + \frac{3x^2}{x^3 - 1} - \frac{2}{2x + 3}\right].$$

Exercises 9-5

Find dy/dx in Exercises 1–14.

1. $y = x^x$
2. $y = x^{x^2}$
3. $y = x^{\sqrt{x}}$
4. $y = x^{1/x}$
5. $y = e^{\sin x}$
6. $y = x^{\cos x}$
7. $y = (x^2)^{\text{Arctan } x}$
8. $y = 10^{\text{Arcsin } x}$
9. $y = (\cos x)^x$
10. $y = (\cos x)^{e^x}$
11. $y = (\tan x)^{\log x}$
12. $y = (\log x)^{\sin x}$
13. $y = e^{x^x}$
14. $y = 2^{x^x}$

Find dy/dx in Exercises 15–20 by using logarithmic differentiation.

15. $y = x^2\sqrt{x + 1}\sqrt{x + 2}$

16. $y = \dfrac{(x^2 + 1)\sqrt{x^2 - 4}}{3x - 2}$

17. $y = \dfrac{\sqrt[3]{1 - x^3}}{(x^2 - 1)(x^2 + 9)}$

18. $y = \dfrac{x^2\sqrt{1 - x^2}}{(x^2 + 1)^3}$

19. $y = \dfrac{x^3 e^{2x}}{\sqrt{1 - 3x^2}}$

20. $y = \dfrac{10^x\sqrt[3]{x^3 - 1}}{x^2}$

21. Solve Exercise 20 of Exercises 6–7 (Chapter 6) by using logarithmic differentiation.

9–8. Exponential Growth

A common natural phenomenon is one in which a variable y, $y > 0$, changes with respect to another variable t at a rate proportional to y; that is, y satisfies the differential equation

$$\frac{dy}{dt} = ky,$$

where k is a constant and may be either positive or negative. This relationship exists in a wide variety of fields. We shall see a few of them in what follows.

This differential equation may be written, separating the variables as in Example 7-15, in the form

$$\frac{dy}{y} = k\,dt.$$

Then

$$\int \frac{dy}{y} = \int k\,dt + C,$$

or

$$\log y = kt + C.$$

Let the initial value of y be y_0; that is,

$$y = y_0 \qquad \text{when } t = 0.$$

Then

$$C = \log y_0 \qquad \text{and} \qquad \log y = kt + \log y_0.$$

Therefore

$$e^{\log y} = e^{(kt + \log y_0)}, \qquad \text{or} \qquad y = y_0\,e^{kt}.$$

Clearly, y is increasing with t if $k > 0$; decreasing, if $k < 0$.

Example 9–12. The amount of active ferment in a culture of yeast grows at a rate proportional to the amount present. If the amount of active ferment doubles in 1 hr, find the amount to be expected in t hours.

If y is the amount of active ferment at time t, we have

$$\frac{dy}{dt} = ky \qquad \text{and} \qquad y = y_0\,e^{kt},$$

where y_0 is the amount present at $t = 0$. When $t = 1$, $y = 2y_0$. Therefore

$$2y_0 = y_0\,e^{k \cdot 1}, \qquad \text{or} \qquad e^k = 2.$$

In order to solve for k we take the natural logarithm of the two members of this equation and obtain

$$k = \log 2.$$

Thus the amount to be expected in t hours is given by

$$y = y_0 \, e^{t \log 2}.$$

Example 9–13. The rate of increase of a certain population is proportional to that population. If it increases from 5000 to 10,000 in 3 months, what will it become in 5 months?

Let y represent the number of individuals in the population at time t. Then

$$\frac{dy}{dt} = ky,$$

from which we obtain, as before,

$$y = y_0 \, e^{kt},$$

where y_0 is the population at $t = 0$; that is,

$$y = 5000e^{kt}.$$

Since $y = 10,000$ when $t = 3$,

$$10,000 = 5000e^{3k}.$$

Thus

$$e^{3k} = 2 \quad \text{or} \quad k = \frac{1}{3} \log 2 \cong 0.23.$$

Hence the general population law takes the form

$$y = 5000e^{0.23t}.$$

When $t = 5$, we have

$$y = 5000e^{1.15} \cong 5000(3.16) = 15,800 \text{ individuals.}$$

Exercises 9-6

1. If $y = 2e^{kt}$ and $y = 10$ when $t = 2$, find k.
2. If $y = 10e^{kt}$ and $y = 3$ when $t = 4$, find k.
3. The population of a given country is estimated to be increasing at a constant rate of 4 percent per year. Assuming that this instantaneous rate can be expected to continue indefinitely, how long will it take for the population to double?

4. The atmospheric pressure is found to vary with respect to the altitude above sea level according to the formula

$$\frac{dp}{dh} = kp.$$

If the pressure (psi) is 14.7 at sea level and 13.0 at 3000-ft elevation, find the formula for p in terms of h.

5. A certain radioactive substance decays at a rate proportional to the amount present. If 1 gram decays to 0.8 gram in 100 years, how long will it take for one-half of it to decay?

6. If bacteria increase at a rate proportional to the number present, how long will it take a given number to double if it will triple in 30 min?

7. If light falling on the surface of water decreases in intensity with respect to depth at a rate proportional to the intensity, and if 6 ft of water absorbs one-fourth of the light falling on its surface, what is the formula for the intensity at depth x? What is the intensity at a depth of 24 ft?

8. Newton's law of cooling asserts that the difference y between the temperature of a body and that of the surrounding medium changes at a rate proportional to this difference. If y changes from 80 deg to 40 deg in 30 min, when will y be 20 deg?

9. A group of 500 construction workers are hired for a five-year project. These workers quit at a rate proportional to the number on the job. If there are 450 workers left at the end of one year, how many will still be on the job at the end of the project?

10. A basic assumption in fish nutrition is that the rate r of consumption of food increases with the concentration of food p at a rate that is proportional to $R - r$, where R is the maximum rate of consumption physiologically possible. Find an expression for r in terms of p and R.

11. The gross national product increased continuously at the rate of 6 percent per year during the years 1961–1966. If it was 740 billion dollars at the end of 1966 and it continues at this same rate, what will be the level of the gross national product at the end of 1975?

12. A firm estimates that its assets appreciate continuously at the rate of 6 percent per year. How long will it take for its assets to triple in value?

13. A firm estimates that the rate of increase in net revenue at the level of x dollars additional investment is given by

$$y = 5^{-x/2a},$$

where a is the initial revenue. Find the equation giving net revenue in terms of additional investment.

14. Insect population studies, where one population parasitizes another by depositing eggs on its members, suggest that

$$\frac{dy}{dx} = \frac{N - y}{N},$$

where N is the original number of hosts, y is the number of parasitized hosts, and x is the number of parasite eggs deposited. On this assumption, find the number of parasitized hosts as a function of parasite eggs deposited.

Chapter 10

METHODS OF INTEGRATION

10-1. Introduction

We have noted in the preceding pages that each differentiation formula may be restated as an integration formula. This *fundamental* set of formulas is collected here for reference.

$$\int u^n \, du = \frac{u^{n+1}}{n+1} + c, \qquad n \neq -1. \tag{10-1}$$

$$\int \frac{du}{u} = \log |u| + c. \tag{10-2}$$

$$\int e^u \, du = e^u + c. \tag{10-3}$$

$$\int a^u \, du = \frac{a^u}{\log a} + c. \tag{10-4}$$

$$\int \sin u \, du = -\cos u + c. \tag{10-5}$$

$$\int \cos u \, du = \sin u + c. \tag{10-6}$$

$$\int \sec^2 u \, du = \tan u + c. \tag{10-7}$$

$$\int \csc^2 u \, du = -\cot u + c. \tag{10-8}$$

$$\int \sec u \tan u \, du = \sec u + c. \tag{10-9}$$

$$\int \csc u \cot u \, du = -\csc u + c. \tag{10-10}$$

$$\int \frac{du}{\sqrt{a^2 - u^2}} = \operatorname{Arcsin} \frac{u}{a} + c, \qquad |u| < |a|. \tag{10-11}$$

$$\int \frac{du}{a^2 + u^2} = \frac{1}{a} \operatorname{Arctan} \frac{u}{a} + c. \tag{10-12}$$

$$\int \frac{du}{u\sqrt{u^2 - a^2}} = \frac{1}{a} \operatorname{Arcsec} \frac{u}{a} + c, \qquad |u| > |a|. \tag{10-13}$$

In this chapter it is our purpose to outline a number of methods of transforming an indefinite integral into one of these fundamental formulas and thereby accomplishing its evaluation. As mentioned in Chapter 7, this is not always possible. An example is

$$\int e^{-x^2}\, dx.$$

That is to say, there is no function f expressible as a finite combination of the functions in (10-1) through (10-13) such that

$$f'(x) = e^{-x^2}.$$

There are many such examples. However, many integrals are reducible to these fundamental formulas and we shall direct our attention to some of them.

10–2. Integration by Parts

Let u and v be two differentiable functions. Then, by the product formula,

$$d(uv) = u\, dv + v\, du$$

or

$$u\, dv = d(uv) - v\, du.$$

Thus

$$\int u\, dv = \int d(uv) - \int v\, du$$

$$= uv - \int v\, du.$$

THEOREM 10–1. If u and v are differentiable functions,

$$\int u\, dv = uv - \int v\, du.$$

In this manner the problem of evaluating $\int u\, dv$ has been transformed into the problem of evaluating $\int v\, du$. If the latter can be accomplished, our transformation has not been in vain, assuming we were in trouble with $\int u\, dv$.

Example 10-1. Evaluate $\int xe^x \, dx$.

This integral does not fall under any of the fundamental formulas. Let us set

$$u = x, \qquad dv = e^x \, dx.$$

Then

$$du = dx, \qquad v = \int e^x \, dx = e^x,$$

where we have ignored the arbitrary constant in the determination of v from dv. Any antiderivative will serve our purpose.

Applying Theorem 10-1, we have

$$\int xe^x \, dx = xe^x - \int e^x \, dx$$

$$= xe^x - e^x + c.$$

Thus the device of *integration by parts* enables us to evaluate the given integral. However, we cannot be careless in the choice of u and dv in the application of Theorem 10-1. Consider

$$u = e^x, \qquad dv = x \, dx.$$

Then

$$du = e^x \, dx, \qquad v = \frac{x^2}{2},$$

and

$$\int xe^x \, dx = \frac{x^2}{2} e^x - \int \frac{x^2}{2} e^x \, dx.$$

While this statement is perfectly true, nothing has been accomplished toward the solution of our problem because the residual integral

$$\int \frac{x^2}{2} e^x \, dx$$

is more complex than the one with which we started. Consequently this choice of u and dv does not provide us with the solution to the given problem.

The third choice we might make is

$$u = xe^x, \qquad dv = dx,$$

so

$$du = (xe^x + e^x) dx, \qquad v = x,$$

and

$$\int xe^x \, dx = x^2 e^x - \int (x^2 e^x + xe^x) \, dx$$

$$= x^2 e^x - \int x^2 e^x \, dx - \int xe^x \, dx,$$

or

$$2 \int xe^x \, dx = x^2 e^x - \int x^2 e^x \, dx.$$

Hence

$$\int xe^x \, dx = \frac{1}{2}\left[x^2 e^x - \int x^2 e^x \, dx \right],$$

which is the same result we obtained with the second choice.

Thus there are three choices of u and dv, only one of which is successful.

Example 10–2. Evaluate $\int \text{Arcsin } x \, dx$.

Let

$$u = \text{Arcsin } x, \qquad dv = dx.$$

Then

$$du = \frac{dx}{\sqrt{1 - x^2}}, \qquad v = x,$$

and

$$\int \text{Arcsin } x \, dx = x \, \text{Arcsin } x - \int \frac{x \, dx}{\sqrt{1 - x^2}}.$$

If we write

$$w = 1 - x^2,$$

we have

$$dw = -2x \, dx,$$

and

$$\int \frac{x \, dx}{\sqrt{1 - x^2}} = \int w^{-1/2}\left(-\frac{dw}{2} \right)$$

$$= -w^{1/2} + c$$

$$= -\sqrt{1 - x^2} + c.$$

Therefore

$$\int \text{Arcsin } x \, dx = x \, \text{Arcsin } x + \sqrt{1 - x^2} + c.$$

Example 10–3. Evaluate $\int e^x \sin x \, dx$.

Let

$$u = e^x, \qquad dv = \sin x \, dx.$$

Then

$$du = e^x \, dx, \qquad v = -\cos x,$$

and

(a) $$\int e^x \sin x \, dx = -e^x \cos x + \int e^x \cos x \, dx.$$

It would appear that we have made no progress, since the two integrals in this relation are equally difficult. Then we try the choice

$$u = \sin x, \qquad dv = e^x \, dx,$$

so

$$du = \cos x \, dx, \qquad v = e^x,$$

and

(b) $$\int e^x \sin x \, dx = e^x \sin x - \int e^x \cos x \, dx.$$

Now let us add (a) and (b) together. We obtain

$$2 \int e^x \sin x \, dx = - e^x \cos x + e^x \sin x + \int e^x \cos x \, dx - \int e^x \cos x \, dx$$

or

$$2 \int e^x \sin x \, dx = e^x(\sin x - \cos x) + c,$$

since arbitrary constants in the two determinations of $\int e^x \cos x \, dx$ may be different. Hence, dividing both members by 2,

$$\int e^x \sin x \, dx = \frac{e^x}{2} (\sin x - \cos x) + c,$$

where we make no distinction between c and $c/2$, since c is arbitrary.

There are many situations in which integration by parts proves to be very effective. We shall see some of these as we proceed.

It may be necessary to apply the process of integration by parts more than once in some of the problems which follow.

Exercises 10-1

Evaluate the following integrals by the method of integration by parts.

1. $\displaystyle\int \log x \, dx$

2. $\displaystyle\int x e^{-x} \, dx$

3. $\displaystyle\int x \sin x \, dx$

4. $\displaystyle\int x \cos 2x \, dx$

5. $\displaystyle\int \text{Arctan } x \, dx$

6. $\displaystyle\int x \log x \, dx$

7. $\displaystyle\int x^2 e^{2x} \, dx$

8. $\displaystyle\int x^2 \log x \, dx$

9. $\displaystyle\int x^2 \sin x \, dx$ **10.** $\displaystyle\int x^2 \cos 2x \, dx$

11. $\displaystyle\int x^3 e^x \, dx$ **12.** $\displaystyle\int x^3 \sin x^2 \, dx$

13. $\displaystyle\int x^3 \sqrt{x^2 + 1} \, dx$ **14.** $\displaystyle\int (\log x)^2 \, dx$

15. $\displaystyle\int e^x \cos 2x \, dx$ **16.** $\displaystyle\int e^{-2x} \sin 3x \, dx$

17. $\displaystyle\int \cos^2 x \, dx$ **18.** $\displaystyle\int (\log \cos x) \sin x \, dx$

10-3. Integrals Involving Trigonometric Functions

In order to cope more efficiently with integrals involving trigonometric functions we shall obtain four more formulas. They are:

$$\int \tan u \, du = -\log|\cos u| + c. \tag{10-14}$$

$$\int \cot u \, du = \log|\sin u| + c. \tag{10-15}$$

$$\int \sec u \, du = \log|\sec u + \tan u| + c. \tag{10-16}$$

$$\int \csc u \, du = -\log|\csc u + \cot u| + c. \tag{10-17}$$

The derivation of (10-14) is very simple. We write

$$\int \tan u \, du = \int \frac{\sin u}{\cos u} \, du,$$

and let

$$w = \cos u.$$

Then

$$dw = -\sin u \, du,$$

and

$$\int \tan u \, du = \int \frac{-dw}{w} = -\log|w| + c$$

$$= -\log|\cos u| + c.$$

An alternate form of this formula is obtained if we recall that

$$\cos u = \frac{1}{\sec u}.$$

Hence we may write

$$\int \tan u \, du = -\log \left| \frac{1}{\sec u} \right| + c$$

$$= -[\log 1 - \log |\sec u|] + c$$

$$= \log |\sec u| + c.$$

We may obtain (10-15) by a similar procedure.

Formula (10-16) results from multiplying numerator and denominator by $\sec u + \tan u$. We have

$$\int \sec u \, du = \int \frac{\sec u(\sec u + \tan u)}{\sec u + \tan u} du$$

$$= \int \frac{\sec u \tan u + \sec^2 u}{\sec u + \tan u} du.$$

Then, if we set

$$w = \sec u + \tan u,$$

we have

$$\int \sec u \, du = \int \frac{dw}{w} = \log |w| + c$$

$$= \log |\sec u + \tan u| + c.$$

Formula (10-17) may be obtained in a like manner.

Now we shall see by examples how (10-1) to (10-17) may be used to evaluate certain types of trigonometric integrals.

Example 10–4. Evaluate $\int \sin^5 x \, dx$.

This problem involves an odd power of the sine function. As we work through this example the student should note that the same general procedure would work for odd powers of the cosine function, and even for products of proper powers of the sine and cosine functions.

First we write

$$I = \int \sin^5 x \, dx = \int \sin^4 x \cdot \sin x \, dx.$$

From (8-6),

$$\sin^2 x = 1 - \cos^2 x,$$

so we have

$$I = \int (1 - \cos^2 x)^2 \sin x \, dx$$

$$= \int (1 - 2\cos^2 x + \cos^4 x)\sin x \, dx$$

$$= \int \sin x \, dx - 2\int \cos^2 x \sin x \, dx + \int \cos^4 x \sin x \, dx.$$

If we set

$$u = \cos x,$$

then

$$du = -\sin x \, dx,$$

and we may write

$$I = -\int du + 2\int u^2 \, du - \int u^4 \, du,$$

and by (10-1),

$$I = -u + 2\frac{u^3}{3} - \frac{u^5}{5} + c.$$

Note that the three arbitrary constants arising from the three integrals have been combined into a single arbitrary constant. This gives us

$$\int \sin^5 x \, dx = -\cos x + \frac{2}{3}\cos^3 x - \frac{1}{5}\cos^5 x + c.$$

Example 10–5. Evaluate $\int \cos^4 x \, dx$.

This problem involves an even power of the cosine function. The method employed will work equally well for even powers of the sine function and for certain products of powers of the sine and cosine functions. An understanding of the method will readily indicate the types of products to which it is applicable.

In problems of this type we rely on (8-10), which we restate in the forms

$$\cos^2 t = \frac{1}{2}(1 + \cos 2t); \tag{8-10a}$$

$$\sin^2 t = \frac{1}{2}(1 - \cos 2t). \tag{8-10b}$$

We write

$$I = \int \cos^4 x \, dx = \int (\cos^2 x)^2 \, dx$$

$$= \int \left[\frac{1 + \cos 2x}{2}\right]^2 dx$$

$$= \frac{1}{4} \int (1 + 2 \cos 2x + \cos^2 2x) \, dx.$$

Then, using (8-10b) once more,

$$I = \frac{1}{4} \int \left(1 + 2 \cos 2x + \frac{1 + \cos 4x}{2}\right) dx$$

$$= \frac{1}{4}\left[\frac{3}{2} \int dx + 2 \int \cos 2x \, dx + \frac{1}{2} \int \cos 4x \, dx\right]$$

$$= \frac{1}{4}\left[\frac{3}{2} x + \sin 2x + \frac{1}{8} \sin 4x\right] + c,$$

where we have used (10-1) and (10-6).

Example 10–6. Evaluate $\int \sin^2 x \cos^3 x \, dx$.

The general method employed in Example 10-4 will work here because of the odd power of the cosine function. We have

$$\int \sin^2 x \cos^3 x \, dx = \int \sin^2 x \cos^2 x \cos x \, dx$$

$$= \int \sin^2 x (1 - \sin^2 x) \cos x \, dx$$

$$= \int \sin^2 x \cos x \, dx - \int \sin^4 x \cos x \, dx$$

$$= \frac{1}{3} \sin^3 x - \frac{1}{5} \sin^5 x + c.$$

Example 10–7. Evaluate $\int \sin^2 x \cos^2 x \, dx$.

Since both powers are even, the method of Examples 10-4 and 10-6 will not work. We make use of (8-10a) and (8-10b) and write

$$I = \int \sin^2 x \cos^2 x \, dx = \int \left[\frac{1 - \cos 2x}{2}\right]\left[\frac{1 + \cos 2x}{2}\right] dx$$

$$= \frac{1}{4} \int (1 - \cos^2 2x) \, dx.$$

Reapplication of (8-10a) gives

$$I = \frac{1}{4} \int \left(1 - \frac{1 + \cos 4x}{2}\right) dx$$

$$= \frac{1}{8} \int (1 - \cos 4x) \, dx$$

$$= \frac{1}{8}\left(x - \frac{1}{4}\sin 4x\right) + c.$$

An alternate solution to this problem is given by making use of (8-9). We have

$$\int \sin^2 x \cos^2 x \, dx = \frac{1}{4} \int \sin^2 2x \, dx$$

$$= \frac{1}{4} \int \frac{1 - \cos 4x}{2} \, dx$$

$$= \frac{1}{8} \int (1 - \cos 4x) \, dx.$$

Example 10-8. Evaluate $\int \tan^5 x \, dx$.

From (8-6)

$$1 + \tan^2 t = \sec^2 t.$$

This is the identity that does the job on this problem. We write

$$I = \int \tan^5 x \, dx = \int \tan^3 x \tan^2 x \, dx$$

$$= \int \tan^3 x(\sec^2 x - 1) \, dx$$

$$= \int \tan^3 x \sec^2 x \, dx - \int \tan^3 x \, dx$$

$$= \int \tan^3 x \sec^2 x \, dx - \int \tan x(\sec^2 x - 1) \, dx$$

$$= \int \tan^3 x \sec^2 x \, dx - \int \tan x \sec^2 x \, dx + \int \tan x \, dx.$$

If we set $u = \tan x$ in the first two integrals,

$$I = \int u^3 \, du - \int u \, du + \int \tan x \, dx$$

$$= \frac{1}{4}u^4 - \frac{1}{2}u^2 - \log|\cos x| + c$$

$$= \frac{1}{4}\tan^4 x - \frac{1}{2}\tan^2 x - \log|\cos x| + c.$$

The student should observe that this method will work on any integral power of the tangent function. The same method coupled with

$$1 + \cot^2 t = \csc^2 t$$

from (8-6) will cope with integral powers of the cotangent function.

Example 10–9. Evaluate $\int \sec^4 x \, dx$.

We have

$$I = \int \sec^4 x \, dx = \int \sec^2 x \sec^2 x \, dx$$

$$= \int (1 + \tan^2 x)\sec^2 x \, dx$$

$$= \int \sec^2 x \, dx + \int \tan^2 x \sec^2 x \, dx.$$

If we set $u = \tan x$,

$$I = \int du + \int u^2 \, du$$

$$= u + \frac{1}{3} u^3 + c$$

$$= \tan x + \frac{1}{3} \tan^3 x + c.$$

This method clearly will work also on even powers of the cosecant function, if we use the proper trigonometric identity.

Example 10–10. Evaluate $\int \csc^3 x \, dx$.

We employ integration by parts to effect the integration of odd powers of the secant and cosecant functions. If we set

$$u = \csc x, \qquad dv = \csc^2 x \, dx,$$

we have

$$du = -\csc x \cot x \, dx, \qquad v = -\cot x,$$

and we may write, by Theorem 10-1,

$$\int \csc^3 x \, dx = -\csc x \cot x - \int \csc x \cot^2 x \, dx$$

$$= -\csc x \cot x - \int \csc x(\csc^2 x - 1) \, dx$$

$$= -\csc x \cot x - \int \csc^3 x \, dx + \int \csc x \, dx.$$

Hence

$$2 \int \csc^3 x \, dx = -\csc x \cot x - \log|\csc x + \cot x| + c,$$

or

$$\int \csc^3 x \, dx = -\frac{1}{2} [\csc x \cot x + \log|\csc x + \cot x|] + c.$$

The principal weapons for attacking simple trigonometric integrals are displayed in the foregoing examples. The student should be able, by means of standard trigonometric identities, to reduce most of the problems he will encounter to a counterpart of one of these examples.

Exercises 10-2

Evaluate the following integrals.

1. $\displaystyle\int \sin^3 2x \, dx$

2. $\displaystyle\int \cos^3 x \, dx$

3. $\displaystyle\int \cos^5 3x \, dx$

4. $\displaystyle\int \sin^2 x \, dx$

5. $\displaystyle\int \cos^2 2x \, dx$

6. $\displaystyle\int \tan^3 \frac{x}{2} \, dx$

7. $\displaystyle\int \cot^3 3x \, dx$

8. $\displaystyle\int \sin^4 2x \, dx$

9. $\displaystyle\int \sin^3 x \cos^2 x \, dx$

10. $\displaystyle\int \cot^4 \frac{x}{2} \, dx$

11. $\displaystyle\int \cot^5 2x \, dx$

12. $\displaystyle\int \cos^4 2x \sin^3 2x \, dx$

13. $\displaystyle\int \cos^4 x \sin^2 x \, dx$

14. $\displaystyle\int \sec^3 5x \, dx$

15. $\displaystyle\int \csc^5 x \, dx$

16. $\displaystyle\int \sec^5 x \tan x \, dx$

17. $\displaystyle\int \cos^2 3x \sin^4 3x \, dx$

18. $\displaystyle\int \csc^4 \frac{x}{2} \cot \frac{x}{2} \, dx$

19. $\displaystyle\int \sqrt{\sin x} \cos^3 x \, dx$

20. $\displaystyle\int \frac{(1 + \sin x) \, dx}{(x - \cos x)^2}$

21. $\displaystyle\int \frac{\sin 4x \, dx}{\cos 2x}$

22. $\displaystyle\int \frac{\tan 3x \, dx}{\sec 3x}$

23. $\displaystyle\int \sqrt{1 - \cos x} \sin x \, dx$

24. $\displaystyle\int \frac{\sin^2 x}{\cos x} \, dx$

10-4. Algebraic Substitution

Many integrals may be transformed to a fundamental formula by an algebraic substitution. We already have used this device on many occasions—in fact, ever since Theorem 7-10 was introduced. In Example 7-2, we set $u = 1 + 2x$ in the integral $\int (1 + 2x)^5 \, dx$, and thereby transformed it to the fundamental form

$$\tfrac{1}{2} \int u^5 \, du.$$

This was an *algebraic substitution*. The student will recognize that we have been making similar transformations from that time.

We do not propose to prove any theorems or develop any general methods, but rather show by example how an algebraic substitution may in some instances save the day.

Example 10-11. Evaluate $\int x\sqrt{1 - x} \, dx$.

This problem may be solved by setting

$$u = \sqrt{1 - x} \qquad \text{or} \qquad u^2 = 1 - x.$$

Then

$$x = 1 - u^2 \qquad \text{and} \qquad dx = -2u \, du.$$

Hence

$$\int x\sqrt{1 - x} \, dx = \int (1 - u^2)(u)(-2u \, du)$$

$$= -2 \int (u^2 - u^4) \, du$$

$$= -2\left(\frac{u^3}{3} - \frac{u^5}{5}\right) + c$$

$$= -2\left[\frac{1}{3}(1 - x)^{3/2} - \frac{1}{5}(1 - x)^{5/2}\right] + c.$$

If we reduce to a common denominator and factor out $(1 - x)^{3/2}$, we obtain

$$\int x\sqrt{1 - x} \, dx = -\frac{2}{15}(1 - x)^{3/2}[5 - 3(1 - x)] + c$$

$$= -\frac{2}{15}(1 - x)^{3/2}(2 + 3x) + c.$$

Example 10-12. Evaluate

$$\int \frac{x^5 \, dx}{\sqrt{x^3 + 1}}.$$

We set

$$u = \sqrt{x^3 + 1}.$$

Then

$$u^2 = x^3 + 1$$

and, differentiating,

$$2u \, du = 3x^2 \, dx \qquad \text{or} \qquad x^2 \, dx = \frac{2}{3} u \, du.$$

Thus, we may write

$$\int \frac{x^5 \, dx}{\sqrt{x^3 + 1}} = \int \frac{x^3 \cdot x^2 \, dx}{\sqrt{x^3 + 1}}$$

$$= \int \frac{(u^2 - 1)(\frac{2}{3}u \, du)}{u}$$

$$= \frac{2}{3} \int (u^2 - 1) \, du$$

$$= \frac{2}{3} \left[\frac{u^3}{3} - u \right] + c$$

$$= \frac{2}{3} \left[\frac{(x^3 + 1)^{3/2}}{3} - (x^3 + 1)^{1/2} \right] + c$$

$$= \frac{2}{9} (x^3 + 1)^{1/2}(x^3 - 2) + c.$$

The student should not get the idea that an algebraic substitution will always effect the immediate solution of such problems. Consider the next example.

Example 10-13. Evaluate

$$\int \frac{x^2 \, dx}{\sqrt{x^2 + 1}}.$$

We try the same sort of substitution that worked in the previous problems. Let

$$u = \sqrt{x^2 + 1}.$$

Then

$$u^2 = x^2 + 1 \quad \text{and} \quad 2u \, du = 2x \, dx.$$

Also

$$x = \sqrt{u^2 - 1},$$

where we take $x \geqslant 0$ for simplicity. Therefore

$$\int \frac{x^2 \, dx}{\sqrt{x^2 + 1}} = \int \frac{\sqrt{u^2 - 1} \, u \, du}{u} = \int \sqrt{u^2 - 1} \, du.$$

While this is a considerable simplification of form over the original, it still does not fit any of the fundamental formulas. Hence we have to look further for the solution. There might be another algebraic substitution which would turn the trick. However, we shall not look further in that direction. In the next section we shall make use of a different type of substitution, which will provide the key to this and many other problems.

10-5. Trigonometric Substitution

In this section we shall see how *trigonometric substitutions* may be used to rationalize certain algebraic forms. Consider the expression $a^2 - u^2$ and set $u = a \sin t$. Then

$$a^2 - u^2 = a^2 - a^2 \sin^2 t = a^2(1 - \sin^2 t) = a^2 \cos^2 t.$$

Hence, under this substitution,

$$(a^2 - u^2)^{n/2} = [(a^2 \cos^2 t)^{1/2}]^n = a^n \cos^n t,$$

which is rational if n is an integer. Therefore we conclude that

$$u = a \sin t \qquad\qquad\qquad (10\text{-}18)$$

will rationalize $(a^2 - u^2)^{n/2}$, n an integer.
   Similarly, making use of

$$1 + \tan^2 t = \sec^2 t,$$

$$u = a \tan t \qquad\qquad\qquad (10\text{-}19)$$

will rationalize $(a^2 + u^2)^{n/2}$, n an integer, and

$$u = a \sec t \qquad\qquad\qquad (10\text{-}20)$$

will rationalize $(u^2 - a^2)^{n/2}$, n an integer.
   These substitutions will convert an algebraic integral containing one of these types of irrationalities into a rational trigonometric integral. If the resulting trigonometric integral is one we can evaluate, we shall have solved our problem.

Example 10–14. Evaluate $\int \sqrt{1 - x^2}\, dx$.

This type of irrationality is eliminated by a substitution of the form (10-18). We set $x = \sin t$. Then

$$dx = \cos t\, dt,$$

and

$$\int \sqrt{1 - x^2}\, dx = \int \sqrt{1 - \sin^2 t}\, \cos t\, dt$$

$$= \int \cos^2 t\, dt$$

$$= \frac{1}{2} \int (1 + \cos 2t)\, dt$$

$$= \frac{1}{2}\left(t + \frac{1}{2} \sin 2t\right) + c.$$

Now our problem becomes that of expressing this result in terms of the original variable x. For this purpose we set up a right triangle with one leg x and the hypotenuse 1 (Fig. 10-1). Then, if t is the acute angle opposite leg x, we have

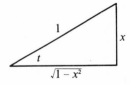

Fig. 10-1

$$\sin t = x,$$

$$\cos t = \sqrt{1 - x^2},$$

$$t = \text{Arcsin } x.$$

Therefore, since

$$\sin 2t = 2 \sin t \cos t,$$

we have

$$\int \sqrt{1 - x^2}\, dx = \frac{1}{2}\,(\text{Arcsin } x + x\sqrt{1 - x^2}) + c.$$

Example 10–15. Evaluate

$$\int \frac{dx}{\sqrt{9x^2 - 16}}.$$

From (10-20) we set $3x = 4 \sec t$. Then

$$dx = \frac{4}{3} \sec t \tan t\, dt,$$

and

$$\int \frac{dx}{\sqrt{9x^2 - 16}} = \int \frac{4/3 \tan t \sec t\, dt}{\sqrt{16 \sec^2 t - 16}} = \frac{1}{3} \int \sec t\, dt$$

$$= \frac{1}{3} \log |\sec t + \tan t| + c.$$

Now we resort to the triangle device again to convert back to the original variable x. This time, since

$$\sec t = \frac{3x}{4},$$

we draw a right triangle with hypotenuse $3x$ and one leg 4 (Fig. 10-2). Then

$$\tan t = \frac{\sqrt{9x^2 - 16}}{4},$$

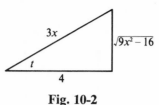

Fig. 10-2

and

$$\int \frac{dx}{\sqrt{9x^2 - 16}} = \frac{1}{3} \log \left| \frac{3x}{4} + \frac{\sqrt{9x^2 - 16}}{4} \right| + c$$

$$= \frac{1}{3} \log |3x + \sqrt{9x^2 - 16}| - \frac{1}{3} \log 4 + c$$

$$= \frac{1}{3} \log |3x + \sqrt{9x^2 - 16}| + c_1,$$

where

$$c_1 = c - \frac{1}{3} \log 4$$

and is arbitrary, since c is arbitrary.

Example 10–16. Evaluate

$$\int \frac{dx}{\sqrt{x^2 + x + 1}}.$$

While this does not appear to fall under the classification of problems currently being considered, we shall soon see that it does. We complete the square in the denominator and obtain

$$\int \frac{dx}{\sqrt{x^2 + x + 1}} = \int \frac{dx}{\sqrt{(x^2 + x + (1/4)) + (3/4)}}$$

$$= \int \frac{dx}{\sqrt{(x + (1/2))^2 + (\sqrt{3/2})^2}}.$$

Then we make the substitution

$$x + \frac{1}{2} = \frac{\sqrt{3}}{2} \tan t,$$

and get

$$\int \frac{dx}{\sqrt{x^2 + x + 1}} = \int \frac{(\sqrt{3}/2)\sec^2 t \, dt}{\sqrt{(3/4)\tan^2 t + (3/4)}}$$

$$= \int \frac{\sec^2 t \, dt}{\sec t}$$

$$= \int \sec t \, dt$$

$$= \log |\sec t + \tan t| + c$$

$$= \log \left| \frac{\sqrt{x^2 + x + 1}}{\sqrt{3}/2} + \frac{x + (1/2)}{\sqrt{3}/2} \right| + c, \qquad \text{(Fig. 10-3)},$$

$$= \log |\sqrt{x^2 + x + 1} + x + (1/2)| + c_1.$$

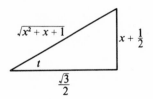

Fig. 10-3

Example 10–17. Evaluate

$$\int_0^3 \frac{dx}{\sqrt{x^2 + 3}}.$$

When working with definite integrals the conversion back to the original variable is not necessary. We transform not only the integrand but also the limits.

We set

$$x = \sqrt{3} \tan t.$$

Then, when

$$x = 0, \qquad \tan t = 0, \qquad \text{or} \quad t = 0;$$

and when

$$x = 3, \qquad \tan t = \sqrt{3}, \qquad \text{or} \quad t = \frac{\pi}{3}.$$

Thus

$$\int_0^3 \frac{dx}{\sqrt{x^2 + 3}} = \int_0^{\pi/3} \frac{\sqrt{3}\, \sec^2 t\, dt}{\sqrt{3 \tan^2 t + 3}}$$

$$= \int_0^{\pi/3} \sec t\, dt$$

$$= \log |\sec t + \tan t| \Big]_0^{\pi/3}$$

$$= \log \left|\sec \frac{\pi}{3} + \tan \frac{\pi}{3}\right| - \log |\sec 0 + \tan 0|$$

$$= \log(2 + \sqrt{3}) - \log 1$$

$$= \log(2 + \sqrt{3}).$$

Now let us return to Example 10-13 and apply a trigonometric substitution to it. If we set $x = \tan t$, we have

$$\int \frac{x^2\, dx}{\sqrt{x^2 + 1}} = \int \frac{\tan^2 t\, \sec^2 t\, dt}{\sqrt{\tan^2 t + 1}}$$

$$= \int \tan^2 t\, \sec t\, dt$$

$$= \int (\sec^2 t - 1)\, \sec t\, dt$$

$$= \int (\sec^3 t - \sec t)\, dt.$$

If we apply the method of Example 10-10, we obtain

$$\int \sec^3 t\, dt = \tfrac{1}{2}[\sec t \tan t + \log |\sec t + \tan t|] + c.$$

This result coupled with (10-16) gives

$$\int (\sec^3 t - \sec t)\, dt = \tfrac{1}{2}[\sec t \tan t + \log |\sec t + \tan t|]$$

$$- \log |\sec t + \tan t| + c$$

$$= \tfrac{1}{2}[\sec t \tan t - \log |\sec t + \tan t|] + c.$$

Then, from Fig. 10-4,

$$\int \frac{x^2\, dx}{\sqrt{x^2 + 1}} = \frac{1}{2} [x\sqrt{1 + x^2} - \log |\sqrt{1 + x^2} + x|] + c.$$

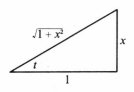

Fig. 10-4

Many problems containing irrationalities of the sort under discussion do not require trigonometric substitution for their solution. The student should be alert for other simpler methods before resorting to this type of transformation.

Exercises 10-3

Evaluate the indefinite integrals in Exercises 1–38.

1. $\int \sqrt{1 - x}\, dx$

2. $\int x\sqrt{x + 2}\, dx$

3. $\int \dfrac{x^2\, dx}{\sqrt{x + 3}}$

4. $\int x\sqrt[3]{1 - x}\, dx$

5. $\int \sqrt[3]{8 + x}\, dx$

6. $\int \dfrac{x\, dx}{9 + x^4}$

7. $\int \dfrac{x\, dx}{\sqrt{1 - x}}$

8. $\int \dfrac{x\, dx}{\sqrt{9 - x^2}}$

9. $\int \dfrac{dx}{\sqrt{9 - x^2}}$

10. $\int \dfrac{x^2\, dx}{\sqrt{9 - x^2}}$

11. $\int \dfrac{dx}{16 + 4x^2}$

12. $\int \dfrac{x\, dx}{16 + 4x^2}$

13. $\int \dfrac{x\, dx}{(x^2 + 1)^2}$

14. $\int \dfrac{x^2\, dx}{\sqrt{4 - x^2}}$

15. $\int \dfrac{x^2\, dx}{\sqrt{4 - x}}$

16. $\int \dfrac{x^3\, dx}{\sqrt{x^2 + 9}}$

17. $\int x^5\sqrt{4 + x^3}\, dx$

18. $\int x\sqrt[3]{(1 + x^2)^2}\, dx$

19. $\int \dfrac{x^3\, dx}{\sqrt{(x^2 + 9)^3}}$

20. $\int \dfrac{x\, dx}{\sqrt{16 - 9x^2}}$

21. $\int \dfrac{dx}{\sqrt{16 - 9x^2}}$

22. $\int \dfrac{x^3\, dx}{\sqrt{16 - 9x^2}}$

23. $\displaystyle\int \frac{\sqrt{x^2 - 4}\,dx}{x}$

24. $\displaystyle\int \sqrt{9 - x^2}\,dx$

25. $\displaystyle\int \sqrt{x^2 + 16}\,dx$

26. $\displaystyle\int \sqrt{x^2 - 9}\,dx$

27. $\displaystyle\int \frac{x^2\,dx}{x^2 + 16}$

28. $\displaystyle\int \frac{x\,dx}{x^2 + 16}$

29. $\displaystyle\int \frac{dx}{x^2 + 16}$

30. $\displaystyle\int x^2 \sqrt{5 - x^2}\,dx$

31. $\displaystyle\int x\sqrt{5 - x}\,dx$

32. $\displaystyle\int \frac{dx}{x^2 + 4x + 13}$

33. $\displaystyle\int \frac{dx}{\sqrt{x^2 + 4x + 13}}$

34. $\displaystyle\int \frac{(2x + 3)\,dx}{\sqrt{9 - 4x^2}}$

35. $\displaystyle\int \frac{dx}{\sqrt{4x - x^2}}$

36. $\displaystyle\int \frac{dx}{9x^2 - 25}$

37. $\displaystyle\int \frac{dx}{(x^2 + 9)^{3/2}}$

38. $\displaystyle\int \frac{dx}{x\sqrt{9x^2 - 16}}$

Evaluate the definite integrals in Exercises 39–44.

39. $\displaystyle\int_{-3}^{3} \sqrt{9 - x^2}\,dx$

40. $\displaystyle\int_{1}^{2} x\sqrt{2 - x}\,dx$

41. $\displaystyle\int_{0}^{1} \frac{x\,dx}{(x^2 + 1)^2}$

42. $\displaystyle\int_{0}^{2} x^2 \sqrt{4 - x^2}\,dx$

43. $\displaystyle\int_{0}^{2} \frac{x^3\,dx}{\sqrt{4 + x^2}}$

44. $\displaystyle\int_{0}^{3} \frac{x^3\,dx}{(9 + x^2)^{5/2}}$

45. Psychologists studying the learning process have found that the probability p of successfully performing a given act and the time t spent in practice appear to satisfy the differential equation

$$\frac{dp}{dt} = k\sqrt{p^3(1 - p)^3}.$$

The solution of this equation calls for the evaluation of

$$\int \frac{dp}{\sqrt{p^3(1 - p)^3}}.$$

Evaluate this integral by means of the substitution $p = \sin^2 x$. Why would you expect such a substitution to help?

10–6. Partial Fractions

As a final device for evaluating integrals, we consider the problem of decomposing a *rational function* (the quotient of two polynomials) into *partial fractions*, that is, into a number of fractions whose sum is the given fraction.
Let the rational function be represented by

$$\frac{p_1(x)}{p_2(x)},$$

where $p_1(x)$ is of lower degree than $p_2(x)$. If this is not so, we divide $p_1(x)$ by $p_2(x)$, obtaining a polynomial $q(x)$ and a remainder $r(x)$, where the latter is a polynomial of degree less than $p_2(x)$. Then we have

$$\frac{p_1(x)}{p_2(x)} = q(x) + \frac{r(x)}{p_2(x)},$$

where the second term in the right member satisfies the hypothesis.

It is a well-known fact from algebra that a polynomial with real coefficients may be decomposed into linear and quadratic factors with real coefficients. Also, the following facts regarding partial fractions are the result of algebraic considerations. We present them here without any attempt at justification.

(1)   Corresponding to each unrepeated linear factor $ax + b$ of $p_2(x)$ there is a partial fraction:

$$\frac{A}{ax + b}.$$

(2)   Corresponding to each linear factor of $p_2(x)$ which is repeated n times, that is, $(ax + b)^n$, there are n partial fractions:

$$\frac{A_1}{ax + b}, \quad \frac{A_2}{(ax + b)^2}, \quad \cdots, \frac{A_n}{(ax + b)^n}; \quad A_n \neq 0.$$

(3)   Corresponding to each unrepeated quadratic factor $ax^2 + bx + c$ of $p_2(x)$ there is a partial fraction:

$$\frac{Ax + B}{ax^2 + bx + c}.$$

(4)   Corresponding to each quadratic factor of $p_2(x)$ which is repeated n times, that is, $(ax^2 + bx + c)^n$, there are n partial fractions:

$$\frac{A_1 x + B_1}{ax^2 + bx + c}, \quad \frac{A_2 x + B_2}{(ax^2 + bx + c)^2}, \quad \cdots, \quad \frac{A_n x + B_n}{(ax^2 + bx + c)^n};$$

where A_n and B_n are not both zero.

The following examples will indicate how these facts help us to evaluate the integrals of rational functions.

Example 10–18. Evaluate

$$\int \frac{(x^2 + 11x - 6)\,dx}{x^3 + x^2 - 6x}.$$

The denominator of this rational function may be broken down into the three factors:

$$x(x - 2)(x + 3).$$

These factors are linear and unrepeated. Therefore, from (1),

$$\frac{x^2 + 11x - 6}{x(x - 2)(x + 3)} = \frac{A}{x} + \frac{B}{x - 2} + \frac{C}{x + 3}$$

$$= \frac{A(x - 2)(x + 3) + Bx(x + 3) + Cx(x - 2)}{x(x - 2)(x + 3)},$$

from which we obtain the identity

(a) $x^2 + 11x - 6 \equiv A(x - 2)(x + 3) + Bx(x + 3) + Cx(x - 2).$

This relationship is true for *all* values of x and therefore true for any particular values we may choose. For reasons that will be obvious, we choose $x = 0, 2, -3$. When these values are chosen, in order, (a) reduces to

$$-6 = -6A;$$

$$20 = 10B;$$

$$-30 = 15C.$$

Hence, $A = 1$, $B = 2$, and $C = -2$, and

$$\int \frac{(x^2 + 11x - 6)\,dx}{x^3 + x^2 - 6x} = \int \left[\frac{1}{x} + \frac{2}{x - 2} - \frac{2}{x + 3}\right] dx$$

$$= \int \frac{dx}{x} + 2\int \frac{dx}{x - 2} - 2\int \frac{dx}{x + 3}$$

$$= \log|x| + 2\log|x - 2| - 2\log|x + 3| + c$$

$$= \log|x| + \log|x - 2|^2 - \log|x + 3|^2 + c$$

$$= \log \frac{|x|(x - 2)^2}{(x + 3)^2} + c.$$

An alternate method for obtaining A, B, and C is to rewrite (a) in the form

$$x^2 + 11x - 6 \equiv (A + B + C)x^2 + (A + 3B - 2C)x - 6A.$$

We equate coefficients of like powers of x in the two members and obtain

$$A + B + C = 1,$$
$$A + 3B - 2C = 11,$$
$$-6A = -6.$$

From the last equation

$$A = 1,$$

which, substituted in the other two, gives

$$B + C = 0,$$
$$3B - 2C = 10.$$

The solution of these two equations yields

$$B = 2, \qquad C = -2,$$

as before.

The first method of obtaining the constants A, B, and C is somewhat simpler in this particular problem. However, it is not so efficient, when quadratic factors are involved.

Example 10–19. Evaluate

$$\int \frac{(3x^2 + 4x + 21)\, dx}{(x + 1)(x^2 + 9)}.$$

Referring to (1) and (3) we write

$$\frac{3x^2 + 4x + 21}{(x + 1)(x^2 + 9)} = \frac{A}{x + 1} + \frac{Bx + C}{x^2 + 9}$$

$$= \frac{A(x^2 + 9) + (Bx + C)(x + 1)}{(x + 1)(x^2 + 9)}$$

$$= \frac{(A + B)x^2 + (C + B)x + (9A + C)}{(x + 1)(x^2 + 9)}.$$

Therefore

$$3x^2 + 4x + 21 \equiv (A + B)x^2 + (C + B)x + (9A + C),$$

from which we obtain the system of equations

$$A + B = 3,$$
$$C + B = 4,$$
$$9A + C = 21.$$

The solution of this system is

$$A = 2, \qquad B = 1, \qquad C = 3.$$

Hence

$$\int \frac{(3x^2 + 4x + 21)\, dx}{(x + 1)(x^2 + 9)} = \int \left[\frac{2}{x + 1} + \frac{x + 3}{x^2 + 9} \right] dx$$

$$= 2 \int \frac{dx}{x + 1} + \int \frac{x\, dx}{x^2 + 9} + 3 \int \frac{dx}{x^2 + 9}$$

$$= 2 \log |x + 1| + \frac{1}{2} \log |x^2 + 9| + 3 \cdot \frac{1}{3} \operatorname{Arctan} \frac{x}{3} + c$$

$$= \log[(x + 1)^2 \sqrt{x^2 + 9}] + \operatorname{Arctan} \frac{x}{3} + c.$$

Example 10–20. Evaluate

$$\int \frac{(4x^2 - 24x + 2)\, dx}{(x - 2)^2 (x^2 + 2x + 2)}.$$

We have, from (2) and (3),

$$\frac{4x^2 - 24x + 2}{(x - 2)^2 (x^2 + 2x + 2)} = \frac{A}{x - 2} + \frac{B}{(x - 2)^2} + \frac{Cx + D}{x^2 + 2x + 2}$$

$$= \frac{A(x - 2)(x^2 + 2x + 2) + B(x^2 + 2x + 2) + (Cx + D)(x - 2)^2}{(x - 2)^2 (x^2 + 2x + 2)}.$$

Therefore

$$4x^2 - 24x + 2 \equiv (A + C)x^3 + (B - 4C + D)x^2$$
$$+ (-2A + 2B + 4C - 4D)x + (-4A + 2B + 4D),$$

from which we obtain

$$A + C = 0,$$
$$B - 4C + D = 4,$$
$$-2A + 2B + 4C - 4D = -24,$$
$$-4A + 2B + 4D = 2.$$

The solution of this system of equations is

$$A = 1, \qquad B = -3, \qquad C = -1, \qquad D = 3.$$

Hence

$$\int \frac{(4x^2 - 24x + 2)\, dx}{(x - 2)^2(x^2 + 2x + 2)} = \int \left[\frac{1}{x - 2} - \frac{3}{(x - 2)^2} - \frac{x - 3}{x^2 + 2x + 2} \right] dx$$

$$= \int \frac{dx}{x - 2} - 3 \int \frac{dx}{(x - 2)^2} - \int \frac{(x + 1 - 4)\, dx}{x^2 + 2x + 2}$$

$$= \log |x - 2| + 3(x - 2)^{-1}$$

$$- \int \frac{(x + 1)\, dx}{x^2 + 2x + 2} + 4 \int \frac{dx}{x^2 + 2x + 2}$$

$$= \log |x - 2| + 3(x - 2)^{-1}$$

$$- \frac{1}{2} \log |x^2 + 2x + 2| + 4 \int \frac{dx}{(x + 1)^2 + 1}$$

$$= \log \frac{|x - 2|}{\sqrt{x^2 + 2x + 2}} + \frac{3}{x - 2} + 4 \operatorname{Arctan}(x + 1) + c.$$

Exercises 10-4

Evaluate the following integrals.

1. $\displaystyle \int \frac{dx}{x^2 - 1}$

2. $\displaystyle \int \frac{x\, dx}{x^2 - x - 2}$

3. $\displaystyle \int \frac{(4x + 2)\, dx}{x^2 + x - 6}$

4. $\displaystyle \int \frac{(2 - x)\, dx}{x^2 + 4x}$

5. $\displaystyle \int \frac{(x^3 - x^2 - 3x + 10)\, dx}{x^2 - 4}$

6. $\displaystyle \int \frac{(2x^2 - 6x - 1)\, dx}{x^2 - 3x}$

7. $\displaystyle \int \frac{(8 - x)\, dx}{x^3 + 4x}$

8. $\displaystyle \int \frac{(5x - 2)\, dx}{x^3 - x^2 + 2x - 2}$

9. $\displaystyle \int \frac{(20 + 7x - x^2)\, dx}{x^3 - 4x^2 - 5x}$

10. $\displaystyle \int \frac{(x + 15)\, dx}{x^4 + 2x^3 + 5x^2}$

11. $\displaystyle \int \frac{dx}{(x^2 + 1)(x^2 + x)}$

12. $\displaystyle \int \frac{(2x^2 + 6x)\, dx}{(x^2 + 1)(x + 1)^2}$

13. $\displaystyle \int \frac{(3x^2 + 15x)\, dx}{(x - 1)^2(x + 2)^2}$

14. $\displaystyle \int \frac{(17x + 6)\, dx}{x^5 - 2x^4 + 3x^3}$

15. $\displaystyle \int \frac{(x - 26)\, dx}{(x - 1)(x^2 + 4)^2}$

16. $\displaystyle \int \frac{dx}{x^6 - 1}$

17. Under certain circumstances the growth of a population may be described by

$$\frac{dx}{dt} = (a - bx)x,$$

where x is the population at time t and a and b are positive constants. What is the ultimate size of such a population? (Hint: Solve this equation by separating the variables as in Example 7-15, and find $\lim_{t \to \infty} x$.)

18. Additional factors may affect population growth in such a way that the sign of the right member of the growth equation in Exercise 17 is changed. If the initial population is x_0, find a general relationship between population size and time for this case.

19. The spread of an infectious disease in a closed community of n individuals, all equally susceptible, is expressed by the equation

$$\frac{dx}{dt} = kx(n - x)$$

where k is a constant and x is the number of infected individuals at time t. If x_0 individuals were infected at time $t = 0$, find the general relationship between the number of infected individuals and time.

10-7. Review

We have studied a number of methods of integration in the preceding pages. Not the least of the skills to be developed in this connection is that of recognizing a particular problem as being a likely subject for a particular method. The following miscellaneous set of problems has been assembled to help the student acquire this skill. They are of varying difficulty but all may be solved by one or more of the methods already studied.

Exercises 10-5

Evaluate the following integrals.

1. $\displaystyle\int \frac{dx}{\sqrt{x^2 - 5}}$

2. $\displaystyle\int \frac{x\,dx}{\sqrt{x^2 - 5}}$

3. $\displaystyle\int \frac{dx}{\sqrt{9 + 4x^2}}$

4. $\displaystyle\int x^{-1} \log x \, dx$

5. $\displaystyle\int x \operatorname{Arctan} x \, dx$

6. $\displaystyle\int \cot^2 3x \, dx$

7. $\displaystyle\int \frac{x^2 \, dx}{\sqrt{x-9}}$

8. $\displaystyle\int \frac{3(x-1) \, dx}{(3x+1)^{3/2}}$

9. $\displaystyle\int x \sin^2 x \, dx$

10. $\displaystyle\int \frac{dx}{x^2 - x}$

11. $\displaystyle\int \sin^3 5x \cos^5 5x \, dx$

12. $\displaystyle\int \frac{\cos x \, dx}{\sin x - 1}$

13. $\displaystyle\int \log^2 2x \, dx$

14. $\displaystyle\int x^3 \cos x \, dx$

15. $\displaystyle\int \csc x \cot^5 x \, dx$

16. $\displaystyle\int \frac{x^2 \, dx}{(x^2 + 9)^2}$

17. $\displaystyle\int x \sec^2 2x \, dx$

18. $\displaystyle\int \cos^3 2x \, dx$

19. $\displaystyle\int \tan^3 2x \, dx$

20. $\displaystyle\int \frac{(2x^2 - 3x - 6) \, dx}{x^2 - 2x}$

21. $\displaystyle\int x(x+1)^{1/3} \, dx$

22. $\displaystyle\int \log^3 x \, dx$

23. $\displaystyle\int x^2 \sqrt{3x^3 + 1} \, dx$

24. $\displaystyle\int \cos^3 2x \sin^2 2x \, dx$

25. $\displaystyle\int x^2 e^{-2x} \, dx$

26. $\displaystyle\int \frac{(x^2 - x) \, dx}{x + 1}$

27. $\displaystyle\int \frac{\cos(\log x) \, dx}{x}$

28. $\displaystyle\int \frac{x^2 \, dx}{\sqrt{1 - 9x^2}}$

29. $\displaystyle\int x(5 - 3x)^{1/4} \, dx$

30. $\displaystyle\int x\sqrt{x^2 + 5} \, dx$

31. $\displaystyle\int e^{2x} \sin x \, dx$

32. $\displaystyle\int \frac{dx}{\sqrt{2x - x^2}}$

33. $\displaystyle\int \frac{dx}{e^x + e^{-x}}$

34. $\displaystyle\int \frac{(\sin x - 1) \, dx}{\cos x}$

35. $\displaystyle\int \frac{(x - 1) \, dx}{x^3 - x^2 - 2x}$

36. $\displaystyle\int \log\sqrt{ex} \, dx$

37. $\displaystyle\int x \sin x \cos x \, dx$

38. $\displaystyle\int \frac{\sqrt{4 - x^2} \, dx}{x^2}$

39. $\displaystyle\int \frac{dx}{4x^2 - 9}$

40. $\displaystyle\int \frac{x^2 \, dx}{(16 - x^2)^{3/2}}$

41. $\displaystyle \int \frac{(1 + x)\, dx}{1 + x^2}$

42. $\displaystyle \int \tan^5 \frac{x}{2} \sec^4 \frac{x}{2}\, dx$

43. $\displaystyle \int \frac{dx}{\sin x - 2 \csc x}$

44. $\displaystyle \int x \, \text{Arcsin } x \, dx$

45. $\displaystyle \int e^{3x} \cos 2x \, dx$

46. $\displaystyle \int \frac{(3x^2 + 2x + 8)\, dx}{x^3 + 4x}$

47. $\displaystyle \int \cos(\log x)\, dx$

48. $\displaystyle \int x\sqrt{x + 5}\, dx$

49. $\displaystyle \int \frac{dx}{\sqrt{16 + 9x^2}}$

50. $\displaystyle \int \frac{(2x^2 - 5x - 1)\, dx}{(x + 2)(x^2 - 4x + 5)}$

10-8. Integral Tables

A collection of integral formulas is called an *integral table* and may consist of a limited number of formulas as in (10-1)–(10-13) which we have called the fundamental set, or it may consist of several hundred formulas as may be found in some mathematical handbooks.

Our efforts at integration so far have been concentrated on transforming a given integral into one or more of the fundamental set. However, in many cases this process may be long and tedious, and, in some instances, may require a considerable degree of ingenuity. This makes it desirable to have a more extensive table of integrals available.

A modest table of integrals appears in Appendix A of this book and we shall use it to gain a little experience in the use of such tables.

The use of an integral table requires finding a formula which may be adjusted to the problem at hand by a proper choice of the constants present and other modifications dictated by the manner in which the variable of integration is involved. Experience with the general techniques of integration is essential in this process.

The problems in Exercises 10-6 may be evaluated by means of the table in Appendix A. The following examples will indicate the general procedure. Practice will do the rest!

Example 10–21. Evaluate $\int x^2 \sqrt{9 - 4x^2}\, dx$.

If we were to evaluate this by reduction to a fundamental form, the substitution $2x = 3 \sin t$ is indicated. However, let us evaluate it by means of the integral table in Appendix A. We seek a formula into which this integral may be transformed. We note that Formula 38 of the table has the left member

$$\int u^2 \sqrt{a^2 - u^2}\, du,$$

which looks promising. If we write $u = 2x$ and $a = 3$, we have

$$\int x^2 \sqrt{9 - 4x^2} \, dx = \int \frac{1}{4}(2x)^2 \sqrt{9 - 4x^2} \cdot \frac{1}{2}(2dx)$$

$$= \frac{1}{8} \int u^2 \sqrt{a^2 - u^2} \, du$$

$$= \frac{1}{8}\left[-\frac{u}{4}(a^2 - u^2)^{3/2} + \frac{a^2}{8} u\sqrt{a^2 - u^2} + \frac{a^4}{8} \, \text{Arcsin} \, \frac{u}{a} \right] + c$$

$$= \frac{1}{8}\left[-\frac{x}{2}(9 - 4x^2)^{3/2} + \frac{9}{4} x\sqrt{9 - 4x^2} + \frac{81}{8} \, \text{Arcsin} \, \frac{2x}{3} \right] + c.$$

Example 10–22. Evaluate $\int \sin^6 2x \, dx$.

If we were to evaluate this integral without the benefit of an integral table, we would follow the procedure outlined in Example 10-5. If we use the table, this problem falls under the *reduction formula*, listed as Formula 69 in Appendix A.

We set $u = 2x$ and obtain

$$\int \sin^6 2x \, dx = \frac{1}{2} \int \sin^6 u \, du$$

$$= \frac{1}{2}\left[-\frac{1}{6} \sin^5 u \cos u + \frac{5}{6} \int \sin^4 u \, du \right].$$

We reapply the same formula to the last term and get

$$\int \sin^6 2x \, dx = \frac{1}{2}\left[-\frac{1}{6} \sin^5 u \cos u + \frac{5}{6}\left\{ -\frac{1}{4} \sin^3 u \cos u + \frac{3}{4} \int \sin^2 u \, du \right\} \right].$$

Then, making use of Formula 65,

$$\int \sin^6 2x \, dx = \frac{1}{2}\left[-\frac{1}{6} \sin^5 u \cos u \right.$$

$$\left. + \frac{5}{6}\left\{ -\frac{1}{4} \sin^3 u \cos u + \frac{3}{4}\left(\frac{u}{2} - \frac{1}{4} \sin 2u \right) \right\} \right] + c$$

$$= \frac{1}{12} \sin^5 u \cos u - \frac{5}{48} \sin^3 u \cos u + \frac{5}{32} u - \frac{5}{64} \sin 2u + c$$

$$= \frac{1}{12} \sin^5 2x \cos 2x - \frac{5}{48} \sin^3 2x \cos 2x + \frac{5}{16} x - \frac{5}{64} \sin 4x + c.$$

Exercises 10-6

Evaluate the following exercises by means of the table of integrals, stating in each case which formula (or formulas) applies.

1. $\displaystyle\int \frac{x\,dx}{1+2x}$

2. $\displaystyle\int \frac{dx}{\sqrt{4x^2+9}}$

3. $\displaystyle\int \sin 2x \sin 3x\,dx$

4. $\displaystyle\int e^x \cos 2x\,dx$

5. $\displaystyle\int \frac{dx}{x\sqrt{3-x}}$

6. $\displaystyle\int \sqrt{(1-x^2)^3}\,dx$

7. $\displaystyle\int \frac{dx}{x\sqrt{3-x^2}}$

8. $\displaystyle\int \cos\frac{x}{2}\sin x\,dx$

9. $\displaystyle\int \operatorname{Arcsin}\frac{2x}{3}\,dx$

10. $\displaystyle\int x^3 \log 2x\,dx$

11. $\displaystyle\int x^2 \cos\frac{x}{2}\,dx$

12. $\displaystyle\int x^2 e^{3x}\,dx$

13. $\displaystyle\int \tan^4 3x\,dx$

14. $\displaystyle\int \cos 5x \cos 3x\,dx$

15. $\displaystyle\int \frac{x^2\,dx}{2+3x}$

16. $\displaystyle\int \sqrt{16-25x^2}\,dx$

17. $\displaystyle\int \frac{x^2\,dx}{\sqrt{9x^2+4}}$

18. $\displaystyle\int \sec^5 3x\,dx$

19. $\displaystyle\int x^2 \sin\frac{x}{2}\,dx$

20. $\displaystyle\int e^{\cos x} \sin x \cos x\,dx$

21. $\displaystyle\int \frac{e^{2x}\,dx}{(1+e^x)^2}$

22. $\displaystyle\int \sqrt{2+\cos x}\,\sin 2x\,dx.$

23. $\displaystyle\int \frac{dx}{x\sqrt{(1-\log^2 x)^3}}$

24. $\displaystyle\int \frac{dx}{x\log^2 x\sqrt{1-\log^2 x}}$

25. $\displaystyle\int \frac{\sec^2 2x\,dx}{\tan 2x\sqrt{1+\tan 2x}}$

26. $\displaystyle\int \frac{(\operatorname{Arctan}(x/2))^2\,dx}{4+x^2}$

27. $\displaystyle\int \sqrt{\frac{\operatorname{Arcsin} 2x}{1-4x^2}}\,dx$

28. $\displaystyle\int \frac{\sqrt{1+x}\,dx}{x}$

29. $\displaystyle\int \frac{x^3\,dx}{\sqrt{(1-x^2)^3}}$

30. $\displaystyle\int e^{2x} \sin e^x\,dx$

Chapter 11

MORE APPLICATIONS

11-1. Volumes of Solids

Let us suppose we have a solid such that the area of cross sections by planes perpendicular to the x-axis may be calculated; that is, it is possible to determine a function A such that the area of the cross section by the plane perpendicular to the x-axis through $(x, 0)$ is given by $A(x)$. Then we may *define* the volume of this solid in a manner analogous to the definition of plane area.

Let the extremes of x between which the solid lies be $a \leqslant x \leqslant b$, and let us subdivide this interval by introducing the values

$$a = x_0 < x_1 < x_2 < \cdots < x_{n-1} < x_n = b,$$

thereby determining

$$(\Delta x)_k = x_k - x_{k-1}, \qquad k = 1, 2, \ldots, n.$$

We also select

$$x_k^*, \qquad k = 1, 2, \ldots, n,$$

such that

$$x_{k-1} \leqslant x_k^* \leqslant x_k,$$

and form the sum

$$\sum_{k=1}^{n} A(x_k^*)(\Delta x)_k.$$

Each term in this sum represents the volume of a right prism with base area $A(x_k^*)$ and altitude $(\Delta x)_k$. This sum therefore represents the volume of a solid which approximates the given solid (Fig. 11-1). Then, following the same line of reasoning as used in Sec. 7-8 to define area, we define the volume of a solid S in the following manner.

DEFINITION 11-1. Let $A(x)$, $a \leqslant x \leqslant b$, be the area of a cross section of a solid S by the plane perpendicular to the x-axis through the point $(x, 0)$. Then the volume of S from $x = a$ to $x = b$ is defined to be

$$V = \lim_{n \to \infty} \sum_{k=1}^{n} A(x_k^*)(\Delta x)_k = \int_a^b A(x)\, dx.$$

272

An *element of volume* is

$$A(x_k^*)(\Delta x)_k,$$

and the limits on x are the extremes between which these elements may be formulated.

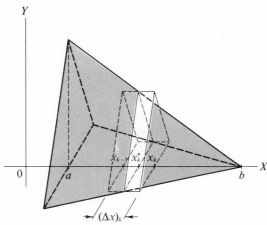

Fig. 11-1

Obviously, there is nothing sacred about the x-axis in this process. It could be set up equally well with respect to the y-axis. We shall not do this formally, but rather leave it to the student to make this adjustment when it seems advisable.

Let us test this definition for consistency with existing formulas by means of the following example.

Example 11-1. Find the volume of a right pyramid with an altitude h units and a square base with side a units.

Let us place the pyramid with respect to a coordinate system as shown in Fig. 11-2. From similar triangles,

$$\frac{BD}{AO} = \frac{DC}{OC},$$

or

$$BD = \frac{(AO)(DC)}{OC}.$$

But

$$AO = \frac{a}{2}, \qquad DC = h - x, \qquad OC = h,$$

so

$$BD = \frac{(a/2)(h - x)}{h}.$$

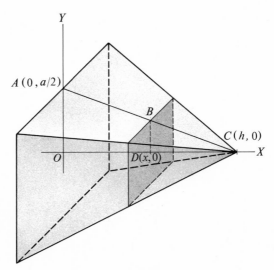

Fig. 11-2

Then

$$A(x) = (2BD)^2 = \left[\frac{a(h-x)}{h}\right]^2,$$

and

$$V = \int_0^h \frac{a^2(h-x)^2}{h^2}\, dx$$

$$= -\frac{a^2}{h^2}\frac{(h-x)^3}{3}\Bigg]_0^h = \frac{1}{3}a^2 h.$$

This agrees with the formula for such a solid as developed in elementary geometry, so we have consistency at least in this case.

Example 11-2. Find the volume of a solid whose base is the area bounded by $y^2 = 2x$ and $x = 2$, and whose cross section by a plane perpendicular to the x-axis is an equilateral triangle.

The cross sections defining A are equilateral triangles whose bases are chords of the parabola perpendicular to its axis (Fig. 11-3). Thus the sides of these triangles are of length $2y$, where y is the ordinate of one end of the chord forming the base. The altitude h satisfies the relationship

$$h^2 + y^2 = 4y^2,$$

from which we obtain

$$h = \sqrt{3}\, y.$$

Therefore

$$A(x) = hy = \sqrt{3}\, y^2 = 2\sqrt{3}\, x,$$

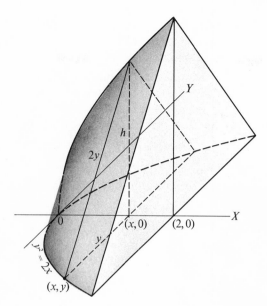

Fig. 11-3

and

$$V = \int_0^2 2\sqrt{3}\,x\,dx = 4\sqrt{3} \text{ cu units.}$$

Exercises 11-1

1. Find the volume of the following solids whose bases are the same as that in Example 11-2, and whose cross sections by planes perpendicular to the x-axis are:
   (a) Squares
   (b) Isosceles right triangles with one of the legs lying in the base
   (c) Isosceles right triangles with the hypotenuse lying in the base

2. Find the volume of the following solids whose bases are a circle of radius 2, and whose cross sections by planes perpendicular to a fixed diameter are:
   (a) Squares
   (b) Equilateral triangles
   (c) Isosceles right triangles with one of the legs a chord of the circle
   (d) Isosceles right triangles with the hypotenuse a chord of the circle

3. The axes of two right-circular cylinders, each of radius 4 in., meet at right angles. Find the volume common to the two cylinders.

4. A variable rectangle moves parallel to a fixed plane with one vertex moving along a fixed line perpendicular to this plane. If one dimension of

the rectangle is four times the distance from the fixed plane and the other dimension is one-half the square of this distance, find the volume swept out by the rectangle as this distance increases from 1 ft to 4 ft.

5. Two circles of radius 6 in. have a common diameter, and their planes are at right angles. A square moves with its vertices on these circles and with its plane perpendicular to the common diameter. Find the volume generated as the square moves from one end of the diameter to the other.

11-2. Volumes of Solids of Revolution—Disks and Washers

The solid swept out by a plane area as it is revolved about some line as an axis is called a *solid of revolution*. For example, the familiar right-circular cylinder is a solid of revolution, since it may be generated by revolving a rectangle about one of its sides. The characteristic property of these solids results from the fact that every point in the area being revolved describes a circle. Hence, areas of cross sections by planes perpendicular to the axis of rotation are bounded by concentric circles. Therefore these areas may be calculated easily, and thus the volumes of solids of revolution may be found by means of Definition 11-1.

Example 11-3. Find the volume of the sphere of radius a.

We choose this example to test again the consistency of Definition 11-1 with formulas established by other methods.

The sphere of radius a may be thought of as the solid of revolution obtained by rotating a semicircle of that radius about its diameter. This semicircle may be taken as the area bounded by

$$x = \sqrt{a^2 - y^2}, \qquad x = 0.$$

This amounts to choosing the y-axis as the axis of rotation (Fig. 11-4). Hence, in the application of Definition 11-1, we take cross sections perpendicular to this axis. If (x, y) is a point on the semicircle, the area of a cross section through that point is πx^2 and the corresponding element of volume is

$$\pi x^2 \, \Delta y = \pi(a^2 - y^2) \, \Delta y.$$

Therefore (Definition 11-1)

$$V = \pi \int_{-a}^{a} (a^2 - y^2) \, dy$$

$$= \pi \left(a^2 y - \frac{y^3}{3} \right) \Big]_{-a}^{a} = \frac{4\pi a^3}{3} \quad \text{cu units.}$$

The student will recognize this result as the well-known formula for the volume of a sphere.

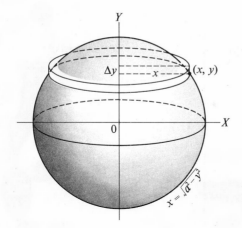

Fig. 11-4

The element of volume used here is the volume of a disk. Hence we often distinguish this procedure from others by calling it the *method of disks*, or by saying we use *disks as elements of volume*.

Example 11–4. Find the volume generated by revolving the area bounded by $x^2 = 4y$, $x = 2$, $y = 0$ about the x-axis.

If (x, y) is a point on the parabola $x^2 = 4y$ (Fig. 11-5), the area of the cross

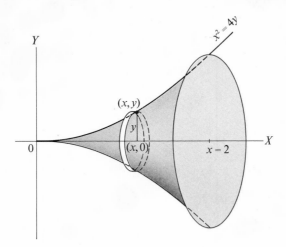

Fig. 11-5

section by a plane through that point and perpendicular to the x-axis (the axis of rotation) is πy^2 and the corresponding element of volume is

$$\pi y^2 \, \Delta x = \frac{\pi x^4}{16} \, \Delta x.$$

Hence the required volume is

$$V = \frac{\pi}{16} \int_0^2 x^4 \, dx = \frac{\pi}{16} \left(\frac{x^5}{5} \right) \Big]_0^2 = \frac{2\pi}{5} \quad \text{cu units.}$$

Example 11–5. Find the volume generated by revolving the area of Example 11-4 about the line $x = 2$.

Cross sections of this solid by planes through (x, y) perpendicular to $x = 2$ have a radius of $2 - x$ (Fig. 11-6). Hence

$$V = \int_0^1 \pi(2 - x)^2 \, dy = \int_0^1 (2 - 2\sqrt{y})^2 \, dy$$

$$= 4\pi \int_0^1 (1 - 2\sqrt{y} + y) \, dy$$

$$= 4\pi \left[y - \frac{4}{3} y^{3/2} + \frac{y^2}{2} \right]_0^1 = \frac{2\pi}{3} \quad \text{cu units.}$$

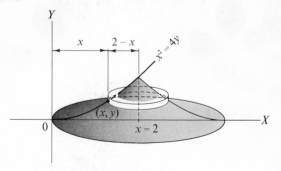

Fig. 11-6

We remind the student that the integral defining V may be expressed in terms of x instead of y. This reduces the work involved in some cases. Since

$$x^2 = 4y.$$

We have

$$2x \, dx = 4dy \quad \text{or} \quad dy = \frac{x}{2} \, dx.$$

Also, $x = 0$ when $y = 0$, and $x = 2$ when $y = 1$. Therefore

$$V = \int_0^1 (2 - x)^2 \, dy = \int_0^2 (2 - x)^2 \frac{x}{2} \, dx.$$

It is a simple matter to verify that the last member gives the same value of V as we obtained previously.

Example 11–6. Find the volume generated by revolving the first quadrant area bounded by the curves $y = \cos x$, $y = \sin x$, $x = 0$, about the x-axis.

When this area is revolved about the x-axis the resulting solid has a hole in it surrounding the axis of rotation (Fig. 11-7). Consequently, a cross

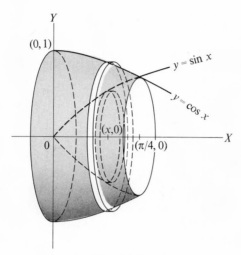

Fig. 11-7

section of it by a plane through $(x, 0)$ perpendicular to the x-axis is a circular ring with inner radius $\sin x$ and outer radius $\cos x$. Hence

$$A(x) = \pi \cos^2 x - \pi \sin^2 x = \pi \cos 2x.$$

Stated otherwise, an element of volume is

$$\pi \cos 2x \, \Delta x.$$

Then

$$V = \pi \int_0^{\pi/4} \cos 2x \, dx = \frac{\pi}{2} \sin 2x \Big]_0^{\pi/4} = \frac{\pi}{2} \quad \text{cu units.}$$

The element of volume in this case is the volume of a *washer*, and we shall refer to it by that name. When we take cross sections of a solid of revolution by planes perpendicular to the axis of rotation, we will always end up with *disks* or *washers*† as elements of volume.

---

† We include in this statement the possibility that a "washer" may consist of a number of concentric rings.

Exercises 11-2

Find the volume of the solid of revolution obtained by revolving the area bounded by the following curves about the indicated axis.

1. $y^2 = 8x$, $y = 0$, $x = 2$, first quadrant; (a) about the x-axis; (b) about the y-axis

2. $y^2 = 8x$, $y = 4$, $x = 0$; (a) about the x-axis; (b) about the y-axis

3. $y = x^3$, $y = 8$, $x = 0$; about the y-axis

4. $y^2 = 3x^3$, $y = 0$, $x = 3$; about the x-axis

5. $8y^2 = 9x^3$, $y = 0$, $x = 2$, first quadrant; (a) about the x-axis; (b) about the y-axis; (c) about $x = 2$

6. $x^2y = 4$, $y = 0$, $x = 1$, $x = 2$; about the x-axis

7. $y = e^{-x}$, $y = 0$, $x = 0$, $x = 1$; about the x-axis

8. $y = e^x$, $y = 0$, $x = 0$, $x = 1$; about $y = e$

9. $y = \log x$, $y = 0$, $x = 3$; about the x-axis

10. $y = x^2$, $y = 0$, $x = 2$, first quadrant; (a) about the x-axis; (b) about the y-axis; (c) about $x = 2$; (d) about $y = 4$

11. $x^2 = 8y$, $y = 0$, $x = 4$; about $x = 4$

12. $y = \sin x$, $y = 0$ from $x = 0$ to $x = \pi$; about the x-axis

13. $y = \sin x$, $y = 0$ from $x = 0$ to $x = \pi$; about $y = 1$

14. The loop of $y^2 = 4x^2 - x^3$; about the x-axis

15. $x^2y^2 + 9y^2 = 12$, $y = 0$, $x = 0$, $x = 3$, $x \geqslant 0$, $y \geqslant 0$; about the x-axis

16. $y = \dfrac{8}{4 + x^2}$, $y = 0$, $x = 0$, $x = 2$; about the x-axis

17. $x^2 + y^2 - 6x + 8 = 0$; about the y-axis

18. $xy = 5$, $x + y = 6$; about the x-axis

19. $x^2 + y^2 = 25$, $x = 3$, $x \geqslant 3$; about $x = 3$

11–3. Volumes of Solids of Revolution—Hollow Cylinders

Let us look at another way to generate an approximating volume for the volume of a solid of revolution. For convenience of discussion let us use A to designate the area to be revolved under $y = f(x)$ from $x = a$ to $x = b$, and let the axis of rotation be the y-axis (Fig. 11-8). Let the interval $a \leqslant x \leqslant b$ be subdivided by the points

$$a = x_0 < x_1 < x_2 < \cdots < x_{n-1} < x_n = b,$$

thereby defining

$$(\Delta x)_k = x_k - x_{k-1}, \qquad k = 1, 2, \ldots, n.$$

Also, let

$$x_k^* = \frac{x_{k-1} + x_k}{2}, \qquad k = 1, 2, \ldots, n,$$

that is, x_k^* is the midpoint of the interval from x_{k-1} to x_k.

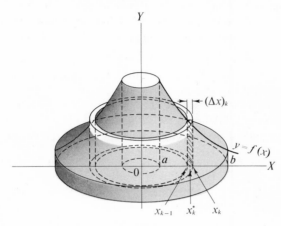

Fig. 11-8

The element of area $f(x_{k-1})(\Delta x)_k$ sweeps out a hollow cylinder of inner radius x_{k-1}, outer radius x_k, and altitude $f(x_{k-1})$ as A is rotated about the y-axis. It is left as an exercise for the student to prove that the volume of this hollow cylinder is

$$V[(\Delta x)_k] = 2\pi x_k^* f(x_{k-1})(\Delta x)_k. \tag{11-1}$$

If we form the sum

$$\sum_{k=1}^{n} V[(\Delta x)_k] = \sum_{k=1}^{n} 2\pi x_k^* f(x_{k-1})(\Delta x)_k,$$

we have the volume of a solid which approximates the given solid of revolution. Then, reasoning as before, we might define the required volume to be

$$V = \lim_{n \to \infty} \sum_{k=1}^{n} 2\pi x_k^* f(x_{k-1})(\Delta x)_k,$$

provided this limit exists. If f is a continuous function, we can appeal to Theorem 7-12 and write

$$V = \lim_{n \to \infty} 2\pi \sum_{k=1}^{n} x_k^* f(x_{k-1})(\Delta x)_k = 2\pi \int_a^b x f(x)\, dx. \tag{11-2}$$

It is not within the scope of this book to prove that the volume as defined by (11-2) is the same as that given by Definition 11-1. However, this can be shown, and we shall assume it to be true from this point on.

It is not feasible to establish formulas like (11-2) to deal with all possible situations in which hollow cylinders may apply. We prefer to set up a general procedure for determining such an element of volume in any particular case. If we examine the element of volume (11-1), we can, taking Theorem 7-12 into account, state the following rule.

The hollow cylindrical element of volume consists of the product of the thickness of the cylinder wall times the curved surface area of a cylinder whose radius lies in the closed range between the inner and outer radii and whose altitude is the same as that of the element.

In order to apply this rule it is first necessary to sketch the solid together with a hollow cylindrical element. Then it is possible to determine the radius, altitude, and wall thickness needed to write down the element of volume and thus obtain the integral defining the volume.

As a first example of this method let us again find the volume of a sphere.

Example 11-7. Find the volume of the sphere of radius a, using the method of hollow cylinders.

Let the area to be rotated be bounded by

$$x = \sqrt{a^2 - y^2}, \qquad x = 0,$$

and let the axis of rotation be the y-axis. We sketch the solid, including an element of volume (Fig. 11-9). From this sketch we observe that the inner

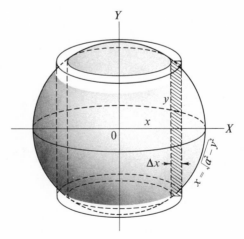

Fig. 11-9

radius is x, the altitude is $2y$, and the wall thickness is Δx, where (x, y) is a point on the semicircle. Therefore the hollow cylindrical element of volume is

$$2\pi x(2y)\,\Delta x,$$

and the volume is given by

$$V = \int_0^a 4\pi xy \, dx = 4\pi \int_0^a x\sqrt{a^2 - x^2} \, dx$$

$$= \frac{4\pi a^3}{3} \quad \text{cu units.}$$

Example II-8. Find the volume generated by revolving the first quadrant area bounded by $x^2 + y^2 = 4$, $y^2 = 3x$, $x = 0$, about the y-axis.

First we solve the equations of the circle and parabola simultaneously, and find the first quadrant point of intersection to be $(1, \sqrt{3})$. Next we sketch the solid of revolution together with a hollow cylinder element of volume (Fig. 11-10). Examination of this sketch shows that if x is the abscissa of a

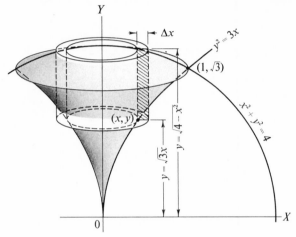

Fig. 11-10

point on either curve, we may use the radius x, the wall thickness Δx, and the altitude $\sqrt{4 - x^2} - \sqrt{3x}$ in the formulation of an element of volume. Hence the element may be written

$$2\pi x(\sqrt{4 - x^2} - \sqrt{3x}) \, \Delta x,$$

and

$$V = 2\pi \int_0^1 x(\sqrt{4 - x^2} - \sqrt{3x}) \, dx$$

$$= 2\pi \left[\int_0^1 x(4 - x^2)^{1/2} \, dx - \sqrt{3} \int_0^1 x^{3/2} \, dx \right]$$

$$= 2\pi \left[-\frac{1}{3}(4 - x^2)^{3/2} - \frac{2\sqrt{3}}{5} x^{5/2} \right]_0^1$$

$$= \frac{2\pi}{15}(40 - 21\sqrt{3}) \quad \text{cu units.}$$

Example 11–9. Find the volume generated by revolving the first quadrant area bounded by $x^2 = 4y$, $y = 1$, $x = 0$, about the line $y = 2$.

From Fig. 11-11 we can write the element of volume:

$$2\pi(2 - y)(2\sqrt{y})\,\Delta y,$$

whence

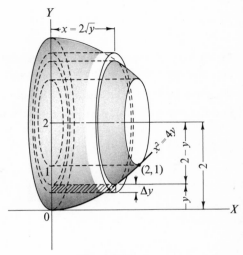

$$V = 4\pi \int_0^1 (2 - y)\sqrt{y}\,dy$$

$$= 4\pi \int_0^1 (2y^{1/2} - y^{3/2})\,dy$$

$$= 4\pi \left[\frac{4y^{3/2}}{3} - \frac{2y^{5/2}}{5}\right]_0^1$$

$$= \frac{56\pi}{15} \quad \text{cu units.}$$

Fig. 11-11

Exercises 11-3

In Exercises 1–18, find the volume of the solid of revolution obtained by revolving the area bounded by the given curves about the indicated axis by using hollow cylindrical elements.

1. $y = 4 - x^2$, $x = 0$, $y = 0$, first quadrant; about $x = 0$

2. Same area as Exercise 1; about $x = 2$

3. $y = 2x^2$, $y = 0$, $x = 1$, $x = 2$; about $x = 0$

4. Same area as Exercise 3; about $x = 3$

5. $y = 4x - x^2$, $y = 0$; about $x = 0$

6. $y = x^3$, $x = 0$, $y = 8$; about $y = 8$

7. $y = x^2 - 6x$, $y = 0$; about $x = 0$

8. Same area as Exercise 7; about $x = -1$

9. $y = x^3 - 1$, $x = 1$, $x = 2$, $y = -1$; about $x = 3$

10. $y = \log x$, $y = 0$, $x = 4$; about $x = -1$

11. $y = \frac{1}{2}(e^x + e^{-x})$, $x = 0$, $x = 1$, $y = 0$; about $x = 0$

12. $y = \frac{8}{4 + x^2}$, $x = 0$, $x = 2$, $y = 0$; about $x = 0$

13. $y^2 = 4x$, $y^2 = 8x - 4$, $y = 0$, first quadrant; about $y = 0$

14. $x^2 + y^2 = 25$, $3y^2 = 16x$, $y = 0$, first quadrant; about $y = 0$

15. $y = \sin x$, $y = 0$, from $x = 0$ to $x = \pi$; about $x = 0$

16. $y = 2x$, $y = x$, $x + y = 6$; about $y = 0$

17. $xy = 4$, $y = 4x$, $x = 2$, $y = 0$; about $y = 0$

18. $x^2 + y^2 = a^2$; about $x = b$, $b > a$

19. A hole 2 in. in diameter is bored through the center of a sphere 4 in. in diameter. What is the volume of material remaining?

20. Prove (11-1).

11-4. Arc Length

We wish to define the length of the curve $y = f(x)$ from $x = a$ to $x = b$. For this purpose let the arc be subdivided by the points (x_k, y_k), $k = 0, 1, 2, \ldots, n$; $a = x_0 < x_1 < x_2 < \cdots < x_{n-1} < x_n = b$; and let successive points be joined by straight-line segments as shown in Fig. 11-12 ($n = 4$). We designate these

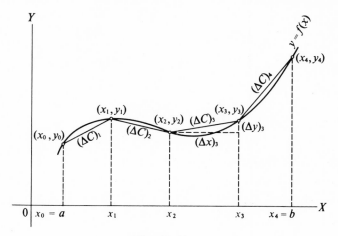

Fig. 11-12

chords by $(\Delta C)_k$, $k = 1, 2, \ldots, n$, and define the given arc length.

DEFINITION 11-2. The length of the arc of $y = f(x)$ from $x = a$ to $x = b$ is

$$S = \lim_{n \to \infty} \sum_{k=1}^{n} (\Delta C)_k, \dagger$$

provided this limit exists.

---

† This limit assumes, like all others of this nature, that $n \to \infty$ in such a manner that max $(\Delta x)_k \to 0$.

We have (Fig. 11-12)

$$[(\Delta C)_k]^2 = [(\Delta x)_k]^2 + [(\Delta y)_k]^2$$

$$= [(\Delta x)_k]^2 + [f(x_k) - f(x_{k-1})]^2.$$

If f is continuous for $a \leqslant x \leqslant b$ and f' exists for $a < x < b$, the mean value theorem (Theorem 6-1) enables us to write

$$f(x_k) - f(x_{k-1}) = f'(x_k^*)(\Delta x)_k,$$

for some x_k^*, $x_{k-1} < x_k^* < x_k$. Therefore

$$\sum_{k=1}^{n} (\Delta C)_k = \sum_{k=1}^{n} \sqrt{[(\Delta x)_k]^2 + [f'(x_k^*)(\Delta x)_k]^2}$$

$$= \sum_{k=1}^{n} \sqrt{1 + [f'(x_k^*)]^2} \,(\Delta x)_k.$$

When f' is continuous, $\sqrt{1 + (f')^2}$ will also be continuous. Hence if f' is continuous for $a \leqslant x \leqslant b$, we may apply Definition 7-2 and Theorem 7-1 to obtain

$$S = \lim_{n \to \infty} \sum_{k=1}^{n} \sqrt{1 + [f'(x_k^*)]^2} \,(\Delta x)_k$$

$$= \int_a^b \sqrt{1 + [f'(x)]^2} \, dx.$$

THEOREM 11-1. If f' is continuous for $a \leqslant x \leqslant b$, the length of arc of $y = f(x)$ from $x = a$ to $x = b$ is

$$S = \int_a^b \sqrt{1 + [f'(x)]^2} \, dx = \int_a^b \sqrt{1 + \left(\frac{dy}{dx}\right)^2} \, dx.$$

Obviously, the roles of x and y could be interchanged in the preceding argument to give

$$S = \int_c^d \sqrt{1 + \left(\frac{dx}{dy}\right)^2} \, dy, \tag{11-3}$$

where (a, c) and (b, d) are the terminal points of the arc. This form may be easier to evaluate in some cases and should be kept in mind for this reason.

The function defined by the relation

$$s(x) = \int_a^x \sqrt{1 + [f'(t)]^2} \, dt$$

is the length of the arc of $y = f(x)$ from $x = a$ to $x = x$. The differential of this *arc length function* is

$$ds = \frac{ds}{dx} dx = \sqrt{1 + [f'(x)]^2} \, dx,$$

or

$$ds = \sqrt{1 + \left(\frac{dy}{dx}\right)^2} \, dx.$$

This may be written in the form

$$ds = \sqrt{(dx)^2 + (dy)^2}. \tag{11-4}$$

This is the fundamental form of the *differential of arc length*. Other forms, as we shall see later, may be developed from this one. In any case

$$S = \int_{x=a}^{x=b} ds.$$

We may also observe from (11-4) that a geometrical interpretation can be placed on ds. We have

$$(ds)^2 = (dx)^2 + (dy)^2,$$

so (Fig. 11-13)

$$ds = PQ.$$

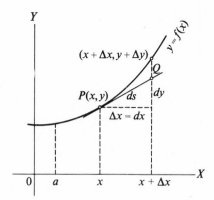

Fig. 11-13

Example II–10. Find the length of the first quadrant arc of the curve $y^2 = x^3$ lying between $x = 0$ and $x = 5$.

We have

$$y = x^{3/2},$$

so

$$\frac{dy}{dx} = \frac{3}{2} x^{1/2}$$

and

$$S = \int_0^5 \sqrt{1 + \left(\frac{3}{2} x^{1/2}\right)^2} \, dx$$

$$= \frac{1}{2} \int_0^5 \sqrt{4 + 9x} \, dx = \frac{335}{27} \text{ unit.}$$

Example 11-11. Find the circumference of the circle $x^2 + y^2 = a^2$.

We shall accomplish this by multiplying by four the length of the arc (Fig. 11-14) of

$$x = \sqrt{a^2 - y^2}$$

from the point $\left(a/\sqrt{2}, \ -a/\sqrt{2}\right)$ to the point $\left(a/\sqrt{2}, \ a/\sqrt{2}\right)$.
We have

$$\frac{dx}{dy} = \frac{-y}{(a^2 - y^2)^{1/2}}.$$

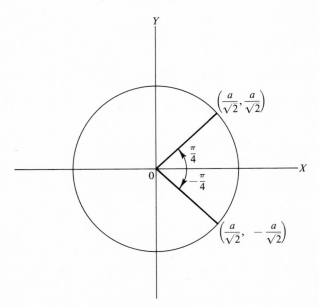

Fig. 11-14

Therefore, from (11-3),

$$S = 4 \int_{-a/\sqrt{2}}^{a/\sqrt{2}} \sqrt{1 + \frac{y^2}{a^2 - y^2}} \, dy$$

$$= 4a \int_{-a/\sqrt{2}}^{a/\sqrt{2}} \frac{dy}{(a^2 - y^2)^{1/2}} \cdot \dagger$$

In order to evaluate this integral we make the trigonometric substitution

$$x = a \sin t$$

and obtain

$$S = 4a \int_{-\pi/4}^{\pi/4} \frac{a \cos t \, dt}{(a^2 - a^2 \sin^2 t)^{1/2}}$$

$$= 4a \int_{-\pi/4}^{\pi/4} dt = 2\pi a,$$

as was to be expected.

Exercises 11-4

Find the length of arc of the following curves between the indicated points.

1. $4y = x^2$; $x = 0$ to $x = 2$

2. $8x = y^2$; first quadrant, $x = 0$ to $x = 2$

3. $y^3 = x^2$; $x = 0$ to $x = 5\sqrt{5}$

4. $3y^2 = (2x + 8)^3$; $x = 0$ to $x = 4$

5. $y = \dfrac{1}{2} a(e^{x/a} + e^{-x/a})$; $x = -a$ to $x = a$

6. $x^{2/3} + y^{2/3} = 1$; first quadrant, $x = 0$ to $x = 1$

7. $y = \dfrac{x^6 + 2}{8x^2}$; $x = 1$ to $x = 2$

8. $y = \log \sin x$; $x = \dfrac{\pi}{2}$ to $x = \dfrac{3\pi}{4}$

9. $y = 12 \log x$; $x = 5$ to $x = 9$

10. $y = e^x$; $x = 0$ to $x = 1$

---

† The particular arc selected in this example was chosen so that this integrand would be defined for all values of y on the closed interval bounded by the limits on this integral $(-a/\sqrt{2} \leq y \leq a/\sqrt{2})$. This is needed to satisfy the hypotheses of Theorem 7-6.

11-5. Area of Surfaces of Revolution

We now consider the *area of the surface of revolution* obtained by revolving the arc of $y = f(x)$ from $x = a$ to $x = b$ around the x-axis. Let this arc be subdivided as in Sec. 11-4 (Fig. 11-12) and designate by $(\Delta A)_k$ the curved area of the frustum of the cone generated by the chord $(\Delta C)_k$ as this revolution takes place (Fig. 11-15).

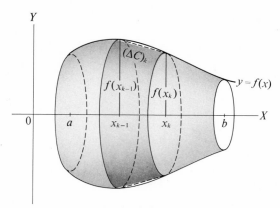

Fig. 11-15

DEFINITION 11-3. The area of the surface of revolution obtained by rotating the arc of $y = f(x)$ from $x = a$ to $x = b$ about the x-axis is

$$A = \lim_{n \to \infty} \sum_{k=1}^{n} (\Delta A)_k,$$

provided this limit exists.

From Fig. 11-14,

$$(\Delta A)_k = 2\pi \left[\frac{f(x_{k-1}) + f(x_k)}{2} \right] (\Delta C)_k .\dagger$$

Let us assume f', and consequently f, is continuous for $a \leqslant x \leqslant b$. Then, since

$$\tfrac{1}{2}[f(x_{k-1}) + f(x_k)]$$

is either equal to or lies between $f(x_{k-1})$ and $f(x_k)$, there exists a number x_k^{**}, $x_{k-1} \leqslant x_k^{**} \leqslant x_k$, such that

$$\tfrac{1}{2}[f(x_{k-1}) + f(x_k)] = f(x_k^{**}).$$

Also under this hypothesis, from Sec. 11-4,

$$(\Delta C)_k = \sqrt{1 + [f'(x_k^*)]^2} \, (\Delta x)_k, \qquad \text{for some } x_k^*, \; x_{k-1} < x_k^* < x_k.$$

Therefore, from Definition 11-3,

$$A = \lim_{n \to \infty} \sum_{k=1}^{n} (\Delta A)_k$$

$$= \lim_{n \to \infty} \sum_{k=1}^{n} 2\pi f(x_k^{**})\sqrt{1 + [f'(x_k^*)]^2} \, (\Delta x)_k$$

$$= 2\pi \int_a^b f(x)\sqrt{1 + [f'(x)]^2} \, dx,$$

by virtue of Theorem 7-12.

THEOREM II-2. If f' is continuous for $a \leqslant x \leqslant b$, the area of the surface of revolution generated by revolving the arc of $y = f(x)$ from $x = a$ to $x = b$ about the x-axis is

$$A = 2\pi \int_a^b f(x)\sqrt{1 + [f'(x)]^2} \, dx.$$

An examination of this formula reveals that the integrand is the product of the circumference of a circle described by a point on the rotating arc, $2\pi f(x)$, by the differential of arc length

$$ds = \sqrt{1 + [f'(x)]^2} \, dx.$$

Hence the integrand, or *element of area*, takes the form

$$2\pi r \, ds. \tag{11-5}$$

By means of this form we may modify Theorem 11-2 so as to find the area of any surface of revolution where the axis of revolution is parallel to one of the coordinate axes. The element of arc length and radius may be expressed in terms of the variable most convenient for computation.

Example II-12. Find the area of the surface generated by revolving the arc of $y = x^3$ from $x = 0$ to $x = 1$ about the x-axis.

Direct application of Theorem 11-2 gives

$$A = 2\pi \int_a^b f(x)\sqrt{1 + [f'(x)]^2} \, dx = 2\pi \int_0^1 x^3 \sqrt{1 + 9x^4} \, dx$$

$$= \frac{\pi}{27}(1 + 9x^4)^{3/2} \Big]_0^1 = \frac{\pi}{27}(10\sqrt{10} - 1) \quad \text{sq units.}$$

---

† We have made use of the formula

$$\text{Area} = 2\pi \left[\frac{r_1 + r_2}{2} \right] l$$

for the lateral surface area of a frustum of a cone with base radii r_1 and r_2, and slant height l.

Example 11–13. Find the area of the sphere of radius a.

Let us consider the sphere as being generated by rotating the semicircle

$$x = \sqrt{a^2 - y^2}$$

about the y-axis. We adapt (11-5) to this situation. We have

$$\frac{dx}{dy} = \frac{-y}{\sqrt{a^2 - y^2}},$$

so we may write

$$ds = \sqrt{1 + \frac{y^2}{a^2 - y^2}}\, dy$$

$$= \frac{a\, dy}{\sqrt{a^2 - y^2}}.$$

Also, from Fig. 11-16, $r = x$. Therefore (11-5) takes the form

$$2\pi r\, ds = (2\pi x)\frac{a\, dy}{\sqrt{a^2 - y^2}} = 2\pi a\, dy,$$

and the required area is given by

$$A = \int_{-a}^{a} 2\pi a\, dy = 4\pi a^2 \text{ sq units.}$$

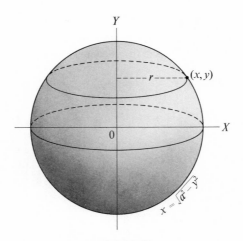

Fig. 11-16

Example 11-14. Find the area of the surface generated by revolving the arc of $4y = x^2$ from $x = 0$ to $x = 2$ about the line $x = 2$.

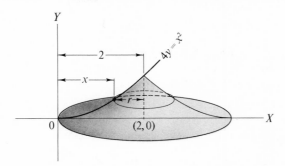

Fig. 11-17

We again make use of (11-5). We have, from Fig. 11-17,

$$r = 2 - x.$$

Also

$$ds = \sqrt{1 + \frac{x^2}{4}}\, dx = \frac{1}{2}\sqrt{4 + x^2}\, dx.$$

Then

$$A = \int_0^2 \pi(2 - x)\sqrt{4 + x^2}\, dx$$

$$= \pi\left[2\int_0^2 \sqrt{4 + x^2}\, dx - \int_0^2 x\sqrt{4 + x^2}\, dx \right].$$

Since these two integrals present two entirely different problems, we shall evaluate them separately. Setting

$$x = 2 \tan t$$

in the first one, we obtain

$$\int_0^2 \sqrt{4 + x^2}\, dx = 4\int_0^{\pi/4} \sec^3 t\, dt$$

$$= 4(\tfrac{1}{2})[\sec t \tan t + \log|\sec t + \tan t|]_0^{\pi/4}$$

$$= 2[\sqrt{2} + \log(\sqrt{2} + 1)].$$

For the second integral, we get

$$\int_0^2 x\sqrt{4 + x^2}\, dx = \tfrac{1}{3}(4 + x^2)^{3/2}\Big]_0^2 = \tfrac{1}{3}(16\sqrt{2} - 8).$$

The combination of these two results yields

$$A = \frac{4\pi}{3} [3 \log(\sqrt{2} + 1) - \sqrt{2} + 2] \quad \text{sq units.}$$

Exercises 11-5

Find the area of the surface of revolution generated by revolving the given arc about the indicated axis. Use integration in all cases.

1. $y = x$ from $x = 0$ to $x = 2$; about the x-axis

2. $y = \frac{1}{2}x + 1$ from $x = 0$ to $x = 2$; about the x-axis

3. Same arc; about the y-axis

4. Same arc; about the line $y = 2$

5. Same arc; about the line $x = 2$

6. $4x = y^2$ from $x = \frac{1}{4}$ to $x = 4$; about the x-axis

7. $4y = x^2$ from $x = 0$ to $x = 2$; about the y-axis

8. $y = \frac{x^6 + 2}{8x^2}$ from $x = 1$ to $x = 2$; about the y-axis

9. $y = \log x$ from $x = 1$ to $x = 2$, about the y-axis

10. The catenary $y = \frac{a}{2}(e^{x/a} + e^{-x/a})$ from $x = 0$ to $x = a$; about the y-axis

11. $y = \frac{a}{2}(e^{x/a} + e^{-x/a})$ from $x = -a$ to $x = a$; about the x-axis

12. The first quadrant arc of the hypocycloid $x^{2/3} + y^{2/3} = a^{2/3}$; about the y-axis

13. $y = x^3$ from $x = 0$ to $x = 1$; about the y-axis

14. $y = \sin x$ from $x = 0$ to $x = \pi$; about the x-axis

15. $y = e^x$ from $x = 0$ to $x = 2$; about the x-axis

11-6. Fluid Pressure

We consider now the force exerted by a liquid on a submerged plane area. For simplicity, we shall consider only areas perpendicular to the surface of the liquid.

The fundamental principal which applies here is that *the pressure (force per unit area) at depth h is wh, where w is the weight per unit volume of liquid.* Further, *this pressure is exerted equally in all directions.*

Let the area A be approximated by a set of *horizontal* elements of area $(\Delta A)_k$, $k = 1, 2, \ldots, n$, one of which is shown in Fig. 11-18. Let $(\Delta F)_k$ be the force exerted by the liquid on $(\Delta A)_k$. Then we define the total force on A.

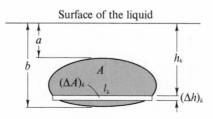

Surface of the liquid

Fig. 11-18

DEFINITION 11–4. The total force F on A is

$$F = \lim_{n \to \infty} \sum_{k=1}^{n} (\Delta F)_k,$$

provided this limit exists.

A little study of the problem will convince the student that this is a reasonable definition.

From Fig. 11-18,

$$(\Delta A)_k = l_k (\Delta h)_k,$$

and the pressure p_k at points in that element satisfies

$$w h_k \leqslant p_k \leqslant w[h_k + (\Delta h)_k].$$

Therefore

$$\sum_{k=1}^{n} w h_k l_k (\Delta h)_k < \sum_{k=1}^{n} (\Delta F)_k < \sum_{k=1}^{n} w[h_k + (\Delta h)_k] l_k (\Delta h)_k;$$

and if l is a continuous function of h, $a \leqslant h \leqslant b$,

$$\lim_{n \to \infty} \sum_{k=1}^{n} w h_k l_k (\Delta h)_k = \lim_{n \to \infty} \sum_{k=1}^{n} w[h_k + (\Delta h)_k] l_k (\Delta h)_k$$

$$= w \int_{a}^{b} h l \, dh,$$

by virtue of Theorem 7-1. Hence we may state the following theorem.

THEOREM 11–3. If the horizontal width l of a vertical submerged plane area A is a continuous function of the depth h, the total force F exerted by the liquid on A is given by

$$F = w \int_{a}^{b} h l \, dh.$$

Example 11–15. A vertical dam has the shape of an isosceles trapezoid with upper base 200 ft and lower base 150 ft. If it is 40 ft high and the water level is 2 ft below the top, find the total force on the dam.

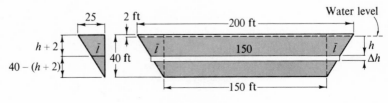

Fig. 11-19

From Fig. 11-19,

$$l = 150 + 2\bar{l},$$

and

$$\frac{l}{25} = \frac{40 - (h + 2)}{40}.$$

Thus

$$\bar{l} = \frac{5(38 - h)}{8},$$

and

$$l = 150 + \frac{5(38 - h)}{4} = \frac{790 - 5h}{4}.$$

Therefore, from Theorem 11-3,

$$F = \int_0^{38} (62.5)h\, \frac{(790 - 5h)}{4}\, dh$$

$$= \frac{62.5}{4} \int_0^{38} (790h - 5h^2)\, dh$$

$$= \frac{62.5}{4} \left(395h^2 - \frac{5h^3}{3} \right) \Big]_0^{38} \cong 3742 \text{ tons.}$$

Example 11–16. A vertical gate in the face of a dam has the shape of a semi-ellipse. If coordinate axes are chosen as shown in Fig. 11-20, it is bounded by the curves $y = 0$ and $y = (4/3)\sqrt{9 - x^2}$, where the units are feet. If the bottom, the straight side, is horizontal and 20 ft below the water surface, what is the total force on the gate?

We have, from Fig. 11-20,

$$l = 2x,$$

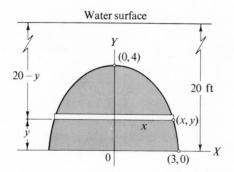

Fig. 11-20

where

$$y = \frac{4}{3}\sqrt{9 - x^2},$$

or

$$x = \frac{3}{4}\sqrt{16 - y^2}.$$

Also

$$h = 20 - y.$$

Therefore

$$F = \int_0^4 62.5(20 - y)\frac{3}{2}\sqrt{16 - y^2}\, dy$$

$$= \frac{3(62.5)}{2}\left[20\int_0^4 \sqrt{16 - y^2}\, dy - \int_0^4 y\sqrt{16 - y^2}\, dy\right]$$

$$= \frac{3(62.5)}{2}\left[20(4\pi) - \frac{64}{3}\right] \cong 10.8 \text{ tons.}$$

As the student has undoubtedly observed from these examples, the principal problem here is to set up a coordinate system and express both l and h in terms of the same variable.

Exercises 11-6

1. A dam has a vertical rectangular gate 6 ft wide and 4 ft high. If the top of the gate is 30 ft below the water surface, what is the total force on it?

2. How far must the water level drop in Exercise 1 to reduce the force on the gate to one-half its original value?

3. The gate in Exercise 1 is constructed to withstand a force of 48 tons. How deep may the water be over the top of the gate and not exceed this safe limit?

4. A water trough has an isosceles right-triangular cross section with the hypotenuse uppermost. If the trough is 4 ft wide, 2 ft deep, and full of water, how much force is exerted against one end?

5. A triangular steel plate has a base of 5 ft and an altitude of 3 ft. If the plate is suspended vertically in water with the base uppermost and 6 ft below and parallel to the water surface, how much force is exerted against one side?

6. Find the total force on one side of the plate of Exercise 5 if it is suspended with the base down and 6 ft below and parallel to the water surface.

7. A dam has a semicircular gate of radius 3 ft. If the gate is vertical, with the straight edge uppermost and 20 ft below the water level, what is the force exerted on it?

8. A dam is in the form of a trapezoid with upper and lower bases 300 ft and 200 ft, respectively, and height 30 ft. If the face is vertical and the water level is at the top of the dam, how much force must it withstand?

9. A water trough has its ends in the shape of a parabolic arc with vertical axis. If the trough is 4 ft wide and 2 ft deep and full of water, how much force is exerted against one end?

10. A square plate 5 ft on a side is suspended in water so that its diagonal is vertical. If the upper corner is 4 ft below the surface of the water, what is the force exerted on one side of the plate?

11. A cylindrical oil drum is 2 ft in diameter and 4 ft tall. It is half-full of oil weighing 50 pcf. If it is standing on end on a level surface, what is the force exerted on the curved sides?

12. The same drum and oil of Exercise 11 is lying on its side on a level surface. What is the total force exerted on the ends?

Chapter 12

PARAMETRIC EQUATIONS—
POLAR COORDINATES

12-1. Parametric Representation

There are many cases in which it is convenient to express the coordinates of points on a curve in terms of a third variable, which we call a *parameter*. The *parametric equations*

$$\begin{aligned} x &= f(t) \\ y &= g(t) \end{aligned}, \qquad t_1 \leqslant t \leqslant t_2, \tag{12-1}$$

define a curve. Each t within the prescribed domain produces, by means of (12-1), a pair of values x and y, which may be used as the coordinates of a point. The totality of these points constitutes the curve. The type of curve will depend on f and g. For example, it is clear that the curve will be continuous if f and g are continuous for $t_1 \leqslant t \leqslant t_2$.

A situation in which such a representation is particularly useful is that in which the position of a particle moving along a curve is a function of time. Then, if t represents time, equations (12-1) not only represent the curve but also give the time when a particular position on the curve is occupied by the particle.

Consider the problem of describing the path of a projectile when air resistance is neglected. Let its initial velocity be V_0, its initial angle of elevation be α, and choose the origin at the initial point of the trajectory (Fig. 12-1).

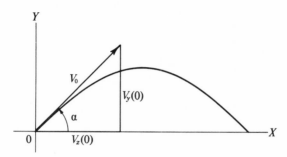

Fig. 12-1

There is no force acting in the horizontal direction. Hence, there is no acceleration in that direction and we have

$$\frac{d^2x}{dt^2} = 0.$$

Thus

$$\frac{dx}{dt} = C_1.$$

This represents the velocity in the x-direction and we write

$$V_x(t) = \frac{dx}{dt} = C_1.$$

From Figure 12-1 we have

$$V_x(0) = V_0 \cos \alpha,$$

so

$$C_1 = V_0 \cos \alpha,$$

and

$$\frac{dx}{dt} = V_0 \cos \alpha.$$

Integrating once more

$$x = (V_0 \cos \alpha)t + C_2.$$

From our choice of coordinate system, $x = 0$ when $t = 0$. Then

$$0 = (V_0 \cos \alpha) \cdot 0 + C_2$$

and

$$x = (V_0 \cos \alpha)t.$$

The force of gravity is accelerating the projectile by the amount $-g$ in the vertical direction. Hence

$$\frac{d^2y}{dt^2} = -g,$$

or, integrating,

$$\frac{dy}{dt} = V_y(t) = -gt + C_3.$$

From Figure 12-1,

$$V_y(0) = V_0 \sin \alpha$$

and we obtain

$$V_0 \sin \alpha = -g \cdot 0 + C_3.$$

Thus

$$\frac{dy}{dt} = -gt + V_0 \sin \alpha.$$

Integrating once more,

$$y = -\tfrac{1}{2}gt^2 + (V_0 \sin \alpha)t + C_4.$$

But $y = 0$ when $t = 0$, whence

$$C_4 = 0,$$

and

$$y = -\tfrac{1}{2}gt^2 + (V_0 \sin \alpha)t.$$

The two equations

$$x = (V_0 \cos \alpha)t, \qquad y = -\tfrac{1}{2}gt^2 + (V_0 \sin \alpha)t,$$

constitute the parametric equations of the required projectile path where the parameter t is time. Thus the parametric approach to this problem gives the solution without difficulty. Other methods do not yield the results so readily.

It will also be found that the conventional representation of some curves by an equation in x and y is so complicated as to be impractical to use. Problems relating to such curves can often be simplified by the use of parametric representation.

If the parameter t is eliminated between the two equations of (12-1), the familiar representation

$$F(x, y) = 0$$

results.

Example 12-1. Discuss the curve represented by

$$x = t \qquad \text{and} \qquad y = t^2.$$

The domain of definition is clearly $-\infty < t < \infty$. Also, the range of x is $-\infty < x < \infty$, but the range of y is $0 \leqslant y < \infty$. If we eliminate t between the two equations, we obtain $y = x^2$, which we recognize as a parabola (Fig. 12-2).

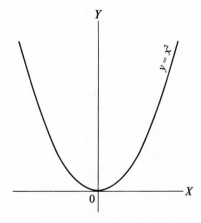

Fig. 12-2

Does this parabola have other parametric representations? The answer to this question is "yes", and we can give numerous examples. Let

$$x = 2t + 1.$$

Then

$$y = (2t + 1)^2,$$

and we have another representation. The student can find many others. However, if we insist on the entire parabola, we have to use a reasonable degree of caution. For example, if we set

$$x = t^2,$$

we have

$$y = t^4.$$

These two equations represent only half the parabola because the range of x is $0 \leqslant x < \infty$.

This difficulty exists in the opposite direction also. The equation obtained by eliminating the parameter may not represent point by point the same curve as the original equations. The following example illustrates this.

Example 12–2. Discuss the curve represented by

$$x = \sin t \quad \text{and} \quad y = 1 - \cos^2 t.$$

The domain of definition is $-\infty < t < \infty$. However, the range of x is $-1 \leqslant x \leqslant 1$ and that of y is $0 \leqslant y \leqslant 1$. If we eliminate t, we obtain $y = x^2$, which is the parabola shown in Fig. 12-2. The curve represented by the original equations is only that portion of this parabola shown in Fig. 12-3.

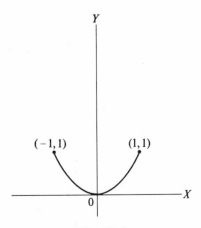

Fig. 12-3

Example 12-3. Sketch the curve represented by

$$x = t^2 + 2t \quad \text{and} \quad y = t + 1.$$

We make a table of values to obtain points on the curve, choosing values of t that will give useful values of x and y. A little experimentation will soon determine what values of t to use.

t	-3	-2	-1	0	1
x	3	0	-1	0	3
y	-2	-1	0	1	2

When we plot the points from this table, we obtain the graph shown in Fig. 12-4.

Next we shall consider two examples of determining a parametric representation from the geometric properties of the curve.

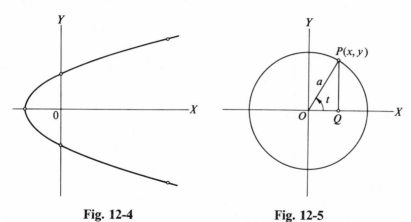

Fig. 12-4   Fig. 12-5

Example 12-4. Derive a parametric representation for the circle of radius a with center at the origin.

Consider any point P on the circle and let its coordinates be (x, y). Join the origin O to P (Fig. 12-5) and let the counterclockwise angle from the positive x-axis to OP be designated by t. Then $OQ = x$ and $QP = y$, and immediately we have from the definitions of the trigonometric functions,

$$x = a \cos t \quad \text{and} \quad y = a \sin t.$$

This particular representation is very useful in many situations.

Example 12-5. A cycloid† is the curve generated by a fixed point on a circle

---

† The cycloid has some very interesting physical properties. For some of these, see R. C. James, "University Mathematics," Wadsworth, Belmont, California (1963), p. 297.

as it rolls without slipping along a straight line. Derive a parametric representation for it.

Let the x-axis be the line along which the circle of radius a rolls; let the origin be chosen as one of the points of contact of the generating point with

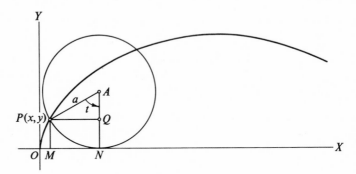

Fig. 12-6

the x-axis; and let t be the radian measure of the angle at the center of the circle between the radius drawn to the generating point P and the radius drawn to the point of tangency N with the x-axis (Fig. 12-6). Then the coordinates of P are

$$x = OM \qquad \text{and} \qquad y = MP.$$

But

$$OM = ON - MN = at - PQ$$

$$= at - a \sin t$$

and

$$MP = NA - QA = a - a \cos t.$$

Therefore a parametric representation of the cycloid is

$$x = a(t - \sin t) \qquad \text{and} \qquad y = a(1 - \cos t).$$

If we should eliminate t between these two equations in order to obtain a single equation in x and y, a very unpleasant equation results.

Exercises 12-1

In Exercises 1–10, sketch the curve from the parametric equations, and then find the equation in x and y by eliminating the parameter.

1. $x = 2t, \ y = 1 - t$

2. $x = \dfrac{w}{2} + 1, \ y = 2w - 1$

3. $x = u^2, \ y = 4u$

4. $x = 1 - 2t, \ y = t^2 - 3$

5. $x = 1 - 3 \cos v$, $y = 2 + 4 \sin v$ **6.** $x = 3 \cos w$, $y = -2 - 2 \sin w$

7. $x = t$, $y = (1 + t)^3$ **8.** $x = u^3$, $y = u^2$

9. $x = \dfrac{1}{v}$, $y = v^2$ **10.** $x = 3 \cos^3 \phi$, $y = 3 \sin^3 \phi$

In Exercises 11–16, find at least two parametric representations for the given curve.

11. $x - 3y = 7$ **12.** $4y^2 = x - 1$

13. $2y - 1 = 3x^2$ **14.** $(x - 1)^2 + 4(y + 2)^2 = 16$

15. $9x^2 + 4y^2 = 36$ **16.** $16(x + 2)^2 = 9(y - 2)$

17. Find parametric equations for the straight line through $P_1(x_1, y_1)$ and $P_2(x_2, y_2)$. *Hint*: Use $t = P_1P/P_1P_2$ as the parameter, where $P(x, y)$ is any point on the line.

12-2. Derivative of a Function Defined by Parametric Equations

First we wish to see that under proper hypotheses the equations (12-1) define y as a function of x; and secondly, that derivatives of y with respect to x may be calculated readily from the parametric form. Both objectives are achieved in the following theorem.

THEOREM 12-1. Let

$$x = f(t) \quad \text{and} \quad y = g(t), \quad t_1 \leqslant t \leqslant t_2,$$

where f and g satisfy the following conditions on the given interval:
(a) f and g are differentiable.
(b) $f'(t) \neq 0$.
(c) f^{-1} exists.
Then y is a function of x and

$$\frac{dy}{dx} = \frac{dy/dt}{dx/dt} = \frac{g'(t)}{f'(t)}.$$

Since f^{-1} exists on $t_1 \leqslant t \leqslant t_2$, we have

$$t = f^{-1}(x),$$

so

$$y = g(f^{-1}(x)) = G(x),$$

and the first conclusion is verified.

From the chain rule and (5-7),

$$\frac{dy}{dx} = \frac{dy}{dt}\frac{dt}{dx} = \frac{dy/dt}{dx/dt}.$$

Thus the theorem is established.

We have not proved, or even stated, an "inverse function theorem." We shall content ourselves with remarking that the conditions of Theorem 12-1 will be met for some interval by those parametric equations which the student will encounter in this book.

Example 12–6. Given

$$x = t + \frac{1}{t}, \qquad y = t - \frac{1}{t},$$

calculate dy/dx.

We have

$$\frac{dx}{dt} = 1 - \frac{1}{t^2}, \qquad \frac{dy}{dt} = 1 + \frac{1}{t^2}.$$

Then, from Theorem 12-1,

$$\frac{dy}{dx} = \frac{dy/dt}{dx/dt} = \frac{1 + (1/t^2)}{1 - (1/t^2)} = \frac{t^2 + 1}{t^2 - 1}, \qquad t \neq 0, \pm 1.$$

This procedure may be extended to calculate higher derivatives. Since

$$\frac{d^2y}{dx^2} = \frac{d}{dx}\left(\frac{dy}{dx}\right) = \frac{dy'}{dx},$$

we may replace y by y' in Theorem 12-1 to obtain

$$\frac{d^2y}{dx^2} = \frac{dy'}{dt} \cdot \frac{dt}{dx} = \frac{dy'/dt}{dx/dt}.$$

Obviously, this may be repeated to obtain still higher-order derivatives.

Example 12–7. Given

$$x = a(t - \sin t), \qquad y = a(1 - \cos t),$$

calculate d^2y/dx^2.

Proceeding as before,

$$\frac{dx}{dt} = a(1 - \cos t), \qquad \frac{dy}{dt} = a \sin t,$$

so

$$y' = \frac{dy}{dx} = \frac{\sin t}{1 - \cos t}.$$

Then

$$\frac{dy'}{dt} = \frac{(1 - \cos t)(\cos t) - (\sin t)(\sin t)}{(1 - \cos t)^2}$$

$$= \frac{\cos t - 1}{(1 - \cos t)^2} = -\frac{1}{1 - \cos t}$$

and

$$\frac{d^2y}{dx^2} = \frac{dy'/dt}{dx/dt} = -\frac{1}{a(1 - \cos t)^2}.$$

Exercises 12-2

In Exercises 1–10, find dy/dx and d^2y/dx^2 without eliminating the parameter.

1. $x = 2t - 1$, $y = t^2 - 3$

2. $x = 3t^2 + 1$, $y = 1 - 2t$

3. $x = t - t^3$, $y = 1 - \dfrac{t^2}{2}$

4. $x = \dfrac{1}{3} u^3$, $y = u^2$

5. $x = \dfrac{1}{v - 1}$, $y = \dfrac{v}{v + 1}$

6. $x = \sin \phi$, $y = \cos 2\phi$

7. $x = 2 - \cos s$, $y = -1 + 2 \sin s$

8. $x = 2 + 3 \cos \theta$, $y = -1 + 3 \sin \theta$

9. $x = t \log t - t$, $y = \log t$

10. $x = 2 \cos^3 t$, $y = 2 \sin^3 t$

In Exercises 11–14, find the maximum and minimum points, the points of inflection, and the ranges of the parameter for which the curve is concave both upward and downward.

11. $x = t^3 - 1$, $y = t^2 + t$

12. $x = r^3 + 1$, $y = 2r^2 - 2r$

13. $x = u^2 + 1$, $y = u^3 - 12u$, $u \geqslant 0$

14. $x = \dfrac{v}{1 + v^2}$, $y = \dfrac{v^2}{1 + v^2}$, $|v| \leqslant 1$

15. Draw the graph of

$$x = 2 \cos^3 t, \qquad y = 2 \sin^3 t,$$

making use of any information available from the derivatives calculated in Exercise 10.

16. If all forces except that of gravity are neglected, the parametric equations of the path of a projectile are (Sec. 12-1)

$$x = (V_0 \cos \alpha)t,$$

$$y = -\frac{1}{2} gt^2 + (V_0 \sin \alpha)t,$$

where α, g, and V_0 are constants. Find the highest point on the path.

17. Show that the path of the projectile in Exercise 16 is a parabola.

12-3. Polar Coordinates

Many difficulties arise when we try to express all problems in terms of
rectangular coordinates. For this reason we find it convenient to develop the
polar coordinate system.

DEFINITION 12-I. Let O be any point and OX the horizontal half-line
extending to the right from O. Let P be any point different from O; let ρ
denote the distance OP; and let α be the counterclockwise angle from OX
to OP (Fig. 12-7). The polar coordinates of P are any of the pairs

$$(r, \theta); \qquad r = \rho, \quad \theta = \alpha \pm 2k\pi, \quad k = 0, 1, 2, \dots ,$$

or

$$(r, \theta); \qquad r = -\rho, \quad \theta = (\alpha + \pi) \pm 2k\pi, \quad k = 0, 1, 2, \dots .$$

The polar coordinates of O are defined to be $(0, \theta)$ where θ is arbitrary.

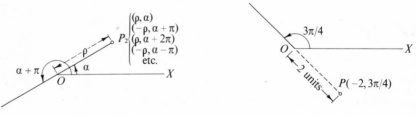

Fig. 12-7   **Fig. 12-8**

Thus the point P is located in terms of its " bearing " and distance from O,
using OX as the reference line. The sign of r distinguishes between the case
where the bearing angle has OP on its terminal side and that where the
terminal side is OP extended through O.

The point O is named the *pole* and OX is called the *polar axis*. The first
member r of the coordinate pair is referred to as the *radius vector*, and the
bearing angle θ is called the *vectorial angle*.

If a set of coordinates is given, say $(-2, (3\pi/4))$, there is a unique point
determined (Fig. 12-8). However, the point so determined has an infinite
number of sets of coordinates, any one of which define it equally well. A
few of them are

$$\left(2, -\frac{\pi}{4}\right), \qquad \left(-2, -\frac{5\pi}{4}\right), \qquad \left(2, \frac{7\pi}{4}\right), \qquad \left(-2, \frac{11\pi}{4}\right).$$

This illustrates a fundamental difference between rectangular and polar
coordinates. Once the rectangular coordinate axes are chosen, there is a "one
to one" correspondence between points and sets of coordinates. In polar
coordinates there corresponds but a single point to each set of coordinates;

however, to any given point there corresponds an infinite number of sets of coordinates.

If we relate the polar and rectangular coordinate systems in a special way, there is a very simple set of equations connecting the two systems. Let us

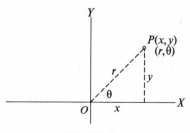

Fig. 12-9

choose them so that the pole and polar axis of the one coincide with the origin and positive x-axis of the other. Then (Fig. 12-9) we have

$$x = r \cos \theta \quad \text{and} \quad y = r \sin \theta, \tag{12-2}$$

or, stated otherwise,

$$x^2 + y^2 = r^2 \quad \text{and} \quad \frac{y}{x} = \tan \theta. \tag{12-3}$$

These pairs of equations enable us to transform from one system to the other. When using (12-3) it is sometimes necessary to exercise the option of choice with care. For example:

Example 12–8. Convert the point $(1, -1)$ in rectangular coordinates to polar coordinates.

Of course we can do this directly without conscious recourse to the preceding equations. However, let us make our point! We have, from (12-3),

$$r = \pm\sqrt{2}, \quad \tan \theta = -1.$$

If we choose θ in the second quadrant, say $\theta = 3\pi/4$, we must choose $r = -\sqrt{2}$. On the other hand, a choice of θ in quadrant IV, say $\theta = -(\pi/4)$, requires us to take $r = \sqrt{2}$.

The locus of an equation $r = f(\theta)$ presents some problems not present when working with rectangular coordinates. The following example illustrates one of these difficulties.

Example 12–9. Sketch the graph of

$$r^2 = \sin \theta.$$

In order to accomplish this, we construct a table of values, $0 \leqslant \theta \leqslant \pi$, from

$$r = \pm\sqrt{\sin \theta}.$$

Only values of θ in the first two quadrants are admissible, since $\sin \theta$ must be positive. We have

θ	0	$\pi/6$	$\pi/3$	$\pi/2$	$2\pi/3$	$5\pi/6$	π
r	0	± 0.7	± 0.9	± 1	± 0.9	± 0.7	0

From this table we obtain the graph in Fig. 12-10.

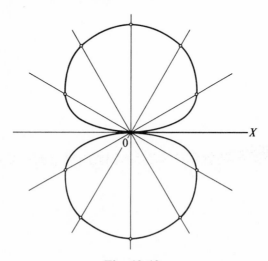

Fig. 12-10

Note that $r = -1$, $\theta = \pi/2$ is a solution of the equation and therefore is a point on the graph. But this point $(-1, \pi/2)$ may be represented also by the coordinates $r = 1$, $\theta = 3\pi/2$, and it is a simple matter to verify that *these values are not a solution of the given equation.* Consequently, it is not enough to try just one set of coordinates to determine if a point is *not* on the graph. If a set of coordinates fails to satisfy the equation, we must make sure that no other representation of this point will satisfy the equation before we conclude that it is not a point on the graph.

DEFINITION 12–2. The graph of $r = f(\theta)$ consists of all points each of which has at least one set of coordinates satisfying the equation.

Exercises 12-3

1. Plot the following points given in polar coordinates. Also write down two more sets of coordinates, one with positive r and one with negative r, for each of them.
   (a) $(2, \pi/2)$ (b) $(-3, -\pi/4)$ (c) $(4, \pi/6)$
   (d) $(-1, 7\pi/3)$ (e) $(2, -5\pi/4)$ (f) $(-5, 2\pi/3)$

2. Convert each of the points in Exercise 1 into rectangular coordinates.

3. Transform each of the following points, given in rectangular coordinates, into polar coordinates. Give four sets of coordinates, two with negative r and two with positive r, for each point.
(a) $(2, 2)$ (b) $(-2, 2)$ (c) $(2, -2)$
(d) $(2, 0)$ (e) $(1, -\sqrt{3})$ (f) $(-1, -\sqrt{3})$

4. Transform the following rectangular equations into polar equations:
(a) $3x - 2y = 5$ (b) $xy = 5$ (c) $x^2 - y^2 = a^2$
(d) $x^2 + y^2 - 4x = 0$

5. Transform the following polar equations into rectangular equations:
(a) $r = a$ (b) $r = 4 \cos \theta$ (c) $r^2 \sin 2\theta = 9$ (d) $r = 2 \sec \theta$

6. Determine whether the following points are on the graph of $r^2 = \cos \theta$:
(a) $(1, 0)$ (b) $(1, \pi)$ (c) $(\sqrt{2}/2, \pi/3)$
(d) $(\sqrt{2}/2, 2\pi/3)$ (e) $(-1, \pi)$ (f) $(1, \pi/2)$

7. Determine whether the following points are on the graph of $r = \sin \theta/2$:
(a) $(0, 0)$ (b) $(1, \pi)$ (c) $(-1, \pi)$ (d) $(1, 2\pi)$
(e) $(\sqrt{2}/2, -\pi/2)$ (f) $(-\sqrt{2}/2, -\pi/2)$

8. Find a formula for the distance from $P_1(r_1, \theta_1)$ to $P_2(r_2, \theta_2)$. *Hint*: Use the law of cosines (8-18).

12–4. Sketching Polar Curves

Proficiency in sketching polar curves is largely a matter of experience and observation. We shall not formulate tests for symmetry or other general properties of polar curves because they are somewhat complicated by the features of polar coordinates which we have already discussed in the preceding section. There is, however, one very simple and useful general rule. It is stated in the following theorem.

THEOREM 12–2. If $f(\theta)$ is continuous for $\theta = \theta_0$ and if $f(\theta_0) = 0$, the line $\theta = \theta_0$ is tangent to the curve $r = f(\theta)$ at the pole.

Since $f(\theta_0) = 0$, the curve passes through the pole. Join the pole to a second point $Q(\Delta r, \theta_0 + \Delta\theta)$ on the curve by a straight line (Fig. 12-11). The limiting position of this line as Q approaches O, if it has one, is, by definition (Sec. 4-3), the tangent to the curve at O. The point Q will approach O when $\Delta r \to 0$. But, due to the continuity of f,

$$\lim_{\Delta\theta \to 0} \Delta r = \lim_{\Delta\theta \to 0} [f(\theta_0 + \Delta\theta) - f(\theta_0)] = 0.$$

Therefore Q will approach O when $\Delta\theta \to 0$. Thus the secant line OQ,

$$\theta = \theta_0 + \Delta\theta,$$

approaches the limiting position

$$\theta = \lim_{\Delta\theta \to 0} (\theta_0 + \Delta\theta) = \theta_0,$$

and the theorem is established.

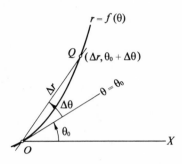

Fig. 12-11

Aside from Theorem 12-2 and some general observations we shall make as we work examples, we shall attack each problem individually. A familiarity with the properties of trigonometric functions is indispensable. Formulas (8-2) and (8-3) are especially recommended for attention.

Example 12–10. Sketch the graph of $r = 4\cos\theta$.

First, since $\cos \pi/2 = 0$, $\theta = \pi/2$ is tangent to the curve at the pole (Theorem 12-2). Other values of θ which give $\cos\theta = 0$ lead to the same line; hence there is no point in listing them.

Also, from (8-2), $\cos(-\theta) = \cos\theta$. Therefore, if (r, θ) is a point on the curve, so is $(r, -\theta)$. This is another way of saying that the curve is symmetric to the polar axis.

We should also note that $\cos\theta$ is negative for θ in the second and third quadrants. Therefore r is negative when θ is in these quadrants. This leads to the conclusion that there are no points on the curve to the left of the tangent $\theta = \pi/2$.

When we put all these results together we observe that we have only to prepare a table of values for $0 \leqslant \theta \leqslant \pi/2$:

θ	0	$\pi/6$	$\pi/3$	$\pi/2$
r	4	3.4	2	0

This curve looks very much like a circle (Fig. 12-12). This fact can be verified by transforming the polar equation into rectangular coordinates. If we multiply both members by r, we obtain

$$r^2 = 4r \cos \theta, \qquad \text{or} \qquad x^2 + y^2 = 4x.$$

This confirms our suspicion that the curve is a circle with center on the polar axis and passing through the pole.

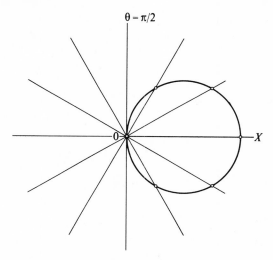

Fig. 12-12

The student should recall at this point that the sine and cosine functions assume the same sequence of values. They are merely out of phase by $\pi/2$ units. Thus we can draw the conclusion that the equations

$$r = \pm 2a \cos \theta \qquad \text{and} \qquad r = \pm 2a \sin \theta$$

are all circles of radius a, passing through the pole and with centers either on the line containing the polar axis or on the line perpendicular to it, passing through the pole. A little experimentation will soon reveal which case it is.

Example 12-11. Sketch the graph of $r = \sin 3\theta$.

Since

$$\sin 3\theta = 0 \qquad \text{when } \theta = 0, \pi/3, 2\pi/3,$$

the lines $\theta = 0$, $\theta = \pi/3$, $\theta = 2\pi/3$ are tangent to the curve at the pole.
From (8-2)

$$\sin 3(-\theta) = -\sin 3\theta.$$

Therefore, if (r, θ) satisfies the equation, so does $(-r, -\theta)$. Graphically, this means that the curve is symmetric to the line through the pole, perpendicular to the polar axis. This in turn means that we have only to calculate a table of values for $-\pi/2 \leqslant \theta \leqslant \pi/2$.

In order to get points reasonably close together we need to take a smaller interval between successive values of θ than we have previously done. This is the result of having to take the multiple 3θ to calculate r.

θ	0°	10°	15°	20°	30°	40°	45°	50°	60°	70°	75°	80°	90°
r	0	0.5	0.7	0.9	1	0.9	0.7	0.5	0	−0.5	−0.7	−0.9	−1

We have no need to write down the table for $-\pi/2 \leqslant \theta \leqslant 0$, since every entry will be opposite in sign to the ones just given. Plotting these points, we obtain Fig. 12-13.

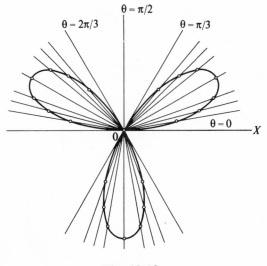

Fig. 12-13

This is called a *three-leaved rose*. The equations

$$r = \pm a \cos n\theta, \quad \text{and} \quad r = \pm a \sin n\theta,$$

n an integer greater than 1, can be shown to give similar curves.

Example 12-12. Sketch the graph of $r = 2(1 - \cos \theta)$.

The only tangent at the pole is $\theta = 0$. It is symmetric to the polar axis for the same reason given in Example 12-10.

A table of values for $0 \leqslant \theta \leqslant \pi$ is

θ	0	$\pi/6$	$\pi/3$	$\pi/2$	$2\pi/3$	$5\pi/6$	π
r	0	0.3	1	2	3	3.7	4

The graph sketched from these values and the known symmetry is shown in Fig. 12-14. This curve is called a *cardioid*. The same reasoning used in Example 12-10 leads us to the conclusion that the graphs of the equations

$$r = a(1 \pm \cos \theta) \qquad \text{and} \qquad r = a(1 \pm \sin \theta)$$

are all cardioids, differing only in their orientation.

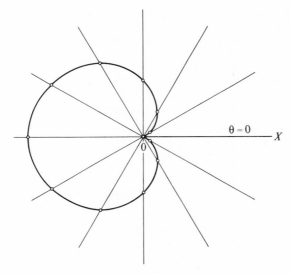

Fig. 12-14

Example 12–13. Sketch the graph of $r = 2(1 - 2 \cos \theta)$.

This curve has tangents $\theta = \pi/3$ and $\theta = -\pi/3$ at the pole, since $1 - 2 \cos \theta = 0$ when $\cos \theta = \frac{1}{2}$.

It is symmetric with respect to the polar axis for the same reason given in Examples 12-10 and 12-12. Now we need only to calculate a table of values for $0 \leqslant \theta \leqslant \pi$.

θ	0	$\pi/6$	$\pi/3$	$5\pi/12$	$\pi/2$	$2\pi/3$	$5\pi/6$	π
r	-2	-1.4	0	0.96	2	4	5.6	6

The graph is shown in Fig. 12-15.

The perceptive student will see that variations may occur in equations of this type.

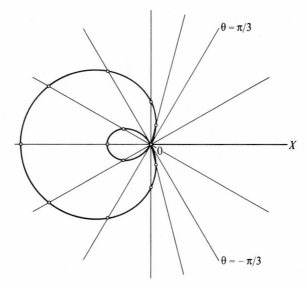

Fig. 12-15

Exercises 12-4

Sketch the graph of each of the following equations.

1. $\theta = \dfrac{\pi}{4}$

2. $r = a$

3. $r = 2 \sec \theta$

4. $r = \frac{1}{2} \csc \theta$

5. $r = 2 \sin \theta$

6. $r = -6 \cos \theta$

7. $r = -\sin \theta$

8. $r = \cos 2\theta$

9. $r = 2 \cos 3\theta$

10. $r = 2 \sin 2\theta$

11. $r = 2(1 + \sin \theta)$

12. $r = 1 + \cos \theta$

13. $r = 3(1 - \sin \theta)$

14. $r = 2(1 + 2 \cos \theta)$

15. $r = 1 - 2 \sin \theta$

16. $r = 2(2 - \sin \theta)$

17. $r = 3 - 2 \cos \theta$

18. $r = \dfrac{4}{1 - \cos \theta}$

19. $r = \dfrac{3}{1 - \sin \theta}$

20. $r = \sin \dfrac{\theta}{2}$

21. $r^2 = \cos \theta$

22. $r^2 = 4 \cos 2\theta$

23. $r^2 = 9 \sin 2\theta$

24. $r = 4\theta$

25. $r = -\theta$

26. $r = \dfrac{2}{\theta}$

12–5. Intersection of Polar Curves

Basically, the problem of finding the points of intersection of two curves is that of solving their equations simultaneously. However, when curves are represented by polar equations, the problem is not always quite this simple. In the following examples we shall see some of the difficulties which may arise. These will be discussed as they occur.

Example 12–14. Find the points of intersection of the two curves $r = 4 \sin 3\theta$ and $r = 2$.

First we draw a sketch (Fig. 12-16) and note that there are six points of

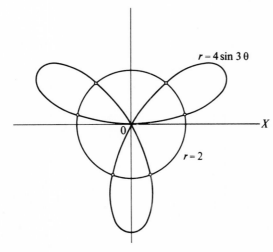

$r = 4 \sin 3\theta$

$r = 2$

Fig. 12-16

intersection. This graphical check is an important part of the process. Next we solve the equations simultaneously and obtain

$$\sin 3\theta = \tfrac{1}{2},$$

whence

$$3\theta = \frac{\pi}{6}, \frac{5\pi}{6}, \frac{\pi}{6} + 2\pi, \frac{5\pi}{6} + 2\pi, \frac{\pi}{6} + 4\pi, \frac{5\pi}{6} + 4\pi;$$

or

$$3\theta = \frac{\pi}{6}, \frac{5\pi}{6}, \frac{13\pi}{6}, \frac{17\pi}{6}, \frac{25\pi}{6}, \frac{29\pi}{6}.$$

Then

$$\theta = \frac{\pi}{18}, \frac{5\pi}{18}, \frac{13\pi}{18}, \frac{17\pi}{18}, \frac{25\pi}{18}, \frac{29\pi}{18}.$$

The student may well ask at this point why we stopped writing values for 3θ where we did, since

$$\sin\left(\frac{\pi}{6} + 2k\pi\right) = \sin\left(\frac{5\pi}{6} + 2k\pi\right) = \frac{1}{2}, \qquad k = 0, \pm1, \pm2, \ldots.$$

The answer to this is that the addition of further multiples of 2π does not lead to new points. For example, if we consider

$$3\theta = \frac{5\pi}{6} + 6\pi,$$

we have

$$\theta = \frac{5\pi}{18} + 2\pi,$$

which has the same terminal side as

$$\theta = \frac{5\pi}{18},$$

and therefore this value of θ does not give a new point.

We conclude that the points of intersection are

$$\left(2, \frac{\pi}{18}\right), \left(2, \frac{5\pi}{18}\right), \left(2, \frac{13\pi}{18}\right), \left(2, \frac{17\pi}{18}\right), \left(2, \frac{25\pi}{18}\right), \left(2, \frac{29\pi}{18}\right).$$

Example 12–15. Find the points of intersection of the curves $r = \sin\theta$ and $r = \cos\theta$.

The sketch of the curves (Fig. 12-17) indicates two intersections.

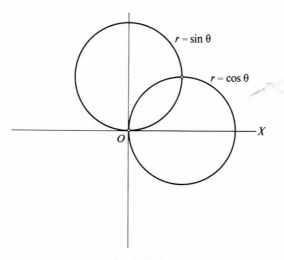

Fig. 12-17

We have

$$\sin \theta = \cos \theta \quad \text{or} \quad \tan \theta = 1.$$

Therefore

$$\theta = \frac{\pi}{4}, \frac{5\pi}{4}.$$

We neglect all other values of θ satisfying this condition, since they yield no new directions from O. We use $\theta = \pi/4$ and get the coordinate set $(\sqrt{2}/2, \pi/4)$; $\theta = 5\pi/4$ gives $(-\sqrt{2}/2, 5\pi/4)$. Both sets of coordinates describe the same point. Consequently, *the simultaneous solution of the two equations does not give the intersection at the pole* shown in Fig. 12-17. This is not surprising, since the vectorial angle of the pole is indeterminate.

In general, intersections at the pole will have to be discovered by other means. *If we have two equations*

$$r = f_1(\theta) \quad \text{and} \quad r = f_2(\theta)$$

and two values θ_1 and θ_2 such that

$$f_1(\theta_1) = f_2(\theta_2) = 0,$$

the two curves intersect at the pole.

Example 12–16. Find the points of intersection of the two curves $r = 1$ and $r = 2 \cos \theta/2$.

The graph of these two equations (Fig. 12-18) indicates four points of

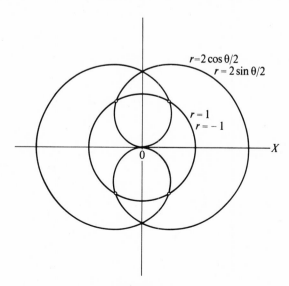

Fig. 12-18

intersection. Solving simultaneously, we have

$$\cos\frac{\theta}{2} = \frac{1}{2},$$

whence

$$\frac{\theta}{2} = \pm\frac{\pi}{3} + 2k\pi, \qquad k = 0, \pm1, \pm2, \ldots.$$

Therefore the solution of the two equations gives

$$r = 1, \qquad \theta = \pm\frac{2\pi}{3} + 4k\pi,$$

that is, nothing more than the two points

$$\left(1, \frac{2\pi}{3}\right), \left(1, -\frac{2\pi}{3}\right).$$

The question then arises: How do we find the other two points of intersection which show in Fig. 12-18?

This difficulty is the result of the multiple representation of a point. *The polar equation of a curve may not be unique* in the customary sense that any particular form of the equation may be obtained from any other form by algebraic manipulation.

Let $P(r_1, \theta_1)$ be a point on the curve defined by

(a) $\qquad\qquad\qquad\qquad\qquad r = f(\theta);$

that is,

(b) $\qquad\qquad\qquad\qquad\qquad r_1 = f(\theta_1).$

Then the point P is on the curve defined by

(c) $\qquad\qquad\qquad\qquad\qquad -r = f(\theta + \pi).$

This can be seen by representing P by the coordinate set $(-r_1, \theta_1 - \pi)$ and substituting in (c). We have

$$-(-r_1) = f[(\theta_1 - \pi) + \pi]$$

or

$$r_1 = f(\theta_1),$$

which is true by virtue of (b). In a similar manner, we can show that any point on (c) is a point on (a). In other words, (a) and (c) represent the same curve.

More generally, this argument can be used to prove that the equations

$$r = f(\theta),$$
$$-r = f[\theta + (2k + 1)\pi], \qquad\qquad\qquad (12\text{-}4)$$
$$r = f(\theta + 2k\pi), \qquad k \text{ any integer,}$$

all represent the same curve.

Thus, from (c), either $r = 1$ or $r = -1$ may be used to represent the circle, and either

$$r = 2 \cos \frac{\theta}{2}$$

or

$$-r = 2 \cos \frac{1}{2}(\theta + \pi) = -2 \sin \frac{\theta}{2}$$

may be used to represent the other curve. Neither of these new equations can be obtained from the original by algebraic operations alone. If we solve simultaneously either of the pairs of equations

$$r = -1 \quad \text{and} \quad r = 2 \cos \frac{\theta}{2},$$

or

$$r = 1 \quad \text{and} \quad r = 2 \sin \frac{\theta}{2},$$

we immediately obtain the missing pair of intersections, $(1, \pi/3)$ and $(1, 5\pi/3)$.

We summarize the problem of finding intersections of polar curves as follows:

Step 1. Sketch the curves to obtain the number and general location of the intersections, and to determine if the pole is one of them.

Step 2. Solve the two equations simultaneously.

Step 3. In case any intersections remain undetermined, replace one of the original equations with an equivalent equation from (12-4).

Exercises 12-5

In Exercises 1–17, find all intersections of the given pairs of curves.

1. $2r = 5, r = 5 \sin \theta$
2. $r = \sqrt{3}, r = 2 \cos \theta$
3. $r = 4(1 + \sin \theta), r \sin \theta = 3$
4. $r = 2 \sin \theta, r \cos \theta = -1$
5. $r = 2(1 - \sin \theta), r = 2(1 - \cos \theta)$
6. $r = 1, r = \cos 2\theta$
7. $r \sin \theta = 1, r = 2 - \sin \theta$
8. $r = 3 \cos \theta, r = 1 + \cos \theta$
9. $r = \sin \theta, r = \sin 2\theta$
10. $r = \sin 2\theta, r = \frac{1}{2}$
11. $r + \cos \theta = 0, 4r + 3 \sec \theta = 0$
12. $r = 2 \cos \theta, 2r(1 + \cos \theta) = 3$
13. $r^2 = 2 \cos \theta, r = 1$
14. $\theta = \frac{\pi}{3}, r = 5$
15. $r = 1, r = \tan \theta$
16. $r = \sin 2\theta, r = \cos 2\theta$
17. $r^2 = 4 \cos 2\theta, r = 2 \cos \theta$

18. Prove that any point P lying on the curve $-r = f(\theta + \pi)$ is also a point on the curve $r = f(\theta)$.

19. Prove that any point P lying on the curve $r = f(\theta)$ is also a point on the curve $r = f(\theta + 2\pi)$.

12-6. Plane Areas

We now consider the plane area bounded by the curve $r = f(\theta)$ and the rays $\theta = a$ and $\theta = b$ (Fig. 12-19). This "sector"-shaped area can be dealt with more effectively by developing a similarly shaped element of area, rather than using the rectangularly shaped element used earlier.

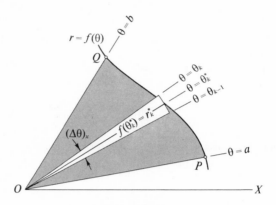

Fig. 12-19

We subdivide the angle POQ by the introduction of rays $\theta = \theta_k$, $k = 1$, $2, \ldots, n - 1$, where

$$a = \theta_0 < \theta_1 < \theta_2 < \cdots < \theta_{n-1} < \theta_n = b,$$

enabling us to define

$$(\Delta\theta)_k = \theta_k - \theta_{k-1}, \qquad k = 1, 2, \ldots, n.$$

Let θ_k^* be any number satisfying

$$\theta_{k-1} \leqslant \theta_k^* \leqslant \theta_k, \qquad k = 1, 2, \ldots, n.$$

The area of the sector with central angle $(\Delta\theta)_k$ and radius $f(\theta_k^*)$ (Fig. 12-19) is given by

$$(\Delta A)_k = \tfrac{1}{2}[f(\theta_k^*)]^2(\Delta\theta)_k,$$

and the total area A is approximated by

$$A \cong \sum_{k=1}^{n} (\Delta A)_k = \sum_{k=1}^{n} \tfrac{1}{2}[f(\theta_k^*)]^2(\Delta\theta)_k.$$

Then, following the precedent we have set on numerous occasions, we define A.

DEFINITION 12–3. The area A bounded by $\theta = a$, $\theta = b$, and $r = f(\theta)$ is given by

$$A = \lim_{n \to \infty} \sum_{k=1}^{n} (\Delta A)_k = \lim_{n \to \infty} \sum_{k=1}^{n} \tfrac{1}{2}[f(\theta_k^*)]^2 (\Delta \theta)_k,$$

provided this limit exists.†

If f is continuous for $a \leqslant \theta \leqslant b$, this limit does exist (Definition 7-2, Theorem 7-1), and we have:

THEOREM 12–3. If f is continuous for $a \leqslant \theta \leqslant b$, the area bounded by $\theta = a$, $\theta = b$, and $r = f(\theta)$ is given by

$$A = \tfrac{1}{2} \int_a^b [f(\theta)]^2 \, d\theta = \tfrac{1}{2} \int_a^b r^2 \, d\theta.$$

In the application of this theorem we use the *polar element of area*

$$(\Delta A)_k = \tfrac{1}{2} r_k^{*2} (\Delta \theta)_k,$$

or, more simply, omitting the subscripts and superscripts,

$$\Delta A = \tfrac{1}{2} r^2 \, \Delta \theta.$$

Example 12–17. Find the total area bounded by $r = 4 \cos \theta$.

From the sketch (Fig. 12-20) we see that the area in question may be

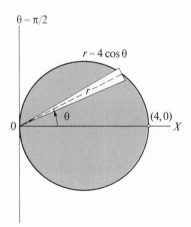

Fig. 12-20

---

† We have already given another definition of plane area in terms of rectangular elements. To be accurate, we should show that these two definitions are consistent. This is true, but we shall not prove it. We leave it to the student's intuition to see that it is at least reasonable to expect the two definitions to lead to the same result.

considered as twice the area bounded by $\theta = 0$, $\theta = \pi/2$, and $r = 4 \cos \theta$. The element of area is

$$\Delta A = \tfrac{1}{2} r^2 \, \Delta\theta = 8 \cos^2 \theta \, \Delta\theta,$$

and

$$A = 2(8) \int_0^{\pi/2} \cos^2 \theta \, d\theta = 8 \left[\frac{1}{2} \sin 2\theta + \theta \right]_0^{\pi/2}$$

$$= 4\pi \quad \text{sq units.}$$

This verifies the consistency of our definitions of area for this case.

Example 12–18. Find the area common to the two circles

$$r = 1 \quad \text{and} \quad r = 2 \cos \theta.$$

First we draw a sketch (Fig. 12-21) and find the points of intersection. They

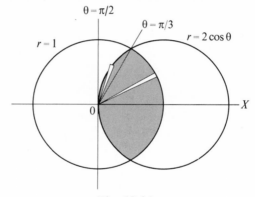

Fig. 12-21

are $(1, \pm\pi/3)$. Let us consider the upper half of the required area. It is made up of A_1 bounded by $\theta = 0$, $\theta = \pi/3$, and $r = 1$; and A_2 bounded by $\theta = \pi/3$, $\theta = \pi/2$, and $r = 2 \cos \theta$. Then

$$A = 2(A_1 + A_2) = 2 \left[\tfrac{1}{2} \int_0^{\pi/3} (1)^2 \, d\theta + \tfrac{1}{2} \int_{\pi/3}^{\pi/2} 4 \cos^2 \theta \, d\theta \right]$$

$$= 2 \left\{ \frac{\pi}{6} + \left(\frac{\sin 2\theta}{2} + \theta \right) \Big]_{\pi/3}^{\pi/2} \right\} = \frac{2\pi}{3} - \frac{\sqrt{3}}{2} \text{ sq units.}$$

Example 12–19. Find the area outside $r = 1$ and inside $r = 1 + \cos \theta$.

The two curves intersect at $(1, \pm\pi/2)$. If we consider the upper half of the required area (Fig. 12-22), we find it to be $A_1 - A_2$, where A_1 is bounded by

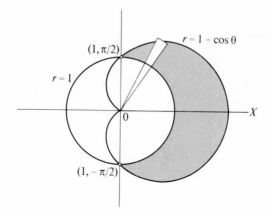

Fig. 12-22

$\theta = 0$, $\theta = \pi/2$, and $r = 1 + \cos \theta$; and A_2 is bounded by $\theta = 0$, $\theta = \pi/2$, and $r = 1$. Then

$$A = 2(A_1 - A_2) = 2\left[\tfrac{1}{2} \int_0^{\pi/2} (1 + \cos \theta)^2 \, d\theta - \tfrac{1}{2} \int_0^{\pi/2} (1)^2 \, d\theta \right]$$

$$= \int_0^{\pi/2} [(1 + 2\cos \theta + \cos^2 \theta) - 1] \, d\theta$$

$$= \int_0^{\pi/2} (2 \cos \theta + \cos^2 \theta) \, d\theta = \frac{8 + \pi}{4} \quad \text{sq units.}$$

Exercises 12-6

1. Find the area inside the cardioid $r = 2(1 - \sin \theta)$.

2. Find the area inside $r = 3 - 2 \sin \theta$.

3. Find the area enclosed by one loop of $r = 2 \sin 2\theta$.

4. Find the area inside the lemniscate $r^2 = 8 \cos 2\theta$.

5. Find the area enclosed by one loop of $r = 2 \cos 3\theta$.

6. Find the larger area bounded by $r = 8 \sin \theta$ and $r = 2 \csc \theta$.

7. Find the larger area bounded by $r = 2 \cos \theta$ and $\theta = \pi/3$.

8. Find the area outside $r = 3$ and inside $r = 6 \sin \theta$.

9. Find the area outside $r = 1$ and inside the lemniscate $r^2 = 2 \cos 2\theta$.

10. Find the area inside $r = 3 \sin \theta$ and outside $r = 1 + \sin \theta$.

11. Find the area inside $r^2 = \sin \theta$.

12. Find the area between the two loops of $r = 1 - 2 \cos \theta$.

12-7. Arc Length

In Sec. 11-4 we developed a formula for arc length when the equation of the curve has the form $y = f(x)$. Now let us assume the curve to be given in parametric form

$$x = g(t),$$

$$y = h(t), \qquad t_1 \leqslant t \leqslant t_2,$$

where g' and h' exist and are continuous on the given interval. Then $(\Delta C)_k$ may be written

$$(\Delta C)_k = \sqrt{[(\Delta x)_k]^2 + [(\Delta y)_k]^2}$$

$$= \sqrt{[g(t_k) - g(t_{k-1})]^2 + [h(t_k) - h(t_{k-1})]^2},$$

where

$$x_k = g(t_k), \qquad k = 0, 1, 2, \ldots, n.\dagger$$

The hypotheses on g and h permit us to use the mean value theorem to write

$$g(t_k) - g(t_{k-1}) = (t_k - t_{k-1})g'(t_k^*)$$

$$= (\Delta t)_k g'(t_k^*),$$

for some t_k^*, $t_{k-1} < t_k^* < t_k$, and

$$h(t_k) - h(t_{k-1}) = (t_k - t_{k-1})h'(t_k^{**})$$

$$= (\Delta t)_k h'(t_k^{**}),$$

for some t_k^{**}, $t_{k-1} < t_k^{**} < t_k$. Thus, from Definition 11-2,

$$S = \lim_{n \to \infty} \sum_{k=1}^{n} (\Delta C)_k = \lim_{n \to \infty} \sum_{k=1}^{n} \sqrt{[g'(t_k^*)]^2 + [h'(t_k^{**})]^2}\,(\Delta t)_k,$$

provided this limit exists.

If it were true that

$$t_k^* = t_k^{**}, \qquad k = 1, 2, 3, \ldots, n,$$

then, from Theorem 7-1,

$$S = \int_{t_1}^{t_2} \sqrt{[g'(t)]^2 + [h'(t)]^2}\,dt.$$

However, there is no assurance that $t_k^* = t_k^{**}$; as a matter of fact, there is little likelihood that it is true. But both numbers are restrained to the interval $t_{k-1} < t < t_k$ and therefore can differ but little in the limiting process. This

---

† We are using the notation of Sec. 11-4, but shall not repeat the definition of it here.

fact, combined with the continuity of g' and h', makes it possible to prove that the limit exists and is the value given. This proof is not within the scope of this book.

THEOREM 12-4. Let the curve C be represented by

$$x = g(t), \qquad y = h(t), \qquad t_1 \leqslant t \leqslant t_2,$$

and let g' and h' be continuous on this interval. Then the length of arc of C from $t = t_1$ to $t = t_2$ is

$$S = \int_{t_1}^{t_2} \sqrt{[g'(t)]^2 + [h'(t)]^2}\, dt = \int_{t_1}^{t_2} \sqrt{\left(\frac{dx}{dt}\right)^2 + \left(\frac{dy}{dt}\right)^2}\, dt.$$

Note that, operating formally, this result follows from (11-4):

$$ds = \sqrt{(dx)^2 + (dy)^2}.$$

We have

$$dx = g'(t)\, dt, \qquad dy = h'(t)\, dt,$$

so

$$ds = \sqrt{[g'(t)]^2 + [h'(t)]^2}\, dt.$$

We can use Theorem 12-4 to obtain the formula for arc length in polar coordinates. Let the equation of C be

$$r = f(\theta).$$

Then, since

$$x = r \cos \theta, \qquad y = r \sin \theta,$$

we may write the parametric equations of C in the form

$$x = f(\theta)\cos \theta \qquad \text{and} \qquad y = f(\theta)\sin \theta.$$

Thus

$$\frac{dx}{d\theta} = -f(\theta)\sin \theta + f'(\theta)\cos \theta,$$

$$\frac{dy}{d\theta} = f(\theta)\cos \theta + f'(\theta)\sin \theta,$$

and

$$\left(\frac{dx}{d\theta}\right)^2 + \left(\frac{dy}{d\theta}\right)^2 = [f(\theta)]^2 + [f'(\theta)]^2.$$

Theorem 12-4 gives the next theorem.

THEOREM 12-5. If f' is continuous for $\theta_1 \leqslant \theta \leqslant \theta_2$, the length of arc of $r = f(\theta)$ from $\theta = \theta_1$ to $\theta = \theta_2$ is

$$S = \int_{\theta_1}^{\theta_2} \sqrt{[f(\theta)]^2 + [f'(\theta)]^2} \, d\theta = \int_{\theta_1}^{\theta_2} \sqrt{r^2 + \left(\frac{dr}{d\theta}\right)^2} \, d\theta.$$

Example 12-20. Find the length of the curve $r = 4 \cos \theta$.

From Fig. 12-23 and Theorem 12-5,

$$S = \int_0^\pi \sqrt{16 \cos^2 \theta + 16 \sin^2 \theta} \, d\theta$$

$$= 4 \int_0^\pi d\theta = 4\pi \quad \text{units.}$$

Note that a point traverses the whole circle as θ ranges from 0 to π.

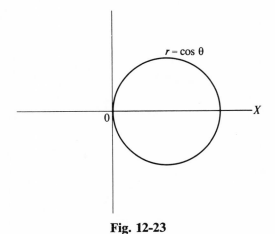

Fig. 12-23

Example 12-21. Find the length of one arch of the cycloid

$$x = a(t - \sin t),$$

$$y = a(1 - \cos t).$$

A point will describe one arch of the cycloid as t varies from $t = 0$ to $t = 2\pi$ because $y = 0$ for both values and for no intermediate value.

We have

$$\frac{dx}{dt} = a(1 - \cos t),$$

$$\frac{dy}{dt} = a \sin t,$$

so, applying Theorem 12-4,

$$S = \int_0^{2\pi} \sqrt{a^2(1 - \cos t)^2 + a^2 \sin^2 t}\, dt$$

$$= a \int_0^{2\pi} \sqrt{2(1 - \cos t)}\, dt$$

$$= 2a \int_0^{2\pi} \sin \frac{t}{2}\, dt = 8a \quad \text{units.}$$

Exercises 12-7

In Exercises 1–7, find the length of the indicated arcs.

1. Total length of the cardioid $r = 2(1 + \cos \theta)$

2. $r = 6 \sin \theta$, from $\theta = \pi/4$ to $\theta = 7\pi/6$

3. $r = 3\theta$, from $\theta = 0$ to $\theta = 2\pi$

4. $r = e^{a\theta}$, from $\theta = 0$ to $\theta = 2\pi$

5. Total length of the hypocycloid $x = a \cos^3 t$, $y = a \sin^3 t$

6. $x = (9/4)t^2$, $y = (9/4)t^3$, from $t = 0$ to $t = 2\sqrt{3}/3$

7. $x = 2t$, $y = 2 \log t$, from $t = 1$ to $t = \sqrt{3}$

In Exercises 8–13, find the area of the surface of revolution generated by revolving the given arc about the indicated axis.

8. The upper half of the cardioid $r = a(1 + \cos \theta)$; about the polar axis

9. The upper half of the lemniscate $r^2 = a^2 \cos 2\theta$; about the polar axis

10. One arch of the cycloid $x = a(t - \sin t)$, $y = a(1 - \cos t)$; about the x-axis

11. $x = 1 - 3t^2$, $y = t - 3t^3$, from $t = 0$ to $t = \sqrt{3}/3$; about the x-axis

12. $x = (\frac{1}{2})t^2 + t$, $y = (\frac{1}{2})t^2 - t$, from $t = 0$ to $t = 1$; about the y-axis

13. The right half of the hypocycloid $x = a \cos^3 t$, $y = a \sin^3 t$; about the y-axis

In Exercises 14–16, find the volume of the solid of revolution generated by revolving the given area about the indicated axis.

14. The area bounded by the right half of the hypocycloid $x = a \cos^3 t$, $y = a \sin^3 t$ and $x = 0$; about the y-axis

15. The area under one arch of the cycloid $x = a(t - \sin t)$, $y = a(1 - \cos t)$; about the x-axis

16. The area bounded by $r = \tan \theta$, $r \cos \theta = \frac{1}{2}$, and the polar axis; about the polar axis

12–8. Tangent Lines to Curves in Polar Coordinates

The concept of the slope of a tangent line to a curve in rectangular coordinates is an important one. We now investigate the similar problem for curves in polar coordinates.

Let α be the angle of inclination of the tangent to the curve $r = f(\theta)$ at the point $P(r, \theta)$; let the tangent be directed in the direction of increasing θ; and let ψ be the counterclockwise angle from the radius vector OP, extended through P, to the directed tangent as indicated in Fig. 12-24. Then

$$\alpha = \theta + \psi + n\pi, \qquad n \text{ an integer.}$$

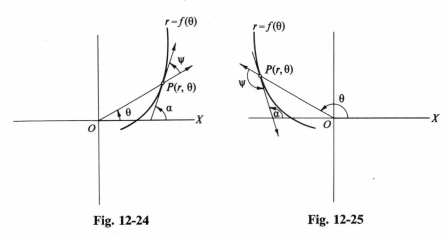

Fig. 12-24   Fig. 12-25

In Fig. 12-24, $n = 0$, but the student can readily verify that $n = -1$ in Fig. 12-25. In any case

(a) $\qquad \tan \alpha = \tan(\theta + \psi + n\pi) = \tan(\theta + \psi).$

The curve may be represented by the parametric equations

$$x = f(\theta)\cos \theta, \qquad y = f(\theta)\sin \theta,$$

so the slope of the tangent line may be written

$$\tan \alpha = \frac{dy}{dx} = \frac{dy/d\theta}{dx/d\theta}, \qquad \frac{dx}{d\theta} \neq 0,$$

$$= \frac{f'(\theta)\sin \theta + f(\theta)\cos \theta}{f'(\theta)\cos \theta - f(\theta)\sin \theta}.$$

Therefore, if $\cos \theta \neq 0$,

$$\tan \alpha = \frac{f'(\theta)\tan \theta + f(\theta)}{f'(\theta) - f(\theta)\tan \theta} = \frac{r + (dr/d\theta)\tan \theta}{(dr/d\theta) - r\tan \theta}. \qquad (12\text{-}5)$$

Thus the formula for slope in polar coordinates is not as simple as its counterpart in rectangular coordinates. We return to (a) in an effort to find a simpler formula which will serve our purposes. We have

$$\tan(\theta + \psi) = \frac{\tan \theta + \tan \psi}{1 - \tan \theta \tan \psi},$$

whence

$$\frac{r + (dr/d\theta)\tan \theta}{(dr/d\theta) - r \tan \theta} = \frac{\tan \theta + \tan \psi}{1 - \tan \theta \tan \psi}.$$

We solve this equation for $\tan \psi$ and obtain

$$\tan \psi = \frac{r}{dr/d\theta}.$$

THEOREM 12–6. Let $P(r, \theta)$ be a point on $r = f(\theta)$, and let the tangent at P be directed toward increasing θ. Then the counterclockwise angle ψ from OP extended through P to the directed tangent satisfies

$$\tan \psi = \frac{r}{dr/d\theta}. \tag{12-6}$$

This is a much simpler formula than (12-5) and can be made to serve many of the same purposes.

Example 12–22. Find the angle between the two curves $r = 1 - \sin \theta$ and $r = \sin \theta$ at their first quadrant point of intersection.

First we obtain this point of intersection. We have

$$\sin \theta = 1 - \sin \theta,$$

from which we get

$$\sin \theta = \tfrac{1}{2}.$$

Thus the first quadrant point of intersection is $P(1/2, \pi/6)$.

Let ψ_1 and ψ_2 be the angles between OP and the directed tangents to the circle and cardioid, respectively (Fig. 12-26). Then, from Theorem 12-6,

$$\tan \psi_1 = \frac{r}{\cos \theta}\bigg]_{(1/2, \pi/6)} = \frac{1}{\sqrt{3}},$$

and

$$\tan \psi_2 = \frac{r}{-\cos \theta}\bigg]_{(1/2, \pi/6)} = -\frac{1}{\sqrt{3}}.$$

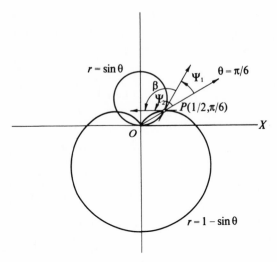

Fig. 12-26

The required angle β is given by

$$\beta = \psi_2 - \psi_1,$$

so

$$\tan \beta = \tan(\psi_2 - \psi_1) = \frac{\tan \psi_2 - \tan \psi_1}{1 + \tan \psi_1 \tan \psi_2}$$

$$= \frac{-(1/\sqrt{3}) - (1/\sqrt{3})}{1 - (1/3)} = -\sqrt{3}.$$

Therefore

$$\beta = \frac{2\pi}{3}.$$

The orthogonality of two curves may be tested by means of $\tan \psi$. Two curves are perpendicular to each other at a point of intersection if the product of the slopes of their tangents at that point is -1. Let two curves have equations

$$r = f_1(\theta), \qquad r = f_2(\theta),$$

and let a point of intersection be (r_1, θ_1). Then, according to (12-5), the condition for orthogonality takes the form

$$\left[\frac{f_1'(\theta_1)\tan \theta_1 + f_1(\theta_1)}{f_1'(\theta_1) - f_1(\theta_1)\tan \theta_1}\right]\left[\frac{f_2'(\theta_1)\tan \theta_1 + f_2(\theta_1)}{f_2'(\theta_1) - f_2(\theta_1)\tan \theta_1}\right] = -1.$$

When we multiply and clear of fractions, this reduces to

$$f_1'(\theta_1)f_2'(\theta_1)[1 + \tan^2 \theta_1] + f_1(\theta_1)f_2(\theta_1)[1 + \tan^2 \theta_1] = 0,$$

whence

$$\frac{f_1(\theta_1)f_2(\theta_1)}{f_1'(\theta_1)f_2'(\theta_1)} = -1.$$

Hence the condition for orthogonality of the two curves is

$$\tan \psi_1 \tan \psi_2 = -1. \tag{12-7}$$

Thus we see that $\tan \psi$ plays much the same role for curves in polar coordinates as $\tan \alpha$ plays for curves in rectangular coordinates.

Exercises 12-8

In Exercises 1–5, find $\tan \psi$ and $\tan \alpha$ at the indicated point on the given curve.

1. $r = 2(3 - 2 \sin \theta)$; $(4, \pi/6)$ **2.** $r = 2(1 - \cos \theta)$; $(1, \pi/3)$

3. $r = 3(1 + \sin \theta)$; $(9/2, \pi/6)$ **4.** $r = 4 \sin 2\theta$; $(4, \pi/4)$

5. $r = 2e^{2\theta}$; $(2, 0)$

6. Find the angle between $r = \sqrt{2} \sin \theta$ and $r^2 = \cos 2\theta$ at their first quadrant point of intersection.

7. Prove that the curves $r = 1 + \sin \theta$ and $r = 1 - \sin \theta$ intersect at right angles.

8. Prove that the curves $r^2 = a^2 \sin 2\theta$ and $r^2 = a^2 \cos 2\theta$ are orthogonal at all points of intersection except the pole.

9. Find the angle between $r = 2$ and $r^2 = 8 \cos 2\theta$ at their first quadrant point of intersection.

10. Find the points at which the tangents to $r = a(1 + \sin \theta)$ are (a) parallel to the polar axis; (b) parallel to $\theta = \pi/2$.

Chapter 13

FUNCTIONS OF SEVERAL VARIABLES

13-1. Fundamental Concepts

There are numerous situations in which one quantity is determined by the independent choice (within possible limitations) of two or more other quantities. For example, the volume of a right-circular cylinder is given by the formula

$$v = \pi r^2 h. \tag{13-1}$$

Thus, v is uniquely determined when r and h are specified, and the validity of the formula is in no way dependent on the values taken. Hence we say that v *is a function of the two independent variables r and h.*

Similarly, if P dollars are invested at $100i$ per cent, compounded annually, they will accumulate to S dollars at the end of t years where (Sec. 9-3)

$$S = P(1 + i)^t. \tag{13-2}$$

This formula exhibits the fact that S is uniquely determined when P, i, and t are given. Therefore we say that it is a function of these three variables.

An example of a function of two variables has already been introduced in Definition 6-5. It defines dy as a function of the two variables x and dx.

Now we wish to discuss the parenthetical expression "within possible limitations," which appears in the first paragraph of this chapter. Consider the function defined by (13-1). No restrictions are imposed on h and r by this equation, but the fact that they are dimensions of a cylinder requires that they be positive quantities. *This is a limitation.* The *volume function of a cylinder* is defined for any pair (h, r) of positive real numbers. Hence, in accord with the terminology used in Chapter 4, we say that *the domain of this function is the set of all ordered pairs of positive real numbers.*

The *accumulation function* defined by (13-2) requires that P and i be positive and that t be a nonnegative integer. Therefore the domain of this function is the set of all ordered triples (P, i, t) which satisfies these conditions. However, prevailing interest rates, opportunities for investment, banking laws, and other practical considerations may impose further limitations on these variables.

The domain of dy is the set of all ordered pairs (x, dx), where x is any number in the domain of f and dx is any real number.

We summarize the foregoing discussion in the following informal definition of functions of two variables.

DEFINITION 13-1. If there is a rule that gives a unique value z when values for x and y are given, we say that z is a function of x and y, and write

$$z = f(x, y).$$

The set of all ordered pairs (x, y) that give a value for z is called the domain of the function.

Let us look at the equation

$$z = \sqrt{1 - x^2 - y^2} \tag{13-3}$$

in the light of this definition. If

$$1 - x^2 - y^2 \geqslant 0,$$

it is clear that (13-3) defines a unique real z and therefore defines z as a function of x and y. Moreover, the domain of this function is the set of all ordered pairs (x, y) such that the above inequality is true. If we consider the pair (x, y) as the coordinates of a point, then we may describe the domain of this function geometrically by saying that it consists of the set of all points interior to and on the unit circle

$$x^2 + y^2 = 1.$$

We shall pursue the geometrical interpretation of functions of two variables further in the next section.

We shall confine ourselves principally to functions of two variables throughout the remainder of this chapter, in the interest of simplification of discussion and notation. However, from time to time, we shall indicate the procedure involved when more variables occur. The student should experience no difficulty in making this extension.

13-2. Three-Dimensional Geometry

This section will be devoted to a brief treatment of three-dimensional geometry in order to attach geometrical significance to some of the concepts of this chapter.

Let us define a three-dimensional rectangular coordinate system by drawing three mutually perpendicular lines through a common point O called the origin. These lines are called *coordinate axes* and each pair of them determines a *coordinate plane*. Thus we have *x*-, *y*-, and *z-axes* and *xy-*, *xz-*, and *yz-planes* (Fig. 13-1). The position of a point P is described by its three directed distances (x, y, z) from each of the three coordinate planes, as

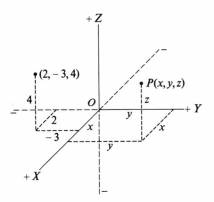

Fig. 13-1

shown in the figure. Positive and negative directions on the axes are assigned as indicated. Thus a point with the coordinates $(2, -3, 4)$ will be located two units in front of the *yz*-plane, three units to the left of the *xz*-plane, and four units above the *xy*-plane.

With a coordinate system defined, we can now talk about graphs of equations in three dimensions.

Suppose we have the equation

$$y = c,$$

where c is a real constant. In three dimensions this says in effect that y is restricted to the value c but that x and z are totally unrestricted. Thus points satisfying this condition are confined to a plane at a directed distance of c units from the *xz*-plane (Fig. 13-2). Similarly

$$x = c, \qquad z = c,$$

represent planes parallel to the *yz*-plane and *xy*-plane, respectively.

Now consider a function of the two variables x and y and set

$$z = f(x, y).$$

Then, to each pair of values (x, y) in the domain of f, there is determined a unique z and the triple of values (x, y, z) may be considered the coordinates of a point in three dimensions. If we use the pair of values (x, y) as coordinates of a point in the *xy*-plane, then the point (x, y, z) is vertically above (or below) it (Fig. 13-3). Let the domain of f be some region† R in the *xy*-plane; that is,

---

† We have not defined the term "region" and do not propose to do so. The intuitive concept conveyed by the word is adequate for our purposes.

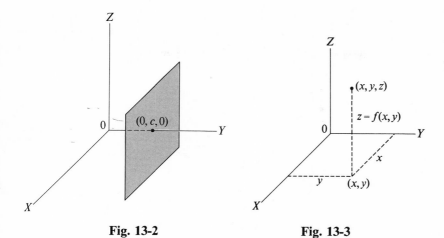

Fig. 13-2   Fig. 13-3

the set of ordered pairs (x, y) for which f is defined, considered as coordinates of points, "fill up" some region in the xy-plane. Then the points (x, y, z) form a "roof" over the region R (Fig. 13-4). We say that this "roof," the

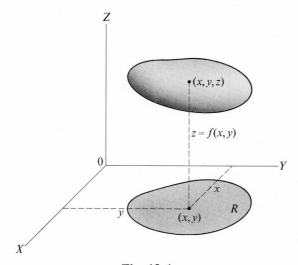

Fig. 13-4

graph of

$$z = f(x, y),$$

is a *surface*. In general, the graph of any equation in x, y, and z is a surface.

A curve in three dimensions may be represented by the equations of two surfaces which intersect in the given curve. For example, the surfaces

$$z = f(x, y) \quad \text{and} \quad y = c$$

intersect in a plane curve "parallel" to the xz-plane, since it lies in the plane $y = c$ (Fig. 13-5). The other characteristics of this curve depend on the nature of the surface $z = f(x, y)$. Curves of this sort are particularly useful in giving a geometrical interpretation to the concepts introduced in the next section.

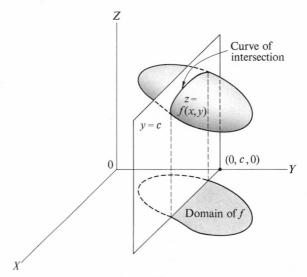

Fig. 13-5

13–3. Partial Differentiation

It often proves useful to study the behavior of a function of several variables when all but one of the variables are held constant. For example, the volume of a gas is a function of the pressure and temperature. We may wish to see what happens to the volume when the pressure is held constant and the temperature allowed to vary.

The study of such situations leads to the process of *partial differentiation.* Suppose we have

$$z = f(x, y),$$

and let us choose a fixed value of y, say $y = y_1$. Then, as long as we retain this value of y,

$$z = f(x, y_1)$$

defines z as a function of x and for a well-behaved function f we can calculate the derivative of z with respect to x. We use the notation,

$$\frac{\partial z}{\partial x} \equiv \frac{\partial f(x, y)}{\partial x} \equiv \frac{\partial f}{\partial x} \equiv z_x \equiv f_x,$$

among others, for this *partial derivative* to indicate that z is only temporarily a function of x alone.

DEFINITION 13-2. Let

$$z = f(x, y).$$

Then

(a) The partial derivative of z (or f) with respect to x is defined by

$$\frac{\partial z}{\partial x} \equiv \frac{\partial f}{\partial x} = \lim_{\Delta x \to 0} \frac{f(x + \Delta x, y) - f(x, y)}{\Delta x};$$

(b) The partial derivative of z (or f) with respect to y is defined by

$$\frac{\partial z}{\partial y} \equiv \frac{\partial f}{\partial y} = \lim_{\Delta y \to 0} \frac{f(x, y + \Delta y) - f(x, y)}{\Delta y};$$

provided these limits exist.

These definitions are readily extended to functions of more than two variables. For example, let

$$w = F(x, y, z).$$

Then

$$\frac{\partial w}{\partial x} = \lim_{\Delta x \to 0} \frac{F(x + \Delta x, y, z) - F(x, y, z)}{\Delta x};$$

$$\frac{\partial w}{\partial y} = \lim_{\Delta y \to 0} \frac{F(x, y + \Delta y, z) - F(x, y, z)}{\Delta y};$$

$$\frac{\partial w}{\partial z} = \lim_{\Delta z \to 0} \frac{F(x, y, z + \Delta z) - F(x, y, z)}{\Delta z}.$$

The calculation of partial derivatives involves no new formulas or techniques when the function is defined in explicit form

$$z = f(x, y).$$

We merely consider all independent variables, except one, as constants and differentiate the resulting function of one variable.

Example 13-1. Calculate $\partial z/\partial x$ and $\partial z/\partial y$ where

$$z = x^2 y + \frac{1}{y}.$$

To obtain $\partial z/\partial x$ we consider y a constant and get

$$\frac{\partial z}{\partial x} = y(2x) + 0 = 2xy.$$

Similarly, for $\partial z/\partial y$, we hold x constant with the result

$$\frac{\partial z}{\partial y} = x^2(1) - \frac{1}{y^2} = x^2 - \frac{1}{y^2}.$$

Example 13–2. Calculate $\partial w/\partial x$, $\partial w/\partial y$, $\partial w/\partial z$, where

$$w = e^x \cos y + yz^2.$$

Holding y and z constant, we obtain

$$\frac{\partial w}{\partial x} = \cos y(e^x) + 0 = e^x \cos y.$$

Also, holding x and z fixed,

$$\frac{\partial w}{\partial y} = e^x(-\sin y) + z^2(1)$$

$$= -e^x \sin y + z^2.$$

If we hold x and y constant, we have

$$\frac{\partial w}{\partial z} = 0 + y(2z) = 2yz.$$

One general interpretation of a partial derivative is obvious from its counterpart in functions of a single variable. It is the rate of change of the function with respect to one independent variable when all others are held fixed. As an illustration of this consider the value of $\partial w/\partial z$ from Example 13-2 for $x = 1$, $y = 2$, $z = 3$. We have

$$\left.\frac{\partial w}{\partial z}\right]_{\substack{x=1 \\ y=2 \\ z=3}} = 2(2)(3) = 12.$$

This simply says that if x and y are held constant at the values 1 and 2, respectively, and z alone is allowed to vary, w is changing 12 units for each unit change in z at the instant z takes the value 3.

We may also put a geometric interpretation on partial derivatives. Let f be a function of two variables and let

$$z = f(x, y).$$

We wish to see what geometric property may be attached to $\partial z/\partial x$ evaluated at $x = x_1$, $y = y_1$. Common notations for this are

$$\left.\frac{\partial z}{\partial x}\right]_{\substack{x=x_1 \\ y=y_1}} \equiv \frac{\partial f(x_1, y_1)}{\partial x} \equiv f_x(x_1, y_1).$$

First of all, this means that y is to be held at the constant value y_1, which means geometrically that points (x, y, z) are constrained to move along the curve of intersection, APB, of the surface

$$z = f(x, y)$$

and the plane

$$y = y_1,$$

(Fig. 13-6).

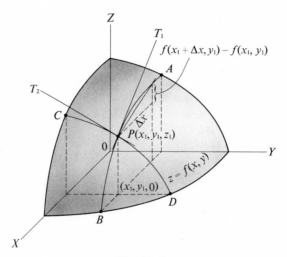

Fig. 13-6

The slope of the tangent line T_1 to APB at (x_1, y_1, z_1), where $z_1 = f(x_1, y_1)$, is, by definition,

$$m = \lim_{\Delta x \to 0} \frac{f(x_1 + \Delta x, y_1) - f(x_1, y_1)}{\Delta x}.$$

But the right member is, by definition, $f_x(x_1, y_1)$. Hence $f_x(x_1, y_1)$ is the *slope of the tangent line T_1 to APB at the point (x_1, y_1, z_1)*.

Similarly, $f_y(x_1, y_1)$ is the *slope of the tangent line T_2 to the curve of intersection, CPD* (Fig. 13-6), *of the surface $z = f(x, y)$ and the plane $x = x_1$ at the point (x_1, y_1, z_1)*.

Since the existence of the derivative implies the continuity of the function, the existence of $\partial z/\partial x$ implies that

$$z = f(x, y_1)$$

is a continuous function of x. A similar statement can be made regarding the continuity of $z = f(x_1, y)$, if $\partial z/\partial y$ exists.

Some comments regarding continuity are essential to the statement of basic theorems on partial differentiation. Hence we shall give a definition of continuity for functions of two variables, but we do not intend to belabor this point. We shall assume that the functions we study from this point on have the necessary properties of continuity. This assumption is necessary in order to avoid developing theorems for testing continuity, a project upon which we have no intention of entering.

DEFINITION 13–3. A function f is said to be continuous for $x = x_0$, $y = y_0$, if, for every $\varepsilon > 0$, there exists a $\delta > 0$ such that

$$|f(x, y) - f(x_0, y_0)| < \varepsilon$$

when both $|x - x_0| < \delta$ and $|y - y_0| < \delta$.

This simply says that we can make $f(x, y)$ as close to $f(x_0, y_0)$ as we please by taking x and y sufficiently close to x_0 and y_0, respectively. Or, stated differently, a small change in x and y produces a small change in $f(x, y)$. The use of the symbol $f(x_0, y_0)$ implies that (x_0, y_0) is in the domain of f.

After *first* partial derivatives are calculated we can repeat the process and obtain *second* partial derivatives. However, there is a new element entering at this point. We may elect to differentiate the second time with respect to y after having first differentiated with respect to x. This gives rise to *mixed* partial derivatives. We write

$$\frac{\partial}{\partial x}\left(\frac{\partial z}{\partial x}\right) = \frac{\partial^2 z}{\partial x^2}; \qquad \frac{\partial}{\partial y}\left(\frac{\partial z}{\partial x}\right) = \frac{\partial^2 z}{\partial y\,\partial x};$$

$$\frac{\partial}{\partial y}\left(\frac{\partial z}{\partial y}\right) = \frac{\partial^2 z}{\partial y^2}; \qquad \frac{\partial}{\partial x}\left(\frac{\partial z}{\partial y}\right) = \frac{\partial^2 z}{\partial x\,\partial y}.$$

Example 13–3. Calculate all second partial derivatives of z where

$$z = x^2 y + \frac{1}{y}.$$

We have, from Example 13-1,

$$\frac{\partial z}{\partial x} = 2xy, \qquad \frac{\partial z}{\partial y} = x^2 - \frac{1}{y^2}.$$

Then

$$\frac{\partial^2 z}{\partial x^2} = \frac{\partial}{\partial x}(2xy) = 2y;$$

$$\frac{\partial^2 z}{\partial y\,\partial x} = \frac{\partial}{\partial y}(2xy) = 2x;$$

$$\frac{\partial^2 z}{\partial y^2} = \frac{\partial}{\partial y}\left(x^2 - \frac{1}{y^2}\right) = \frac{2}{y^3};$$

$$\frac{\partial^2 z}{\partial x\,\partial y} = \frac{\partial}{\partial x}\left(x^2 - \frac{1}{y^2}\right) = 2x.$$

It should be noted that, in this case at least,

$$\frac{\partial^2 z}{\partial y \, \partial x} = \frac{\partial^2 z}{\partial x \, \partial y};$$

that is, the order of differentiation is immaterial. In this connection, we state the following theorem without proof.

THEOREM 13-1. Let $z = f(x, y)$. If f and its first and second partial derivatives are defined for x and y close to x_0 and y_0, respectively, and if both

$$\frac{\partial^2 z}{\partial x \, \partial y}, \qquad \frac{\partial^2 z}{\partial y \, \partial x},$$

are continuous for $x = x_0$, $y = y_0$, then

$$\frac{\partial^2 z}{\partial x \, \partial y} = \frac{\partial^2 z}{\partial y \, \partial x}.$$

This theorem can be extended to higher-order mixed derivatives. Consider the case in which we calculate the third partial derivative of f by differentiating twice with respect to y and once with respect to x. There are three ways of doing this, as indicated by the symbols

$$\frac{\partial^3 f}{\partial x \, \partial y^2}, \qquad \frac{\partial^3 f}{\partial y \, \partial x \, \partial y}, \qquad \frac{\partial^3 f}{\partial y^2 \, \partial x}.$$

The extension of Theorem 13-1 asserts that with proper conditions of continuity, which we assume, these are all equal.

Example 13-4. Verify that

$$\frac{\partial^3 f}{\partial x^2 \, \partial y} = \frac{\partial^3 f}{\partial y \, \partial x^2},$$

if

$$f(x, y, z) = ze^{xy} + x \cos y.$$

Although f is a function of the three variables x, y, and z, we are not asked to differentiate with respect to z. Hence it will be considered a constant throughout the calculation.

We have, for the left member,

$$\frac{\partial f}{\partial y} = zxe^{xy} - x \sin y;$$

$$\frac{\partial^2 f}{\partial x \, \partial y} = \frac{\partial}{\partial x}\left(\frac{\partial f}{\partial y}\right) = z(e^{xy} + xye^{xy}) - \sin y;$$

$$\frac{\partial^3 f}{\partial x^2 \, \partial y} = \frac{\partial}{\partial x}\left(\frac{\partial^2 f}{\partial x \, \partial y}\right) = z(ye^{xy} + ye^{xy} + xy^2 e^{xy})$$

$$= ze^{xy}(2y + xy^2).$$

For the right member,

$$\frac{\partial f}{\partial x} = zye^{xy} + \cos y;$$

$$\frac{\partial^2 f}{\partial x^2} = \frac{\partial}{\partial x}\left(\frac{\partial f}{\partial x}\right) = zy^2 e^{xy};$$

$$\frac{\partial^3 f}{\partial y\,\partial x^2} = \frac{\partial}{\partial y}\left(\frac{\partial^2 f}{\partial x^2}\right) = z(2ye^{xy} + y^2 xe^{xy})$$

$$= ze^{xy}(2y + xy^2).$$

Thus the equality is verified.

Exercises 13-1

Find all first and second partial derivatives of the functions given in Exercises 1–8.

1. $f(x, y) = x^2 + xy + y^2$ 2. $F(s, t) = (s^2 - t^2)(s + t)$

3. $h(x, y) = \sqrt{x - y}$ 4. $g(u, v) = \sin uv$

5. $G(r, s) = \log \dfrac{r}{s}$ 6. $H(x, y) = x \log y$

7. $f(u, v) = \cos(u - v)$ 8. $g(r, \theta) = r(\sin \theta - r)$

Find all first partial derivatives of the functions given in Exercises 9–12.

9. $F(x, y, z) = \sqrt{16 - x^2 - y^2 - z^2}$ 10. $W(x, y, z) = \dfrac{x - y}{z}$

11. $U(r, s, t) = \text{Arctan } rst$ 12. $G(r, \theta, z) = z - r^2 \cos 2\theta$

13. Suppose the annual profits P of an enterprise are expressible in the form

$$P = 50{,}000 - (30 - E)^2 - (40 - I)^2$$

where E is the number of employees and I is the inventory in tens of thousands of dollars. Find $\partial P/\partial E$ and $\partial P/\partial I$ and interpret.

14. The drawing power P of a shopping center is directly proportional to the size M and inversely proportional to the square of the distance D. Find $\partial P/\partial M$ and $\partial P/\partial D$ and interpret.

15. Find the slope of the tangent at the point $(3, 2, -3)$ to the curve of intersection of the surfaces (a) $z = x^2 - 2xy$ and $y = 2$; (b) $z = x^2 - 2xy$ and $x = 3$.

16. The volume V of a gas is given in terms of temperature T and pressure P by $V = KT/P$, where K is a constant. Find $\partial V/\partial T$ and $\partial V/\partial P$ and interpret.

Find all second partial derivatives of the functions given in Exercises 17–20.

17. $g(x, y, z) = e^{xyz}$ **18.** $f(u, v, w) = uv^2 \cos w$

19. $F(x, y, z) = \log(x - y + 2z)$ **20.** $G(r, \theta) = r^2 - 4 \cos 2\theta$

In Exercises 21–24, verify that

$$\frac{\partial^2 z}{\partial x \, \partial y} = \frac{\partial^2 z}{\partial y \, \partial x}.$$

21. $z = \sqrt{x^2 + y^2}$ **22.** $z = \text{Arctan} \dfrac{y}{x}$

23. $z = e^x \log xy$ **24.** $z = \dfrac{x + y}{x - y}$

25. Verify that

$$\frac{\partial^3 z}{\partial x \, \partial y^2} = \frac{\partial^3 z}{\partial y \, \partial x \, \partial y} = \frac{\partial^3 z}{\partial y^2 \, \partial x},$$

if

$$z = y \sin x + x \cos y.$$

26. Verify that

$$\frac{\partial^2 z}{\partial x^2} - \frac{\partial^2 z}{\partial y^2} = 0,$$

if $z = \sin(x + y) + \cos(x - y)$.

27. Verify that

$$9\left(\frac{\partial^2 z}{\partial x^2}\right) - \frac{\partial^2 z}{\partial y^2} = 0,$$

if $z = \log(x - 3y) + e^{x+3y}$.

28. Verify that

$$x\left(\frac{\partial z}{\partial x}\right) + y\left(\frac{\partial z}{\partial y}\right) = 0,$$

if $z = e^{y/x}$.

29. Verify that

$$x^2\left(\frac{\partial z}{\partial x}\right) + y^2\left(\frac{\partial z}{\partial y}\right) = 0,$$

if $z = (x - y)/xy$.

13–4. Fundamental Increment Formula

Let f be a function of the two independent variables x and y. We seek a formula for the increment Δf produced by the increments Δx and Δy of the independent variables. We have

$$\Delta f = f(x + \Delta x, y + \Delta y) - f(x, y),$$

which we may put in the form

$$\Delta f = f(x + \Delta x, y + \Delta y) - f(x, y + \Delta y) + f(x, y + \Delta y) - f(x, y)$$

by adding and subtracting $f(x, y + \Delta y)$.

Consider the first pair of function values: f is evaluated at the two points $(x + \Delta x, y + \Delta y)$ and $(x, y + \Delta y)$, so the y value is the same in both. Momentarily then, we may think of f as a function of x alone and, if proper hypotheses are satisfied, make use of the mean value theorem to write

$$f(x + \Delta x, y + \Delta y) - f(x, y + \Delta y) = \Delta x \, \frac{\partial f(x + \theta_1 \, \Delta x, y + \Delta y)}{\partial x}, \qquad 0 < \theta_1 < 1.$$

Similar reasoning applies to the second pair of function values, and we obtain

$$f(x, y + \Delta y) - f(x, y) = \Delta y \, \frac{\partial f(x, y + \theta_2 \, \Delta y)}{\partial y}, \qquad 0 < \theta_2 < 1.$$

These two statements require that f be continuous for $x_0 \leqslant x \leqslant x_1$, $y_0 \leqslant y \leqslant y_1$, and that the partial derivatives exist on the interior of these intervals.

Now let us also assume that $\partial f / \partial x$, $\partial f / \partial y$ are continuous in the same domain as f. Then, from the basic property of continuity,

$$\frac{\partial f(x + \theta_1 \, \Delta x, y + \Delta y)}{\partial x} - \frac{\partial f(x, y)}{\partial x} = \varepsilon_1$$

and

$$\frac{\partial f(x, y + \theta_2 \, \Delta y)}{\partial y} - \frac{\partial f(x, y)}{\partial y} = \varepsilon_2,$$

where†

$$\lim_{\substack{\Delta x \to 0 \\ \Delta y \to 0}} \varepsilon_1 = 0, \qquad \lim_{\substack{\Delta x \to 0 \\ \Delta y \to 0}} \varepsilon_2 = 0,$$

provided x, $x + \Delta x$, y, $y + \Delta y$ are values on the two intervals. Combining these results, we may state the following theorem.

---

† We leave the meaning of such limits to the students' intuition. They occur but seldom in this book.

THEOREM 13–2. Let f be a function of the independent variables x and y, and let f, $\partial f/\partial x$, $\partial f/\partial y$ be continuous for $x_0 \leqslant x \leqslant x_1$, $y_0 \leqslant y \leqslant y_1$. Then the total increment Δf produced by the increments Δx and Δy, for x, $x + \Delta x$, y, $y + \Delta y$ on these intervals, is given by

$$\Delta f = \Delta x \, \frac{\partial f(x, y)}{\partial x} + \Delta y \, \frac{\partial f(x, y)}{\partial y} + \varepsilon_1 \, \Delta x + \varepsilon_2 \, \Delta y,$$

where

$$\lim_{\substack{\Delta x \to 0 \\ \Delta y \to 0}} \varepsilon_1 = 0, \qquad \lim_{\substack{\Delta x \to 0 \\ \Delta y \to 0}} \varepsilon_2 = 0.$$

Obviously, there is no problem in extending this theorem to more than two independent variables.

13–5. Total Differential

The formula for Δf in Theorem 13-2 is analogous to the formula (6-5),

$$\Delta y = f'(x) \, \Delta x + \eta \Delta x,$$

for the increment of a function of one variable. We used this relation to define dy (or df) by discarding the term $\eta \, \Delta x$, which is the product of two numbers, each of which approaches zero as $\Delta x \to 0$. We proceed similarly for functions of two variables.

DEFINITION 13–4. Let f be a function of the two independent variables x and y, and let f, $\partial f/\partial x$, $\partial f/\partial y$ be continuous in some domain. Then we define dx, dy, and df by:

$$dx = \Delta x, \qquad dy = \Delta y,$$

$$df = \frac{\partial f}{\partial x} \, \Delta x + \frac{\partial f}{\partial y} \, \Delta y = \frac{\partial f}{\partial x} \, dx + \frac{\partial f}{\partial y} \, dy,$$

where df is called the total differential, to distinguish it from the partial differentials

$$\left(\frac{\partial f}{\partial x}\right) dx, \qquad \left(\frac{\partial f}{\partial y}\right) dy.$$

It can be shown, making use of Theorem 13-2, that

$$\lim_{\substack{\Delta x \to 0 \\ \Delta y \to 0}} \frac{df - \Delta f}{|\Delta x| + |\Delta y|} = 0.$$

Thus, as in functions of one variable, df serves as an approximation to Δf. This is convenient because df is usually easier to calculate than Δf.

The modifications to be made in Definition 13-4 for functions of more than two independent variables are obvious.

Example 13–5. Given

$$f(x, y) = x^2 + 3xy,$$

find Δf, df, and $\Delta f - df$.

$$\Delta f = (x + \Delta x)^2 + 3(x + \Delta x)(y + \Delta y) - (x^2 + 3xy)$$

$$= (2x + 3y)\, \Delta x + 3x\, \Delta y + (\Delta x)^2 + 3\Delta x\, \Delta y.$$

Note that ε_1, ε_2 of Theorem 13-2 are, in this problem,

$$\varepsilon_1 = \Delta x, \qquad \varepsilon_2 = 3\Delta x.$$

$$df = \frac{\partial(x^2 + 3xy)}{\partial x}\, \Delta x + \frac{\partial(x^2 + 3xy)}{\partial y}\, \Delta y$$

$$= (2x + 3y)\, \Delta x + 3x\, \Delta y$$

$$= (2x + 3y)\, dx + 3x\, dy.$$

Thus

$$\Delta f - df = (\Delta x)^2 + 3\Delta x\, \Delta y.$$

Example 13–6. Given

$$z = \sqrt{x^2 + 9y^2}.$$

If x changes from 5 to 4.98 and y changes from 4 to 4.03, find approximately the change in z.

$$\Delta z \cong dz = \frac{\partial}{\partial x}(x^2 + 9y^2)^{1/2}\, dx + \frac{\partial}{\partial y}(x^2 + 9y^2)^{1/2}\, dy$$

$$= \frac{x}{(x^2 + 9y^2)^{1/2}}\, dx + \frac{9y}{(x^2 + 9y^2)^{1/2}}\, dy$$

$$= \frac{1}{\sqrt{x^2 + 9y^2}}(x\, dx + 9y\, dy).$$

Now let

$$x = 5, \qquad dx = -0.02, \qquad y = 4, \qquad dy = 0.03.$$

Then

$$\Delta z \cong dz = \frac{1}{13}[5(-0.02) + 36(0.03)]$$

$$= \frac{0.98}{13} \cong 0.075.$$

Example 13–7. A city water tank is in the form of a hemisphere surmounted by a cylinder of equal radius. The altitude and radius of the cylinder are reported as being 20 ft and 10 ft, respectively. What is the approximate possible error in the computed capacity of the tank, if these dimensions are subject to an error of 1 percent? What is the approximate possible percent error in the computed capacity?

Let the altitude and radius be represented by h and r, respectively. Then, from Definition 6-6,

$$\frac{\Delta r}{r} = \frac{\Delta r}{10} = \pm 0.01; \qquad \frac{\Delta h}{h} = \frac{\Delta h}{20} = \pm 0.01.$$

Thus

$$\Delta r \equiv dr = \pm 0.1; \qquad \Delta h \equiv dh = \pm 0.2.$$

Also, if V is the capacity,

$$V = \frac{2}{3}\pi r^3 + \pi r^2 h,$$

whence

$$\Delta V \cong dV = (2\pi r^2 + 2\pi rh)\, dr + \pi r^2\, dh$$

$$= \pi[2(100 + 200)(\pm 0.1) + 100(\pm 0.2)]$$

$$= \pi(\pm 60 \pm 20).$$

Therefore the approximate maximum possible error in the computed capacity is

$$dV = 80\pi \text{ cu ft} \cong 1880 \text{ gal.}$$

From the given dimensions,

$$V = \tfrac{2}{3}\pi(1000) + \pi(2000)$$

$$= \frac{8000\pi}{3} \quad \text{cu ft.}$$

Hence the maximum percent error in the computed capacity is approximately

$$\frac{dV}{V} = \frac{80\pi}{8000\pi/3} = 0.03 = 3 \text{ percent.}$$

Exercises 13-2

In Exercises 1–4, compare the total increment and the total differential of the given functions for the values stated.

1. $f(x, y) = \dfrac{1}{x + y}$; $x = 1$, $y = 2$, $\Delta x = 0.01$, $\Delta y = 0.02$

2. $F(s, t) = t^2 - 2ts$; $t = 4$, $s = -3$, $\Delta t = 0.02$, $\Delta s = 0.03$

3. $g(u, v) = \log(uv)$; $u = 1$, $v = 1$, $\Delta u = 0.01$, $\Delta v = -0.01$

4. $h(x, y, z) = x + y^2 - 3xz$; $x = 2$, $y = 3$, $z = 5$, $\Delta x = 0.02$, $\Delta y = -0.01$, $\Delta z = 0.03$

In Exercises 5–10, find the total differential of the given function.

5. $f(x, y) = \sqrt{xy + y^2}$ 6. $h(r, \theta) = r^2 \cos 2\theta$

7. $g(r, s) = \log \dfrac{s}{\sqrt{s^2 + r^2}}$ 8. $F(u, v) = e^v \sin 2u$

9. $H(x, y, z) = x^2 + y^2 - 2xyz$ 10. $f(t, u, v) = \dfrac{uv}{t^2 - u^2}$

11. Find the approximate change in the volume of a right-circular cylinder if the radius is increased from 5 in. to 5.02 in. and the altitude is decreased from 10 in. to 9.96 in.

12. Find approximately how much the volume of a right-circular cone will change if the altitude is increased from 10 in. to 10.03 in. and the diameter of the base is decreased from 4 in. to 3.99 in.

13. Compute the approximate volume of a rectangular solid of dimensions 3.01 in., 7.98 in., 10.03 in. by means of differentials. Compare this approximate volume with the actual volume.

14. Find approximately the volume of metal in a closed cylindrical can 6 in. high and 5 in. in diameter, if the metal is 0.015 in. thick.

15. Find the approximate error in the calculated area of a circular sector of radius 6 in. and central angle 0.5 rad, if errors of 0.1 in. and 0.005 rad are made in taking these measurements.

16. A runner is timed at 10 sec for the 100-yd dash. If the time may be in error by 0.1 sec and the distance in error by 6 in., what possible approximate error could result in his computed velocity?

17. What is the possible approximate percent error in the computed volume of a right circular cylinder caused by percent errors of 0.5% in the radius and height?

18. If the dimensions of a rectangular box are measured with a possible error of 1%, what is the approximate possible percent error in the computed volume? If these dimensions are measured as 2, 3, and 4 units, compare dv and Δv.

19. The volume V of a gas is given in terms of the pressure P and temperature T by $V = kT/P$, where k is a constant. If there is initially 10 cu ft of gas, find the approximate change in the volume when the temperature is increased 1 percent and the pressure decreased 0.5 percent.

20. The index of refraction is given by

$$\mu = \frac{\sin i}{\sin r},$$

where i is the angle of incidence and r is the angle of refraction. What is the approximate maximum error in μ if $i = 45$ deg, $r = 40$ deg, these angular measurements being subject to 0.5 percent error? What is the approximate maximum percent error in μ?

21. It is known that the time P for one complete oscillation of a simple pendulum of length l is

$$P = 2\pi \sqrt{\frac{l}{g}},$$

where g is the acceleration of gravity. The value of g is to be determined from this relation. If l and P can be measured with a maximum error of 0.1 %, what is the maximum percent error in the calculated value of g? (*Hint*: Logarithmic differentiation will simplify the algebra in this problem.)

22. The area of a triangular plot of ground is to be calculated by measuring two sides and the included angle. If the linear and angular measurements are subject to errors of 0.04 percent and 0.06 percent, respectively, and the angle is found to be 60 deg, what is the approximate percent error possible in the calculated area?

13-6. The Chain Rule

If f is a function of the variables x and y, it may be that x and y are, in turn, functions of a third variable t. For example, let

$$f(x, y) = x^2 + y^2,$$

where

$$x = 2t, \qquad y = \frac{1}{t^2}.$$

Then we may write

$$f(x, y) = f\left(2t, \frac{1}{t^2}\right)$$

$$= 4t^2 + \frac{1}{t^4} = F(t).$$

Thus, in such a circumstance, we are dealing with a function of the single independent variable t. We propose to develop a means by which we can calculate the derivative of this function without the necessity of replacing x and y by the corresponding functions of t, as done above. This results in a direct extension of the chain rule stated in Theorem 5-7.

We shall prove the following theorem.

THEOREM 13-3. Let f, $\partial f/\partial x$, $\partial f/\partial y$ be continuous functions of x and y in some domain, and let

$$x = x(t), \qquad y = y(t)$$

be differentiable functions of t in a corresponding domain. Then†

$$\frac{df}{dt} = \frac{\partial f}{\partial x}\frac{dx}{dt} + \frac{\partial f}{\partial y}\frac{dy}{dt}.$$

The result in Theorem 13-3 is also called the chain rule.

From Theorem 13-2,

$$\Delta f = \frac{\partial f}{\partial x}\Delta x + \frac{\partial f}{\partial y}\Delta y + \varepsilon_1\,\Delta x + \varepsilon_2\,\Delta y,$$

whence

$$\frac{\Delta f}{\Delta t} = \frac{\partial f}{\partial x}\frac{\Delta x}{\Delta t} + \frac{\partial f}{\partial y}\frac{\Delta y}{\Delta t} + \varepsilon_1\frac{\Delta x}{\Delta t} + \varepsilon_2\frac{\Delta y}{\Delta t}.$$

Thus

(a)
$$\frac{df}{dt} = \lim_{\Delta t \to 0}\frac{\Delta f}{\Delta t}$$

$$= \frac{\partial f}{\partial x}\lim_{\Delta t \to 0}\frac{\Delta x}{\Delta t} + \frac{\partial f}{\partial y}\lim_{\Delta t \to 0}\frac{\Delta y}{\Delta t} + \lim_{\Delta t \to 0}\varepsilon_1\frac{\Delta x}{\Delta t} + \lim_{\Delta t \to 0}\varepsilon_2\frac{\Delta y}{\Delta t}.$$

Let us examine the last term in the right member of (a). Since $x = x(t)$, $y = y(t)$ are differentiable, they are also continuous. Hence $\Delta x \to 0$, $\Delta y \to 0$ when $\Delta t \to 0$. Therefore we may write

$$\lim_{\Delta t \to 0}\varepsilon_2\frac{\Delta y}{\Delta t} = \lim_{\Delta t \to 0}\varepsilon_2\lim_{\Delta t \to 0}\frac{\Delta y}{\Delta t}$$

$$= \lim_{\substack{\Delta x \to 0 \\ \Delta y \to 0}}\varepsilon_2\lim_{\Delta t \to 0}\frac{\Delta y}{\Delta t} = 0\cdot\frac{dy}{dt} = 0.$$

---

† It is to be understood that there is an interval $t_0 \leqslant t \leqslant t_1$ for which $x = x(t)$, $y = y(t)$ are values of x and y for which f and its partial derivatives are continuous. We omit this from the statement of the theorem in the interest of simplicity.

Similarly,

$$\lim_{\Delta t \to 0} \varepsilon_1 \frac{\Delta x}{\Delta t} = 0,$$

and the theorem follows from (a).

Example 13-8. Given

$$f(x, y) = x^2 + y^2,$$

where

$$x = 2t, \qquad y = \frac{1}{t^2},$$

find df/dt.

We shall do this by direct substitution and by the chain rule. From above

(b)
$$\frac{df}{dt} = \frac{d}{dt}\left(4t^2 + \frac{1}{t^4}\right) = 8t - \frac{4}{t^5}.$$

If we use the chain rule, we obtain

$$\frac{df}{dt} = \frac{\partial}{\partial x}(x^2 + y^2)\frac{d}{dt}(2t) + \frac{\partial}{\partial y}(x^2 + y^2)\frac{d}{dt}\left(\frac{1}{t^2}\right)$$

$$= (2x)(2) + (2y)\left(-\frac{2}{t^3}\right)$$

$$= 4x - \frac{4y}{t^3}.$$

We easily verify that this gives the same result as (b) by substituting

$$x = 2t, \qquad y = \frac{1}{t^2}.$$

If f is a function of three variables x, y, and z, and these in turn are all differentiable functions of a fourth variable t, the chain rule of Theorem 13-3 takes the form

$$\frac{df}{dt} = \frac{\partial f}{\partial x}\frac{dx}{dt} + \frac{\partial f}{\partial y}\frac{dy}{dt} + \frac{\partial f}{\partial z}\frac{dz}{dt}. \tag{13-1}$$

Example 13-9. Given

$$f(x, y, z) = x^2 y - \log z,$$

where

$$x = t^2, \qquad y = e^t, \qquad z = \frac{1}{t},$$

calculate df/dt.

From (13-1),

$$\frac{df}{dt} = \frac{\partial}{\partial x}(x^2y - \log z)\frac{d}{dt}(t^2) + \frac{\partial}{\partial y}(x^2y - \log z)\frac{d}{dt}(e^t)$$

$$+ \frac{\partial}{\partial z}(x^2y - \log z)\frac{d}{dt}\left(\frac{1}{t}\right)$$

$$= (2xy)(2t) + (x^2)(e^t) + \left(-\frac{1}{z}\right)\left(-\frac{1}{t^2}\right)$$

$$= 4xyt + x^2e^t + \frac{1}{zt^2}.$$

If it appears desirable, this result is readily expressible in terms of t alone.
   A case of particular interest appears in the following example.

Example 13-10. Given

$$F(x, y, z) = x^2 + y^2 - z^2$$

where

$$y = e^x, \qquad z = \log x,$$

find dF/dx.

Note that the independent variable here is one of those appearing in the
definition of F. We can accomplish our purpose by utilizing (13-1). We have

$$\frac{dF}{dx} = \frac{\partial F}{\partial x}\frac{dx}{dx} + \frac{\partial F}{\partial y}\frac{dy}{dx} + \frac{\partial F}{\partial z}\frac{dz}{dx}$$

$$= 2x + 2ye^x - \frac{2z}{x}.$$

The student should note the difference between the symbols

$$\frac{dF}{dx}, \qquad \frac{\partial F}{\partial x},$$

appearing above. We have

$$\frac{dF}{dx} = \frac{d}{dx}F(x, e^x, \log x) = \frac{d}{dx}(x^2 + e^{2x} - \log^2 x)$$

and

$$\frac{\partial F}{\partial x} = \frac{\partial}{\partial x}F(x, y, z) = \frac{\partial}{\partial x}(x^2 + y^2 - z^2).$$

Consider next the case where f is a function of the two variables x and y,
and

$$x = x(r, s), \qquad y = y(r, s).$$

Thus the independent variables are r and s, while x and y are only *intermediate* variables.

For example, this situation arises in transforming from one coordinate system to another. If we wish to transform from rectangular coordinates to polar coordinates, we set

$$x = r \cos \theta, \qquad y = r \sin \theta,$$

thereby changing from x and y as independent variables to r and θ.

From Theorem 13-2,

$$\Delta f = \frac{\partial f}{\partial x} \Delta x + \frac{\partial f}{\partial y} \Delta y + \varepsilon_1 \Delta x + \varepsilon_2 \Delta y.$$

Now let us hold s fixed and permit r to assume an increment Δr. Then we may write

$$\frac{\Delta f}{\Delta r} = \frac{\partial f}{\partial x} \frac{\Delta x}{\Delta r} + \frac{\partial f}{\partial y} \frac{\Delta y}{\Delta r} + \varepsilon_1 \frac{\Delta x}{\Delta r} + \varepsilon_2 \frac{\Delta y}{\Delta r}$$

and

(c) $$\lim_{\Delta r \to 0} \frac{\Delta f}{\Delta r} = \frac{\partial f}{\partial x} \lim_{\Delta r \to 0} \frac{\Delta x}{\Delta r} + \frac{\partial f}{\partial y} \lim_{\Delta r \to 0} \frac{\Delta y}{\Delta r} + \lim_{\Delta r \to 0} \left[\varepsilon_1 \frac{\Delta x}{\Delta r} + \varepsilon_2 \frac{\Delta y}{\Delta r} \right].$$

Methods similar to those used in connection with Theorem 13-3 may be easily adapted, under proper hypotheses, to show

$$\lim_{\Delta r \to 0} \left[\varepsilon_1 \frac{\Delta x}{\Delta r} + \varepsilon_2 \frac{\Delta y}{\Delta r} \right] = 0.$$

Hence, from (c),

$$\frac{\partial f}{\partial r} = \frac{\partial f}{\partial x} \frac{\partial x}{\partial r} + \frac{\partial f}{\partial y} \frac{\partial y}{\partial r}.$$

The same general procedure will give

$$\frac{\partial f}{\partial s} = \frac{\partial f}{\partial x} \frac{\partial x}{\partial s} + \frac{\partial f}{\partial y} \frac{\partial y}{\partial s}.$$

THEOREM 13-4. Let f, $\partial f / \partial x$, $\partial f / \partial y$ be continuous functions of x and y in some domain, and let

$$x = x(r, s), \qquad y = y(r, s)$$

be differentiable functions of r and s in a corresponding domain. Then

$$\frac{\partial f}{\partial r} = \frac{\partial f}{\partial x} \frac{\partial x}{\partial r} + \frac{\partial f}{\partial y} \frac{\partial y}{\partial r};$$

$$\frac{\partial f}{\partial s} = \frac{\partial f}{\partial x} \frac{\partial x}{\partial s} + \frac{\partial f}{\partial y} \frac{\partial y}{\partial s}.$$

This result is also called the chain rule. The extension to more variables is obvious.

Example 13–11. Given

$$f(x, y) = x^2 - y^2,$$

where

$$x = r \cos \theta, \qquad y = r \sin \theta,$$

calculate

$$\frac{\partial f}{\partial r}, \quad \frac{\partial f}{\partial \theta}.$$

From Theorem 13-4,

$$\frac{\partial f}{\partial r} = \frac{\partial}{\partial x}(x^2 - y^2)\frac{\partial}{\partial r}(r \cos \theta) + \frac{\partial}{\partial y}(x^2 - y^2)\frac{\partial}{\partial r}(r \sin \theta)$$

$$= 2x \cos \theta - 2y \sin \theta.$$

Similarly,

$$\frac{\partial f}{\partial \theta} = \frac{\partial}{\partial x}(x^2 - y^2)\frac{\partial}{\partial \theta}(r \cos \theta) + \frac{\partial}{\partial y}(x^2 - y^2)\frac{\partial}{\partial \theta}(r \sin \theta)$$

$$= -2xr \sin \theta - 2yr \cos \theta.$$

Exercises 13-3

In Exercises 1–14, find the derivative of the given function.

1. $f(x, y) = 2x^2 + xy + y^2; x = \dfrac{1}{t}, y = t^3$

2. $F(u, v) = \dfrac{u}{u^2 + v^2}; u = \sin s, v = \cos s$

3. $g(r, s) = \sqrt{r^2 + s^2}; r = 1 - t, s = 1 + 2t$

4. $h(w, t) = (1 + w)(1 - t); w = x^2, t = x^3$

5. $G(x, y) = e^x \cos y; x = \log r, y = \pi r$

6. $F(s, t) = e^{st}; s = \sin u, t = \cot u$

7. $F(r, w) = \log(r^2 - w^2); r = te^t, w = e^t$

8. $H(x, y, z) = xy^2 + x^2z + yz^2; x = t, y = \dfrac{1}{t}, z = \tfrac{1}{2}t^2$

9. $g(u, v, w) = \text{Arctan } \dfrac{uv}{w}; u = \sin t, v = \cos t, w = t^2$

10. $f(x, y, z) = \log(x^2 + y^2 + z^2); \; y = x^2, z = e^{-x}$

11. $F(r, s, t) = \dfrac{r}{s} + \dfrac{s}{t} + \dfrac{t}{r}; \; r = 2t^2, s = \dfrac{1}{2t}$

12. $h(x, y, t) = \sqrt{x^2 + y^2 - t^2}; \; x = e^t, y = \sin t$

13. $G(r, \theta, z) = \dfrac{r^2 - \sin 2\theta}{z}; \; r = 1 - z^2, \theta = \text{Arctan } z$

14. $g(x, y, z) = e^{xy} \sin z; \; y = \log x, z = x \log x$

In Exercises 15–20, find the first partial derivatives of the given function with respect to each of the independent variables.

15. $f(x, y) = \sqrt{x^2 - y^2}; \; x = \dfrac{r}{s}, y = rs$

16. $g(u, v) = \dfrac{uv}{u^2 + v^2}; \; u = r \sin \theta, v = r \cos \theta$

17. $h(w, t) = \dfrac{\cos w}{t^2}; \; w = x - z, t = x + z$

18. $H(r, \theta) = e^{-r^2} \cos 2\theta; \; r = \sqrt{x^2 + y^2}, \theta = \text{Arctan } \dfrac{y}{x}$

19. $F(x, y, z) = \sqrt{x^2 + y^2 - z^2}; \; x = r \cos t, y = r \sin t, z = t^2$

20. $G(u, v, w) = \log\sqrt{u^2 + v^2 + w^2}; \; u = se^t, v = se^{-t}, w = e^t$

21. Given f is a function of x and y. If we change to polar coordinates by setting $x = r \cos \theta, y = r \sin \theta$, show that:

(a) $\dfrac{\partial f}{\partial r} = \dfrac{\partial f}{\partial x} \cos \theta + \dfrac{\partial f}{\partial y} \sin \theta$

(b) $\dfrac{1}{r} \dfrac{\partial f}{\partial \theta} = -\dfrac{\partial f}{\partial x} \sin \theta + \dfrac{\partial f}{\partial y} \cos \theta$

(c) $\dfrac{\partial f}{\partial x} = \dfrac{\partial f}{\partial r} \cos \theta - \dfrac{\partial f}{\partial \theta} \dfrac{\sin \theta}{r}$

(d) $\dfrac{\partial f}{\partial y} = \dfrac{\partial f}{\partial r} \sin \theta + \dfrac{\partial f}{\partial \theta} \dfrac{\cos \theta}{r}$

13–7. Higher Derivatives

We have already briefly discussed partial derivatives of higher order in Sec. 13-3. In this section we shall see how the chain rule may be used in this connection. The following examples illustrate the process.

Example 13-12. Calculate $\partial^2 f/\partial r^2$ for

$$f(x, y) = x^2 - 4xy; \qquad x = se^r, \quad y = se^{-r}.$$

From the chain rule,

$$\frac{\partial f}{\partial r} = \frac{\partial}{\partial x}(x^2 - 4xy)\frac{\partial}{\partial r}(se^r) + \frac{\partial}{\partial y}(x^2 - 4xy)\frac{\partial}{\partial r}(se^{-r})$$

$$= (2x - 4y)(se^r) + (4x)(se^{-r}).$$

Then

$$\frac{\partial^2 f}{\partial r^2} = \frac{\partial}{\partial r}\left(\frac{\partial f}{\partial r}\right) = \frac{\partial}{\partial r}[(2x - 4y)(se^r)] + \frac{\partial}{\partial r}[(4x)(se^{-r})]$$

$$= se^r\frac{\partial}{\partial r}(2x - 4y) + (2x - 4y)\frac{\partial}{\partial r}(se^r)$$

$$+ se^{-r}\frac{\partial}{\partial r}(4x) + (4x)\frac{\partial}{\partial r}(se^{-r}),$$

making use in the last step of the product rule of differentiation.
   In order to compute

$$\frac{\partial}{\partial r}(2x - 4y), \qquad \frac{\partial}{\partial r}(4x),$$

we apply the chain rule once more and obtain

$$\frac{\partial}{\partial r}(2x - 4y) \quad \frac{\partial}{\partial x}(2x - 4y)\frac{\partial}{\partial r}(se^r) + \frac{\partial}{\partial y}(2x - 4y)\frac{\partial}{\partial r}(se^{-r})$$

$$= 2se^r + 4se^{-r},$$

and

$$\frac{\partial}{\partial r}(4x) = \frac{\partial}{\partial x}(4x)\frac{\partial}{\partial r}(se^r) + \frac{\partial}{\partial y}(4x)\frac{\partial}{\partial r}(se^{-r})$$

$$= 4se^r + (0)(se^{-r}) = 4se^r.$$

Hence

$$\frac{\partial^2 f}{\partial r^2} = se^r(2se^r + 4se^{-r}) + (2x - 4y)(se^r)$$

$$+ se^{-r}(4se^r) + 4x(-se^{-r})$$

$$= 2s[s(e^{2r} + 4) + (x - 2y)e^r - 2xe^{-r}].$$

The general case of Example 13-12 is the following example.

Example 13-13. Calculate $\partial^2 f/\partial r^2$, where f is a function of x and y and

$$x = g(r, s), \qquad y = h(r, s).$$

From the chain rule,

$$\frac{\partial f}{\partial r} = \frac{\partial f}{\partial x}\frac{\partial x}{\partial r} + \frac{\partial f}{\partial y}\frac{\partial y}{\partial r}.$$

Then

(a)
$$\frac{\partial^2 f}{\partial r^2} = \frac{\partial}{\partial r}\left(\frac{\partial f}{\partial r}\right) = \frac{\partial}{\partial r}\left(\frac{\partial f}{\partial x}\frac{\partial x}{\partial r}\right) + \frac{\partial}{\partial r}\left(\frac{\partial f}{\partial y}\frac{\partial y}{\partial r}\right)$$

$$= \frac{\partial x}{\partial r}\frac{\partial}{\partial r}\left(\frac{\partial f}{\partial x}\right) + \frac{\partial f}{\partial x}\frac{\partial}{\partial r}\left(\frac{\partial x}{\partial r}\right) + \frac{\partial y}{\partial r}\frac{\partial}{\partial r}\left(\frac{\partial f}{\partial y}\right) + \frac{\partial f}{\partial y}\frac{\partial}{\partial r}\left(\frac{\partial y}{\partial r}\right).$$

In order to calculate

$$\frac{\partial}{\partial r}\left(\frac{\partial f}{\partial x}\right), \qquad \frac{\partial}{\partial r}\left(\frac{\partial f}{\partial y}\right),$$

we apply the chain rule once more and obtain

$$\frac{\partial}{\partial r}\left(\frac{\partial f}{\partial x}\right) = \frac{\partial}{\partial x}\left(\frac{\partial f}{\partial x}\right)\frac{\partial x}{\partial r} + \frac{\partial}{\partial y}\left(\frac{\partial f}{\partial x}\right)\frac{\partial y}{\partial r}$$

$$= \frac{\partial^2 f}{\partial x^2}\frac{\partial x}{\partial r} + \frac{\partial^2 f}{\partial y\,\partial x}\frac{\partial y}{\partial r},$$

and

$$\frac{\partial}{\partial r}\left(\frac{\partial f}{\partial y}\right) = \frac{\partial}{\partial x}\left(\frac{\partial f}{\partial y}\right)\frac{\partial x}{\partial r} + \frac{\partial}{\partial y}\left(\frac{\partial f}{\partial y}\right)\frac{\partial y}{\partial r}$$

$$= \frac{\partial^2 f}{\partial x\,\partial y}\frac{\partial x}{\partial r} + \frac{\partial^2 f}{\partial y^2}\frac{\partial y}{\partial r}.$$

Substituting these expressions in (a), we get

$$\frac{\partial^2 f}{\partial r^2} = \left[\frac{\partial^2 f}{\partial x^2}\frac{\partial x}{\partial r} + \frac{\partial^2 f}{\partial y\,\partial x}\frac{\partial y}{\partial r}\right]\frac{\partial x}{\partial r} + \frac{\partial f}{\partial x}\frac{\partial^2 x}{\partial r^2}$$

$$+ \left[\frac{\partial^2 f}{\partial x\,\partial y}\frac{\partial x}{\partial r} + \frac{\partial^2 f}{\partial y^2}\frac{\partial y}{\partial r}\right]\frac{\partial y}{\partial r} + \frac{\partial f}{\partial y}\frac{\partial^2 y}{\partial r^2}$$

$$= \frac{\partial^2 f}{\partial x^2}\left(\frac{\partial x}{\partial r}\right)^2 + 2\frac{\partial^2 f}{\partial x\,\partial y}\frac{\partial x}{\partial r}\frac{\partial y}{\partial r} + \frac{\partial^2 f}{\partial y^2}\left(\frac{\partial y}{\partial r}\right)^2$$

$$+ \frac{\partial f}{\partial x}\frac{\partial^2 x}{\partial r^2} + \frac{\partial f}{\partial y}\frac{\partial^2 y}{\partial r^2}.$$

This is not presented for use as a formula but rather as an example of a general process. The same basic steps may be performed to obtain

$$\frac{\partial^2 f}{\partial s^2}, \qquad \frac{\partial^2 f}{\partial r\,\partial s}.$$

These are left for exercises.

Exercises 13-4

1. Given

$$f(x, y) = 2xy + y^2,$$

where

$$x = r \cos \theta, \qquad y = r \sin \theta,$$

find $\partial^2 f / \partial r^2$, $\partial^2 f / \partial r \, \partial \theta$, $\partial^2 f / \partial \theta^2$ in terms of x, y, r, θ by means of the chain rule.

2. Given

$$f(x, y) = e^{-x} \cos y,$$

where

$$x = r - t, \qquad y = r + t,$$

find $\partial^2 f / \partial r^2$, $\partial^2 f / \partial r \, \partial t$, $\partial^2 f / \partial t^2$ in terms of x, y, r, t by means of the chain rule.

3. If $f(x, y) = f_1(x - ay) + f_2(x + ay)$, show that

$$a^2 \frac{\partial^2 f}{\partial x^2} = \frac{\partial^2 f}{\partial y^2}.$$

(*Hint*: Set $x - ay = r$, $x + ay = s$ and use the chain rule.)

4. Given f is a function of x and y and

$$x = g(t), \qquad y = h(t),$$

show that

$$\frac{d^2 f}{dt^2} = \frac{\partial^2 f}{\partial x^2} \left(\frac{dx}{dt}\right)^2 + 2 \frac{\partial^2 f}{\partial x \, \partial y} \frac{dx}{dt} \frac{dy}{dt} + \frac{\partial^2 f}{\partial y^2} \left(\frac{dy}{dt}\right)^2 + \frac{\partial f}{\partial x} \frac{d^2 x}{dt^2} + \frac{\partial f}{\partial y} \frac{d^2 y}{dt^2}.$$

5. Given f is a function of x and y and

$$x = g(r, s), \qquad y = h(r, s),$$

show that

(a) $\dfrac{\partial^2 f}{\partial s^2} = \dfrac{\partial^2 f}{\partial x^2} \left(\dfrac{\partial x}{\partial s}\right)^2 + 2 \dfrac{\partial^2 f}{\partial x \, \partial y} \dfrac{\partial x}{\partial s} \dfrac{\partial y}{\partial s} + \dfrac{\partial^2 f}{\partial y^2} \left(\dfrac{\partial y}{\partial s}\right)^2 + \dfrac{\partial f}{\partial x} \dfrac{\partial^2 x}{\partial s^2} + \dfrac{\partial f}{\partial y} \dfrac{\partial^2 y}{\partial s^2};$

(b) $\dfrac{\partial^2 f}{\partial r \, \partial s} = \dfrac{\partial^2 f}{\partial x^2} \dfrac{\partial x}{\partial r} \dfrac{\partial x}{\partial s} + \dfrac{\partial^2 f}{\partial x \, \partial y} \left(\dfrac{\partial y}{\partial r} \dfrac{\partial x}{\partial s} + \dfrac{\partial x}{\partial r} \dfrac{\partial y}{\partial s}\right) + \dfrac{\partial^2 f}{\partial y^2} \dfrac{\partial y}{\partial r} \dfrac{\partial y}{\partial s} + \dfrac{\partial f}{\partial x} \dfrac{\partial^2 x}{\partial r \, \partial s}$

$\qquad + \dfrac{\partial f}{\partial y} \dfrac{\partial^2 y}{\partial r \, \partial s}.$

13-8. Implicit Differentiation

The methods of partial differentiation provide a means of calculating derivatives when the function is defined in implicit form. Let us assume that the relation

(a) $$F(x, y) = 0$$

defines y as a function of x. For values satisfying (a),

$$\Delta F = F(x + \Delta x, y + \Delta y) - F(x, y) = 0,$$

since each term in the difference is zero. Hence, from Theorem 13-2,

$$\Delta F = \frac{\partial F}{\partial x} \Delta x + \frac{\partial F}{\partial y} \Delta y + \varepsilon_1 \Delta x + \varepsilon_2 \Delta y = 0$$

or, dividing both members by Δx,

$$\frac{\partial F}{\partial x} + \frac{\partial F}{\partial y} \frac{\Delta y}{\Delta x} + \varepsilon_1 + \varepsilon_2 \frac{\Delta y}{\Delta x} = 0.$$

From this we obtain, taking the limit of both members as $\Delta x \to 0$,

(b) $$\frac{\partial F}{\partial x} + \frac{\partial F}{\partial y} \lim_{\Delta x \to 0} \frac{\Delta y}{\Delta x} + \lim_{\Delta x \to 0} \left[\varepsilon_1 + \varepsilon_2 \frac{\Delta y}{\Delta x} \right] = 0.$$

Since

$$\lim_{\Delta x \to 0} \frac{\Delta y}{\Delta x} = \frac{dy}{dx}, \qquad \lim_{\Delta x \to 0} \left[\varepsilon_1 + \varepsilon_2 \frac{\Delta y}{\Delta x} \right] = 0,$$

equation (b) reduces to

$$\frac{\partial F}{\partial x} + \frac{\partial F}{\partial y} \frac{dy}{dx} = 0.$$

Thus, we are able to state the following theorem.

THEOREM 13-5. Let y be a differentiable function of x defined by the relation

$$F(x, y) = 0.$$

Then

$$\frac{dy}{dx} = -\frac{\partial F / \partial x}{\partial F / \partial y}, \qquad \frac{\partial F}{\partial y} \neq 0.$$

Example 13–14. Calculate dy/dx if $x^2 + xy + y^2 = 1$.

We write

$$F(x, y) = x^2 + xy + y^2 - 1 = 0.$$

Then

$$\frac{\partial F}{\partial x} = 2x + y, \qquad \frac{\partial F}{\partial y} = x + 2y,$$

and, from Theorem 13-5,

$$\frac{dy}{dx} = -\frac{2x + y}{x + 2y}.$$

To confirm this result, we proceed according to the methods developed in Sec. 5-7, and obtain

$$2x + y + x\frac{dy}{dx} + 2y\frac{dy}{dx} = 0,$$

whence

$$(x + 2y)\frac{dy}{dx} = -2x - y$$

or

$$\frac{dy}{dx} = -\frac{2x + y}{x + 2y}.$$

Let us assume now that the relation

$$F(x, y, z) = 0$$

defines z as a function of the variables x and y. Then, reasoning as above,

$$\Delta F = \frac{\partial F}{\partial x}\Delta x + \frac{\partial F}{\partial y}\Delta y + \frac{\partial F}{\partial z}\Delta z + \varepsilon_1\,\Delta x + \varepsilon_2\,\Delta y + \varepsilon_3\,\Delta z = 0.$$

Let us assume further that y is held constant, that is, $\Delta y = 0$. This gives, after dividing both members by Δx,

$$\frac{\partial F}{\partial x} + \frac{\partial F}{\partial z}\frac{\Delta z}{\Delta x} + \varepsilon_1 + \varepsilon_3\frac{\Delta z}{\Delta x} = 0,$$

or, taking the limit of both members as $\Delta x \to 0$,

(c) $$\frac{\partial F}{\partial x} + \frac{\partial F}{\partial z}\lim_{\Delta x \to 0}\frac{\Delta z}{\Delta x} + \lim_{\Delta x \to 0}\left[\varepsilon_1 + \varepsilon_3\frac{\Delta z}{\Delta x}\right] = 0.$$

Since we are considering the limit as $\Delta x \to 0$ with y held constant, we have

$$\lim_{\Delta x \to 0} \frac{\Delta z}{\Delta x} = \frac{\partial z}{\partial x}.$$

Therefore (c) results in

$$\frac{\partial F}{\partial x} + \frac{\partial F}{\partial z} \frac{\partial z}{\partial x} = 0.$$

Similar reasoning, holding x constant, gives

$$\frac{\partial F}{\partial y} + \frac{\partial F}{\partial z} \frac{\partial z}{\partial y} = 0,$$

and we may state the next theorem.

THEOREM 13-6. Let z be a differentiable function of x and y defined by the relation

$$F(x, y, z) = 0.$$

Then

$$\frac{\partial z}{\partial x} = -\frac{\partial F/\partial x}{\partial F/\partial z},$$

and

$$\frac{\partial z}{\partial y} = -\frac{\partial F/\partial y}{\partial F/\partial z}, \qquad \frac{\partial F}{\partial z} \neq 0.$$

Example 13-15. Calculate $\partial z/\partial x$, $\partial z/\partial y$, where

$$\sin(2x + y) + \cos(y + 2z) = 0.$$

Let

$$F(x, y, z) = \sin(2x + y) + \cos(y + 2z).$$

Then

$$\frac{\partial F}{\partial x} = 2 \cos(2x + y),$$

$$\frac{\partial F}{\partial y} = \cos(2x + y) - \sin(y + 2z),$$

$$\frac{\partial F}{\partial z} = -2 \sin(y + 2z).$$

Hence

$$\frac{\partial z}{\partial x} = -\frac{2\cos(2x+y)}{-2\sin(y+2z)} = \frac{\cos(2x+y)}{\sin(y+2z)},$$

and

$$\frac{\partial z}{\partial y} = -\frac{\cos(2x+y) - \sin(y+2z)}{-2\sin(y+2z)}$$

$$= \frac{1}{2}\left[\frac{\cos(2x+y)}{\sin(y+2z)} - 1\right].$$

Exercises 13-5

In Exercises 1–8, find dy/dx in two ways.

1. $x^2 + 2xy - 2y^2 = 0$ **2.** $3x^3 - xy + y^3 = 0$

3. $x^{1/2} + y^{1/2} = 1$ **4.** $x\cos y + y\sin x = 1$

5. $\text{Arctan}\,\dfrac{y}{x} = \sqrt{x^2 + y^2}$ **6.** $e^x\cos y + e^{-x}\sin y = 0$

7. $\log\left(xy - \dfrac{x}{y}\right) = 0$ **8.** $e^{xy}\cot y = \tan x$

In Exercises 9–16, find $\partial z/\partial x$, $\partial z/\partial y$.

9. $x^2 + y^2 - z^2 = 1$ **10.** $\dfrac{x^2}{y} + \dfrac{y^2}{z} + \dfrac{z^2}{x} = 1$

11. $\dfrac{y^2}{x^2 - z^2} = x$ **12.** $e^{xy}\cos z = \sin xy$

13. $\text{Arctan}\,\dfrac{y}{x} = \text{Arcsin}\,\dfrac{z}{y}$ **14.** $x^{2/3} + y^{2/3} + z^{2/3} = 1$

15. $\sin(x+y) + \sin(y+z) + \sin(x+z) = 0$

16. $\log\left(xyz + \dfrac{y}{z}\right) = x$

Chapter 14

INFINITE SERIES

14-1. Introduction

If we should follow the first method used in solving Example 7-2 to solve Example 7-3, we would need to expand

$$\sqrt{1 + 3x^2} = (1 + 3x^2)^{1/2}$$

by the binomial theorem. This would give

$$1 + \frac{1}{2}(3x^2) + \frac{\frac{1}{2}(-\frac{1}{2})}{2}(3x^2)^2 + \frac{\frac{1}{2}(-\frac{1}{2})(-\frac{3}{2})}{3\cdot2}(3x^2)^3 + \cdots.$$

There is no place to stop! The terms go on indefinitely. This is what we call an *infinite series*.

Infinite sums present problems that finite sums do not. A finite sum can be calculated simply by applying the process of addition to the terms. This cannot be done with an infinite sum because, regardless of how many terms are included in the process of addition, there are always some left over. One way of dealing with this problem is indicated in the following example.

Example 14-1. A specially constructed ball dropped from a height h will bounce to a·height $0.9h$. If we neglect the diameter of the ball and drop it from a height of 10 ft, how far will it travel after it strikes the floor the first time?

The first bounce will be $(0.9)(10)$ ft high. Consequently the ball will travel twice that distance, $2(0.9)(10)$ ft, on the first bounce. The second bounce will be $(0.9)[(0.9)(10)]$ ft, or $(0.9)^2(10)$ ft high. Hence the distance traveled on that bounce will be $2(0.9)^2(10)$ ft. Similarly, it will travel $2(0.9)^3(10)$ ft on the third bounce, and so on.

Since the diameter of the ball is assumed to be zero, it will never stop bouncing! Thus, if we are to talk about the total distance traveled by our special ball, we have to attach some meaning to the infinite sum,

$$2(0.9)(10) + 2(0.9)^2(10) + 2(0.9)^3(10) + \cdots.$$

Suppose we find the sum of the first n terms. This is easily accomplished because these n terms,

$$2(0.9)(10) + 2(0.9)^2(10) + 2(0.9)^3(10) + \cdots + 2(0.9)^n(10),$$

form a finite geometric sum; that is, any term may be obtained by multiplying the preceding one by a fixed number, 0.9 in this case. Then, applying the formula† for the sum of a finite geometric progression and denoting this sum by S_n, we have

$$S_n = \frac{2(10)(0.9)[1 - (0.9)^n]}{1 - 0.9}$$

$$= 180 - 180(0.9)^n.$$

What happens to this sum as we include more and more terms? This question is answered by considering

$$\lim_{n \to \infty} S_n = 180 - 180 \lim_{n \to \infty} (0.9)^n$$

$$= 180 \text{ ft.}$$

Thus, as time goes on, the distance traveled by the ball gets closer and closer to 180 ft and eventually approximates 180 ft as closely as we please. Hence we say that the total distance traveled, or the sum of the infinite series, is 180 ft.

Many situations in a wide variety of fields are expressible in the form of nonending sums such as those above. Therefore in this final chapter we shall introduce the student to a few of the basic concepts and elementary theorems which relate to this topic.

14-2. Fundamental Concepts

DEFINITION 14-1. Let $\{u_n\}$ be an infinite sequence (Sec. 4-4). Then the indicated sum

$$\sum_{n=1}^{\infty} u_n = u_1 + u_2 + u_3 + \cdots + u_n + \cdots$$

is called an infinite series.

---

† The formula for the sum of the geometric progression

$$a + ar + ar^2 + \cdots + ar^{n-1}$$

is easily obtained. We set

$$S = a + ar + ar^2 + \cdots + ar^{n-1}.$$

Then

$$rS = ar + ar^2 + ar^3 + \cdots + ar^n,$$

so

$$S - rS = a - ar^n.$$

We solve this equation for S and get

$$S = \frac{a - ar^n}{1 - r} = \frac{a(1 - r^n)}{1 - r}.$$

As noted in the preceding section, such a sum has no meaning in the usual sense. However, we shall see (as in Example 14-1) that it is possible to attach meaning to some of them through the medium of limits of sequences.

Returning to Example 14-1 we see that the first term, u_1, is $2(10)(0.9)$; the second term, u_2, is $2(10)(0.9)^2$, and so forth. The nth term, u_n, also called the *general term*, or the *law of formation*, is $2(10)(0.9)^n$. The general term may be used to write any particular term, and thus it actually defines the series. For example the 1000th term is $2(10)(0.9)^{1000}$. No infinite series is completely defined unless the general term is known.

DEFINITION 14–2. The sum of the first n terms of an infinite series is called the nth partial sum and is indicated by S_n.

Some partial sums from Example 14-1 are

$$S_1 = 2(10)(0.9); \qquad S_2 = 2(10)(0.9) + 2(10)(0.9)^2;$$
$$S_n = 2(10)(0.9) + 2(10)(0.9)^2 + \cdots + 2(10)(0.9)^n.$$

Since these partial sums have a finite number of terms, they have meaning as a sum and may be calculated. Hence, associated with each infinite series is an infinite sequence,

$$\{S_n\} \equiv S_1, S_2, S_3, \ldots, S_n, \ldots,$$

of partial sums.

DEFINITION 14–3. Given the infinite series

$$\sum_{n=1}^{\infty} u_n = u_1 + u_2 + u_3 + \cdots + u_n + \cdots.$$

If the infinite sequence $\{S_n\}$ of partial sums has a finite limit S, that is, if

$$\lim_{n \to \infty} S_n = \lim_{n \to \infty} (u_1 + u_2 + u_3 + \cdots + u_n) = S, \qquad |S| < \infty,$$

the series is said to converge to the sum S.

A series that converges to a sum S is said to be a *convergent* series. This is a reasonable definition to give to the sum of an infinite number of terms. It merely says that a convergent series may have its sum approximated to any desired degree of accuracy by taking the sum of a sufficiently large number of terms. Thus, at least for numerical work, a convergent infinite series may be replaced by one of its partial sums, the particular partial sum being dictated by the accuracy required in the problem.

According to this definition, the series in Example 14-1 converges to the sum 180.

DEFINITION 14–4. An infinite series which does not converge is called a divergent series.

Since a divergent series does not have a sum† in the sense we are using, it becomes important to us to determine whether a particular series converges or diverges. Most of what we have to say from here on will be directed to this problem.

14–3. General Tests

It is not always possible to write down a convenient expression for S_n such as that in Example 14-1. Hence it is usually difficult to apply the definition directly to an infinite series to settle the question of convergence or divergence. For this reason we shall develop some general theorems which will help determine these facts.

First, we shall determine a *necessary condition* for convergence, that is, a condition which must be satisfied if a series is to converge.

Let the series

$$\sum_{n=1}^{\infty} u_n = u_1 + u_2 + \cdots + u_n + \cdots$$

converge to the sum S. Then, by definition,

$$\lim_{n \to \infty} S_{n-1} = \lim_{n \to \infty} S_n = S.$$

Also

$$u_n = S_n - S_{n-1}.$$

Hence

$$\lim_{n \to \infty} u_n = \lim_{n \to \infty} (S_n - S_{n-1})$$

$$= \lim_{n \to \infty} S_n - \lim_{n \to \infty} S_{n-1} = 0.$$

THEOREM 14–1. A necessary condition for the series

$$\sum_{n=1}^{\infty} u_n$$

to converge is that

$$\lim_{n \to \infty} u_n = 0.$$

Note that this condition was satisfied in Example 14-1. In that series,

$$u_n = 2(10)(0.9)^n,$$

so

$$\lim_{n \to \infty} u_n = \lim_{n \to \infty} 2(10)(0.9)^n = 0.$$

---

† In advanced mathematics, some different types of sums are associated with infinite series in such a way that a divergent series may have a sum in this new sense. However, at this point, we shall consider a divergent series to have no sum.

However, it should also be noted that Theorem 14-1 states a *necessary* condition, not a *sufficient* condition. We shall show the latter by the following useful example.

Example 14-2. Show that the series

$$\sum_{n=1}^{\infty} \frac{1}{n} = 1 + \frac{1}{2} + \frac{1}{3} + \cdots + \frac{1}{n} + \cdots$$

diverges.

First we observe that this series satisfies the necessary condition of Theorem 14-1. We have

$$\lim_{n \to \infty} u_n = \lim_{n \to \infty} \frac{1}{n} = 0.$$

Hence we cannot use this theorem to establish divergence.

Let us group the terms in this series as follows:

$$\sum_{n=1}^{\infty} \frac{1}{n} = 1 + \frac{1}{2} + \left(\frac{1}{3} + \frac{1}{4}\right) + \left(\frac{1}{5} + \frac{1}{6} + \frac{1}{7} + \frac{1}{8}\right)$$

$$+ \left(\frac{1}{9} + \frac{1}{10} + \frac{1}{11} + \frac{1}{12} + \frac{1}{13} + \frac{1}{14} + \frac{1}{15} + \frac{1}{16}\right) + \cdots,$$

where each group after the second one has twice as many terms as the preceding one. Each of these groups has a sum greater than or equal to $\frac{1}{2}$. Since there is an inexhaustible supply of terms, we can take enough so that their sum will exceed $k(\frac{1}{2})$, where k is arbitrarily large. Hence, by definition, the series diverges.

This series is known as the *harmonic series* and we shall find it useful in establishing the divergence of other series.

Example 14-3. Show that the series

$$\sum_{n=1}^{\infty} (-1)^{n-1} = 1 - 1 + 1 - 1 + 1 - \cdots$$

diverges.

We shall do this in two ways. First,

$$S_{2n-1} = 1, \qquad S_{2n} = 0,$$

so the sequence of partial sums,

$$\{S_n\} = 1, 0, 1, 0, 1, \ldots,$$

fails to have a limit. Therefore the series diverges.
Second,

$$u_n = (-1)^n;$$

consequently

$$\lim_{n \to \infty} u_n = \lim_{n \to \infty} (-1)^n$$

fails to exist and equal zero. Thus the necessary condition for convergence is not fulfilled and the series diverges.

Our next project is to develop a general theorem on the convergence of an infinite series of positive terms. In this endeavor we shall have need of a theorem on sequences, which has not been stated previously in this book. We present it here without proof.

THEOREM 14–2. If the sequence $\{S_n\}$ has the properties

(a) $$S_{n+1} \geqslant S_n, \qquad \text{all } n,$$

and

(b) $$S_n \leqslant M, \qquad M \text{ a constant and for all } n,$$

then

$$\lim_{n \to \infty} S_n = S \leqslant M.$$

That is, if we have a sequence whose terms are continually increasing (unless they are constant, in which case the theorem is obvious) but which are at the same time bounded, it must approach a limit less than or equal to the bound. A little reflection will convince the student that this is a reasonable conclusion.

Let

$$\sum_{n=1}^{\infty} u_n = u_1 + u_2 + \cdots + u_n + \cdots$$

be a series of positive terms. Then

$$S_{n+1} > S_n, \qquad \text{all } n.$$

Suppose further that

$$S_n < M, \qquad \text{all } n.$$

Under these hypotheses, the sequence of partial sums, $\{S_n\}$, satisfies the requirements of Theorem 14-2, so

$$\lim_{n \to \infty} S_n = S \leqslant M.$$

Thus we may state the following theorem.

THEOREM 14–3. Let

$$\sum_{n=1}^{\infty} u_n = u_1 + u_2 + \cdots + u_n + \cdots$$

be a series of positive terms. If $S_n < M$ for all n, M a constant, the series converges to a sum S, $S \leqslant M$.

This theorem is not too useful for investigating particular series, but it is productive of other theorems which do serve this purpose.

Exercises 14-1

In Exercises 1–6, write out the first five terms of the given series.

1. $\displaystyle\sum_{n=1}^{\infty} \frac{1}{n^2}$

2. $\displaystyle\sum_{n=1}^{\infty} \frac{2n}{n^2 + 1}$

3. $\displaystyle\sum_{n=1}^{\infty} (-1)^n \frac{2n}{(n+1)(n+2)}$

4. $\displaystyle\sum_{n=1}^{\infty} \frac{1}{(n+1)(n^2+1)}$

5. $\displaystyle\sum_{n=1}^{\infty} (-1)^{n-1} \frac{n}{\sqrt{n}(n+1)}$

6. $\displaystyle\sum_{n=1}^{\infty} \frac{1}{n^{2n}}$

In Exercises 7–12, find the nth term of the infinite series for which the first five terms are given.

7. $1 + \dfrac{1}{8} + \dfrac{1}{27} + \dfrac{1}{64} + \dfrac{1}{125} + \cdots$

8. $1 - \dfrac{1}{2} + \dfrac{1}{4} - \dfrac{1}{8} + \dfrac{1}{16} - \cdots$

9. $\dfrac{3}{1\cdot2} + \dfrac{4}{2\cdot3} + \dfrac{5}{3\cdot4} + \dfrac{6}{4\cdot5} + \dfrac{7}{5\cdot6} + \cdots$

10. $1 - \dfrac{1}{2} + \dfrac{1}{3} - \dfrac{1}{4} + \dfrac{1}{5} - \cdots$

11. $\dfrac{1}{2\cdot3} - \dfrac{3}{3\cdot4} + \dfrac{5}{4\cdot5} - \dfrac{7}{5\cdot6} + \dfrac{9}{6\cdot7} - \cdots$

12. $\dfrac{1}{2} + \dfrac{1}{3} + \dfrac{1}{5} + \dfrac{1}{9} + \dfrac{1}{17} + \cdots$

In Exercises 13–16, determine whether Theorem 14-1 is satisfied. If a statement regarding convergence or divergence can be made, do so.

13. $\dfrac{1}{2} - \dfrac{2}{9} + \dfrac{3}{28} - \dfrac{4}{65} + \dfrac{5}{126} - \cdots$

14. $\dfrac{1}{3} + \dfrac{2}{5} + \dfrac{3}{7} + \dfrac{4}{9} + \dfrac{5}{11} + \cdots$

15. $\dfrac{1}{6} + \dfrac{4}{21} + \dfrac{9}{46} + \dfrac{16}{81} + \dfrac{25}{126} + \cdots$

16. $\dfrac{1}{2\cdot3} - \dfrac{2}{3\cdot5} + \dfrac{3}{4\cdot7} - \dfrac{4}{5\cdot9} + \dfrac{5}{6\cdot11} - \cdots$

14-4. The Geometric Series

The infinite series

$$\sum_{n=1}^{\infty} ar^{n-1} = a + ar + ar^2 + \cdots + ar^{n-1} + \cdots$$

is, for obvious reasons, called the geometric series. We have

$$S_n = a + ar + ar^2 + \cdots + ar^{n-1}$$

$$= \frac{a(1 - r^n)}{1 - r} = \frac{a}{1 - r} - \frac{ar^n}{1 - r}, \qquad r \neq 1,$$

whence

$$\lim_{n \to \infty} S_n = \frac{a}{1 - r} - \frac{a}{1 - r} \lim_{n \to \infty} r^n, \qquad r \neq 1.$$

If $r = 1$,

$$S_n = na,$$

so

$$\lim_{n \to \infty} S_n = \lim_{n \to \infty} na = \pm \infty.$$

If $r \neq 1$, we have the following three cases:

CASE I $(-1 < r < 1)$:

$$\lim_{n \to \infty} r^n = 0.$$

CASE II $(r > 1)$:

$$\lim_{n \to \infty} r^n = \infty.$$

CASE III $(r \leqslant -1)$:

$$\lim_{n \to \infty} r^n \text{ fails to exist.}$$

The limit in Case III does not exist because successive values of r^n alternate between positive and negative numbers greater than or equal to 1 in magnitude.
  Hence

$$\lim_{n \to \infty} S_n = \frac{a}{1 - r}, \qquad |r| < 1,$$

and

$$\lim_{n \to \infty} S_n = \pm \infty \text{ (or fails to exist)}, \qquad |r| \geqslant 1.$$

This completes the proof of the following theorem.

THEOREM 14–4. The geometric series

$$\sum_{n=1}^{\infty} ar^{n-1}$$

converges to the sum $a/(1-r)$ if $|r| < 1$ and diverges if $|r| \geqslant 1$.

Example 14–4. Determine if the series

$$3 + 2 + \frac{4}{3} + \frac{8}{9} + \frac{16}{27} + \cdots$$

converges, and if so, to what sum?

We observe that each term is obtained from the preceding one by multiplying by 2/3. Hence, this is a geometric series with ratio 2/3 and (by Theorem 14-4) converges to the sum

$$S = \frac{3}{1 - \frac{2}{3}} = 9.$$

Example 14–5. Determine if the series

$$1 - \frac{3}{2} + \frac{9}{4} - \frac{27}{8} + \frac{81}{16} - \cdots$$

converges, and if so, to what sum?

This is a geometric series with $r = -(3/2)$. Hence, by Theorem 14-4, it diverges. This result could also be obtained by noting that u_n does not approach zero as $n \to \infty$.

14–5. The p-Series

We now investigate the convergence of the series

$$\sum_{n=1}^{\infty} \frac{1}{n^p} = 1 + \frac{1}{2^p} + \frac{1}{3^p} + \cdots + \frac{1}{n^p} + \cdots,$$

which is usually referred to as the p-series, or the *hyperharmonic series*. We shall prove the following theorem.

THEOREM 14–5. The p-series

$$\sum_{n=1}^{\infty} \frac{1}{n^p}$$

converges when $p > 1$ and diverges when $p \leqslant 1$.

CASE I ($p = 1$): This is the case discussed in Example 14-2.

CASE II ($p > 1$): We group the terms in the series in the following manner:

$$\sum_{n=1}^{\infty} \frac{1}{n^p} = 1 + \left(\frac{1}{2^p} + \frac{1}{3^p}\right) + \left(\frac{1}{4^p} + \frac{1}{5^p} + \frac{1}{6^p} + \frac{1}{7^p}\right) + \cdots$$

where each group has double the number of terms in the preceding group. We have, since $p > 1$,

$$\frac{1}{2^p} + \frac{1}{3^p} < \frac{1}{2^p} + \frac{1}{2^p} = \frac{2}{2^p} = \frac{1}{2^{p-1}},$$

$$\frac{1}{4^p} + \frac{1}{5^p} + \frac{1}{6^p} + \frac{1}{7^p} < \frac{1}{4^p} + \frac{1}{4^p} + \frac{1}{4^p} + \frac{1}{4^p} = \frac{4}{4^p} = \frac{1}{2^{2(p-1)}},$$

. . . .

Therefore each group is less than or equal to the corresponding term in the series

$$1 + \frac{1}{2^{p-1}} + \frac{1}{2^{2(p-1)}} + \frac{1}{2^{3(p-1)}} + \cdots.$$

This is a geometric series with $r = 1/2^{p-1} < 1$, and thus converges to the sum

$$S = \frac{1}{1 - (1/2^{p-1})} = \frac{2^{p-1}}{2^{p-1} - 1}.$$

Hence the partial sums of the p-series are all less than S, and it converges by virtue of Theorem 14-3.

CASE III ($p < 1$): In this case, the partial sums of

$$\sum_{n=1}^{\infty} \frac{1}{n^p}$$

are all greater than the corresponding partial sums of the divergent series

$$\sum_{n=1}^{\infty} \frac{1}{n}.$$

Hence the p-series, $p < 1$, diverges.

Example 14–6. Investigate the convergence of

$$1 + \frac{1}{\sqrt{2}} + \frac{1}{\sqrt{3}} + \frac{1}{\sqrt{4}} + \frac{1}{\sqrt{5}} + \cdots.$$

This is the p-series with $p = 1/2$. Hence it diverges.

Exercises 14-2

Use Theorems 14-1 through 14-5 to investigate the convergence of the following infinite series. Whenever possible, determine the sum.

1. $1 + \dfrac{1}{3} + \dfrac{1}{9} + \dfrac{1}{27} + \cdots$

2. $1 + \dfrac{1}{2^2} + \dfrac{1}{3^2} + \dfrac{1}{4^2} + \cdots$

3. $1 + \dfrac{1}{\sqrt[3]{2}} + \dfrac{1}{\sqrt[3]{3}} + \dfrac{1}{\sqrt[3]{4}} + \cdots$

4. $\dfrac{1}{2} + \dfrac{2}{3} + \dfrac{3}{4} + \dfrac{4}{5} + \cdots$

5. $1 + \dfrac{1}{\sqrt{8}} + \dfrac{1}{\sqrt{27}} + \dfrac{1}{\sqrt{64}} + \cdots$

6. $1 + 1.1 + 1.21 + 1.331 + \cdots$

7. $1 + \dfrac{1}{\sqrt[3]{4}} + \dfrac{1}{\sqrt[3]{9}} + \dfrac{1}{\sqrt[3]{16}} + \cdots$

8. $1 + 0.1 + 0.01 + 0.001 + \cdots$

9. $1 + \dfrac{5}{6} + \dfrac{25}{36} + \dfrac{125}{216} + \cdots$

10. $1 + \dfrac{1}{\sqrt[3]{16}} + \dfrac{1}{\sqrt[3]{81}} + \dfrac{1}{\sqrt[3]{256}} + \cdots$

11. A machine costs $100,000. In any year it depreciates by 10 percent of its value at the beginning of the year. What is the annual depreciation in each of the first five years? What is the sum of the annual depreciation costs as the number of years n gets large, that is, as $n \to \infty$?

12. The original cost of an item is $8000 and it depreciates annually 10 percent of its value at the beginning of the year. Write an expression for the depreciation cost for the nth year. Does the infinite series, of which this is the general term, converge? If so, to what sum? Interpret your result.

14-6. The Comparison Test

One of the most important tests for series of positive terms is that stated in the next two theorems.

THEOREM 14-6. If

$$\sum_{n=1}^{\infty} a_n = a_1 + a_2 + \cdots + a_n + \cdots$$

is a convergent series of positive terms, and if

$$0 \leqslant u_n \leqslant a_n$$

for all n, the series

$$\sum_{n=1}^{\infty} u_n = u_1 + u_2 + \cdots + u_n + \cdots$$

converges.

If S is the sum of the convergent series $\sum_{n=1}^{\infty} a_n$, and if

$$\bar{S}_n = a_1 + a_2 + \cdots + a_n$$

and

$$S_n = u_1 + u_2 + \cdots + u_n,$$

we have

$$S_n \leqslant \bar{S}_n < S$$

for all n. The theorem follows from Theorem 14-3.

The companion theorem is as follows.

THEOREM 14–7. If

$$\sum_{n=1}^{\infty} a_n = a_1 + a_2 + \cdots + a_n + \cdots$$

is a divergent series of positive terms, and if

$$u_n \geqslant a_n \qquad \text{for all } n,$$

the series

$$\sum_{n=1}^{\infty} u_n = u_1 + u_2 + \cdots + u_n + \cdots$$

diverges.

Using the same notation as in the preceding argument, we have

$$\lim_{n \to \infty} \bar{S}_n = \infty \qquad \text{and} \quad S_n \geqslant \bar{S}_n \qquad \text{for all } n.$$

Hence

$$\lim_{n \to \infty} S_n = \infty,$$

and the series diverges.

Theorems 14-6 and 14-7 together constitute the *comparison test*. It enables us to determine the convergence properties of series by comparing them with series of known convergence or divergence.

Example 14–7. Investigate the convergence of the series

$$\frac{1}{3} + \frac{1}{5} + \frac{1}{9} + \frac{1}{17} + \frac{1}{33} + \cdots.$$

The general term of this series is

$$u_n = \frac{1}{2^n + 1}.$$

Let us compare this series with

$$\frac{1}{2} + \frac{1}{2^2} + \frac{1}{2^3} + \frac{1}{2^4} + \cdots,$$

for which the general term is

$$a_n = \frac{1}{2^n}.$$

This geometric series has $r = 1/2$, and therefore converges. Moreover,

$$u_n = \frac{1}{2^n + 1} < \frac{1}{2^n} = a_n \qquad \text{for all } n.$$

Consequently, by Theorem 14-6, the given series converges.

Example 14–8. Investigate the convergence of the series

$$1 + \frac{1}{\sqrt{2} - 1} + \frac{1}{\sqrt{3} - 1} + \frac{1}{\sqrt{4} - 1} + \cdots.$$

The general term is

$$u_n = \frac{1}{\sqrt{n} - 1}.$$

Let us compare this series with

$$1 + \frac{1}{\sqrt{2}} + \frac{1}{\sqrt{3}} + \frac{1}{\sqrt{4}} + \cdots,$$

for which the general term is

$$a_n = \frac{1}{\sqrt{n}} = \frac{1}{n^{1/2}}.$$

This is a p-series, $p = 1/2$, and therefore diverges. Also,

$$u_n = \frac{1}{\sqrt{n} - 1} > \frac{1}{\sqrt{n}} = a_n,$$

so, by Theorem 14-7, the given series diverges.

It should be noted that nothing is accomplished by showing that a given series is term by term greater than a convergent series or term by term less than a divergent series. The student should make a guess, based on observation and experience, as to which of Theorems 14-6 or 14-7 is most appropriate. Then he should attempt to confirm his conjecture, always being ready to reverse himself if it appears that he made a poor first guess.

It should also be observed that a finite number of terms do not affect the convergence of a series. For example, if we discard the first ten terms of a series, this will in no way affect its convergence or divergence. If it converged before, it will still converge. The same remarks apply to divergence. However, *it will affect the sum.*

With this in mind, it is sometimes convenient to discard a few terms at the beginning of a series in order to apply the comparison test.

Example 14–9. Investigate the convergence of

$$\frac{\sqrt{2}}{3} + \frac{\sqrt{3}}{4} + \frac{\sqrt{4}}{5} + \frac{\sqrt{5}}{6} + \cdots.$$

Let us compare this with the divergent harmonic series

$$1 + \frac{1}{2} + \frac{1}{3} + \cdots + \frac{1}{n} + \cdots,$$

or rather let us use the comparison series

$$\frac{1}{3} + \frac{1}{4} + \frac{1}{5} + \cdots,$$

which results from dropping the first two terms of the harmonic series. Then, comparing the general terms of these two series, we have

$$\frac{\sqrt{n+1}}{n+2} > \frac{1}{n+2} \qquad \text{for all } n,$$

with the result that the given series diverges.

Exercises 14-3

Test the following series for convergence by means of the comparison test, using the geometric series, the harmonic series, and the *p*-series as sources of comparison series.

1. $1 + \dfrac{1}{2} + \dfrac{1}{5} + \dfrac{1}{10} + \cdots$ **2.** $\sqrt{2} + \dfrac{\sqrt{3}}{2} + \dfrac{\sqrt{4}}{3} + \dfrac{\sqrt{5}}{4} + \cdots$

3. $\dfrac{1}{\log 2} + \dfrac{1}{\log 3} + \dfrac{1}{\log 4} + \dfrac{1}{\log 5} + \cdots$

4. $\dfrac{1}{1 \cdot 2} + \dfrac{1}{2 \cdot 3} + \dfrac{1}{3 \cdot 4} + \dfrac{1}{4 \cdot 5} + \cdots$

5. $\dfrac{1}{\sqrt{1 \cdot 2}} + \dfrac{1}{\sqrt{2 \cdot 3}} + \dfrac{1}{\sqrt{3 \cdot 4}} + \dfrac{1}{\sqrt{4 \cdot 5}} + \cdots$

6. $\dfrac{1}{1 \cdot 2 \cdot 3} + \dfrac{1}{2 \cdot 3 \cdot 4} + \dfrac{1}{3 \cdot 4 \cdot 5} + \dfrac{1}{4 \cdot 5 \cdot 6} + \cdots$

7. $\dfrac{1}{1\cdot 2}+\dfrac{1}{3\cdot 4}+\dfrac{1}{5\cdot 6}+\dfrac{1}{7\cdot 8}+\cdots$

8. $\dfrac{1}{3}+\dfrac{1}{4^2}+\dfrac{1}{5^3}+\dfrac{1}{6^4}+\cdots$

9. $\dfrac{1}{2}+\dfrac{2}{9}+\dfrac{3}{28}+\dfrac{4}{65}+\cdots$

10. $\dfrac{1}{2}+\dfrac{1}{2\cdot 4}+\dfrac{1}{3\cdot 8}+\dfrac{1}{4\cdot 16}+\cdots$

11. $\dfrac{1\cdot 2}{2\cdot 3}+\dfrac{2\cdot 4}{3\cdot 9}+\dfrac{3\cdot 8}{4\cdot 27}+\dfrac{4\cdot 16}{5\cdot 81}+\cdots$

12. $2+\dfrac{3}{4}+\dfrac{4}{9}+\dfrac{5}{16}+\dfrac{6}{25}+\cdots$

14–7. The Ratio Test

Consider the series of positive terms

(a)
$$\sum_{n=1}^{\infty} u_n,$$

and let

$$\lim_{n\to\infty}\frac{u_{n+1}}{u_n}=\rho.$$

Since $u_n > 0$ for all n, it follows that $\rho \geqslant 0$. There are three cases to consider.

CASE I ($\rho < 1$): In this case we can choose an r satisfying $\rho < r < 1$. Moreover, from the definition of a limit, there exists an m such that

$$\frac{u_{n+1}}{u_n} < r, \qquad n \geqslant m.$$

Hence

$$\frac{u_{m+1}}{u_m} < r,$$

$$\frac{u_{m+2}}{u_{m+1}} < r,$$

$$\cdot \quad \cdot \quad \cdot \quad \cdot \quad \cdot \quad ,$$

$$\frac{u_{m+p}}{u_{m+p-1}} < r,$$

$$\cdot \quad \cdot \quad \cdot \quad \cdot \quad \cdot \quad \cdot$$

Thus

$$u_{m+1} < ru_m,$$

$$u_{m+2} < ru_{m+1} < r^2 u_m,$$

$$\cdot \quad \cdot \quad \cdot \quad \cdot \quad \cdot \quad \cdot \quad \cdot \quad ,$$

$$u_{m+p} < r^p u_m$$

$$\cdot \quad \cdot \quad \cdot \quad \cdot \quad \cdot \quad \cdot \quad \cdot$$

Therefore the series

(b) $$u_m + u_{m+1} + u_{m+2} + \cdots + u_{m+p} + \cdots$$

is term by term less than the series

$$u_m + u_m r + u_m r^2 + \cdots + u_m r^p + \cdots,$$

which is convergent, since $r < 1$. Consequently, by the comparison test, (b) converges. The convergence of (b) implies the convergence of (a) because they are identical except for $m - 1$ terms.

CASE II $(\rho > 1)$: Here we can choose an r satisfying $1 < r < \rho$. Also there exists an m such that

$$\frac{u_{n+1}}{u_n} > r, \qquad n \geq m.$$

Hence

$$u_{n+1} > ru_n > u_m > 0, \qquad n \geq m.$$

Consequently

$$\lim_{n \to \infty} u_n \neq 0,$$

and (a) diverges (Theorem 14-1).

CASE III $(\rho = 1)$: Nothing can be determined from this test in this case. Consider the two series

$$\sum_{n=1}^{\infty} \frac{1}{n}, \qquad \sum_{n=1}^{\infty} \frac{1}{n^2}.$$

We have

$$\lim_{n \to \infty} \frac{u_{n+1}}{u_n} = 1$$

for each of them. However, one of them converges and the other diverges. Therefore $\rho = 1$ gives no information regarding convergence.

THEOREM 14-8. Given the series of positive terms

$$\sum_{n=1}^{\infty} u_n \qquad \text{and} \qquad \lim_{n \to \infty} \frac{u_{n+1}}{u_n} = \rho.$$

If $\rho < 1$, the series converges; if $\rho > 1$, the series diverges; if $\rho = 1$, this test provides no information on convergence.

This test is called the *ratio test*. Its advantage is that it provides a straightforward method of approaching a convergence problem. Its disadvantage is that many series have $\rho = 1$.

Before we look at an example of the ratio test, let us define the *factorial notation.*

DEFINITION 14–5. We define the symbol $n!$, n a nonnegative integer to mean

$$n! = 1 \cdot 2 \cdot 3 \cdot 4 \cdots (n-1)(n), \qquad n > 0,$$

and

$$0! = 1.$$

Thus

$$5! = 1 \cdot 2 \cdot 3 \cdot 4 \cdot 5,$$

and is read "five factorial."

Example 14–10. Use the ratio test to examine

$$1 + \frac{2!}{3} + \frac{3!}{3^2} + \frac{4!}{3^3} + \cdots$$

for convergence.

The general term may be written

$$u_n = \frac{n!}{3^{n-1}},$$

whence

$$\frac{u_{n+1}}{u_n} = \frac{\dfrac{(n+1)!}{3^n}}{\dfrac{n!}{3^{n-1}}} = \frac{(n+1)!}{3^n} \cdot \frac{3^{n-1}}{n!}$$

$$= \frac{n+1}{3}.$$

Hence

$$\lim_{n \to \infty} \frac{u_{n+1}}{u_n} = \lim_{n \to \infty} \frac{n+1}{3} = \infty,$$

so we conclude the series diverges.

Example 14–11. Use the ratio test to examine

$$\frac{1}{1 \cdot 2} + \frac{1}{2 \cdot 3} + \frac{1}{3 \cdot 4} + \frac{1}{4 \cdot 5} + \cdots$$

for convergence.

The general term is

$$u_n = \frac{1}{n(n+1)},$$

so

$$\frac{u_{n+1}}{u_n} = \frac{\dfrac{1}{(n+1)(n+2)}}{\dfrac{1}{n(n+1)}} = \frac{n}{n+2}.$$

Thus

$$\lim_{n \to \infty} \frac{u_{n+1}}{u_n} = \lim_{n \to \infty} \frac{n}{n+2} = 1,$$

and we conclude that the ratio test fails to discriminate between convergence and divergence for this series. However, comparison with the p-series, $p = 2$, readily shows that it converges.

Example 14–12. Use the ratio test to examine the series

$$1 + \frac{1}{3} + \frac{2}{3^2} + \frac{3}{3^3} + \frac{4}{3^4} + \cdots$$

for convergence.

We have, neglecting the first term,

$$u_n = \frac{n}{3^n},$$

whence

$$\frac{u_{n+1}}{u_n} = \frac{(n+1)/3^{(n+1)}}{n/3^n} = \frac{n+1}{3n}.$$

Therefore

$$\lim_{n \to \infty} \frac{u_{n+1}}{u_n} = \lim_{n \to \infty} \frac{n+1}{3n} = \frac{1}{3},$$

which indicates that the series converges.

Exercises 14-4

Investigate the convergence of the following series by the ratio test. If this test fails, use another test.

1. $1 + \dfrac{1}{3} + \dfrac{1}{9} + \dfrac{1}{27} + \cdots$

2. $1 + \dfrac{1!}{3} + \dfrac{3!}{3^3} + \dfrac{5!}{3^5} + \cdots$

3. $\dfrac{1 \cdot 2}{3} + \dfrac{2 \cdot 3}{3^2} + \dfrac{3 \cdot 4}{3^3} + \dfrac{4 \cdot 5}{3^4} + \cdots$

4. $\dfrac{1}{2 \cdot 3 \cdot 4} + \dfrac{2}{3 \cdot 4 \cdot 5} + \dfrac{3}{4 \cdot 5 \cdot 6} + \dfrac{4}{5 \cdot 6 \cdot 7} + \cdots$

5. $\dfrac{3}{1} + \dfrac{3^2}{2^2} + \dfrac{3^3}{3^2} + \dfrac{3^4}{4^2} + \cdots$ **6.** $1 + \dfrac{2^3}{2^2} + \dfrac{3^3}{2^3} + \dfrac{4^3}{2^4} + \cdots$

7. $\dfrac{1}{3} + \dfrac{3}{3^2} + \dfrac{5}{3^3} + \dfrac{7}{3^4} + \cdots$ **8.** $1 + \dfrac{2}{\sqrt{3}} + \dfrac{3}{\sqrt{5}} + \dfrac{4}{\sqrt{7}} + \cdots$

9. $2 + \dfrac{5}{2} + \dfrac{10}{2^2} + \dfrac{17}{2^3} + \cdots$

10. $\dfrac{1}{1!} + \dfrac{1 \cdot 3}{2!} + \dfrac{1 \cdot 3 \cdot 5}{3!} + \dfrac{1 \cdot 3 \cdot 5 \cdot 7}{4!} + \cdots$

11. $1 + \dfrac{10^2}{2!} + \dfrac{10^3}{3!} + \dfrac{10^4}{4!} + \cdots$

12. $\dfrac{3}{1^{1.1}} + \dfrac{3}{2^{1.1}} + \dfrac{3}{3^{1.1}} + \dfrac{3}{4^{1.1}} + \cdots$

13. $\dfrac{1!}{1^2 \cdot 2^2} + \dfrac{2!}{2^2 \cdot 3^2} + \dfrac{3!}{3^2 \cdot 4^2} + \dfrac{4!}{4^2 \cdot 5^2} + \cdots$

14. $2^2 + \dfrac{4^2}{5} + \dfrac{6^2}{5^2} + \dfrac{8^2}{5^3} + \cdots$

15. $\dfrac{2 + \sqrt{1}}{1!} + \dfrac{2 + \sqrt{2}}{2!} + \dfrac{2 + \sqrt{3}}{3!} + \dfrac{2 + \sqrt{4}}{4!} + \cdots$

16. $\dfrac{1^{10}}{e} + \dfrac{2^{10}}{e^2} + \dfrac{3^{10}}{e^3} + \dfrac{4^{10}}{e^4} + \cdots$

14-8. Alternating Series

Except for the geometric series we have confined our attention to series of positive terms. In this section we shall consider series which contain both positive and negative terms. In particular we shall investigate the convergence of *alternating series*.

DEFINITION 14-6. A series whose terms are alternately of opposite sign is called an alternating series.

Consider the series

$$\sum_{n=1}^{\infty} (-1)^{n-1} u_n = u_1 - u_2 + u_3 - u_4 + \cdots,$$

where $u_n > 0$, for all n, and let us assume

(a) $u_{n+1} \leqslant u_n,$ for all $n,$

and

(b) $\lim_{n \to \infty} u_n = 0.$

Then

$$S_{2n} = (u_1 - u_2) + (u_3 - u_4) + \cdots + (u_{2n-1} - u_{2n})$$

is increasing because each quantity in parentheses is nonnegative. But we may also write

$$S_{2n} = u_1 - (u_2 - u_3) - (u_4 - u_5) - \cdots - (u_{2n-2} - u_{2n-1}) - u_{2n},$$

which indicates that

$$S_{2n} < u_1.$$

Hence, by Theorem 14-2,

$$\lim_{n \to \infty} S_{2n} = S \leqslant u_1.$$

Moreover,

$$S_{2n+1} = S_{2n} + u_{2n+1}$$

whence

$$\lim_{n \to \infty} S_{2n+1} = \lim_{n \to \infty} S_{2n} + \lim_{n \to \infty} u_{2n+1}$$

$$= S,$$

by virtue of hypothesis (b). Hence the partial sums have the same limit S whether their index is even or odd, and we conclude that the alternating series converges under the conditions specified.

If condition (a) is satisfied for all but a finite number of terms, the series will still converge.

THEOREM 14-9. Given the series

$$\sum_{n=1}^{\infty} (-1)^{n-1} u_n = u_1 - u_2 + u_3 - u_4 + \cdots.$$

If $0 < u_{n+1} \leqslant u_n$, $n > N$, and $\lim_{n \to \infty} u_n = 0$, the series converges.

This theorem is stated for $u_n > 0$, $n > N$. The purpose of this requirement is to ensure the alternating character of the series for $n > N$. No difficulties would be introduced if $u_n < 0$ in this range. However, for the sake of definiteness, we shall retain the original hypothesis.

It is easy to obtain an upper bound for the difference of S and S_n for a convergent alternating series. We have

$$S - S_{2n} = u_{2n+1} - (u_{2n+2} - u_{2n+3}) - \cdots,$$

whence we obtain

$$S - S_{2n} < u_{2n+1};$$

and

$$S_{2n+1} - S = u_{2n+2} - (u_{2n+3} - u_{2n+4}) - \cdots,$$

so we can conclude that

$$S_{2n+1} - S < u_{2n+2}.$$

In any case, n even or odd,

$$|S - S_n| < u_{n+1}.$$

THEOREM 14-10. If a series satisfies the hypotheses of Theorem 14-9,

$$|S - S_n| < u_{n+1}.$$

In other words, the difference between the sum of a convergent alternating series and any partial sum does not exceed in magnitude the numerical value of the first term in the series omitted from the partial sum.

Example 14-13. Use the alternating series test to prove the convergence of

$$\sum_{n=1}^{\infty} (-1)^{n-1} \frac{1}{n} = 1 - \frac{1}{2} + \frac{1}{3} - \frac{1}{4} + \cdots.$$

We have

$$u_n = \frac{1}{n},$$

so

$$\lim_{n \to \infty} u_n = \lim_{n \to \infty} \frac{1}{n} = 0 \quad \text{and} \quad \frac{1}{n+1} < \frac{1}{n}, \quad \text{for all } n > 0.$$

Therefore both hypotheses of Theorem 14-9 are satisfied and the series converges.

Also from Theorem 14-10,

$$|S - S_n| < \frac{1}{n+1}.$$

For example,

$$|S - S_3| < 0.25;$$
$$|S - S_7| < 0.125;$$

and so forth.

Exercises 14-5

In Exercises 1–10, test the series for convergence by means of the alternating series test.

1. $\dfrac{1}{3} - \dfrac{1}{5} + \dfrac{1}{7} - \dfrac{1}{9} + \cdots$

2. $\dfrac{1}{9} - \dfrac{1}{25} + \dfrac{1}{49} - \dfrac{1}{81} + \cdots$

3. $\dfrac{1}{2^2} - \dfrac{2}{2^3} + \dfrac{3}{2^4} - \dfrac{4}{2^5} + \cdots$

4. $\dfrac{3}{2} - \dfrac{4}{3} + \dfrac{5}{4} - \dfrac{6}{5} + \cdots$

5. $\dfrac{1}{1 \cdot 2} - \dfrac{1}{2 \cdot 3} + \dfrac{1}{3 \cdot 4} - \dfrac{1}{4 \cdot 5} + \cdots$

6. $1 - \dfrac{3}{4} + \dfrac{5}{8} - \dfrac{9}{16} + \dfrac{17}{32} - \cdots$

7. $1 - \dfrac{1}{\sqrt{3}} + \dfrac{1}{\sqrt{5}} - \dfrac{1}{\sqrt{7}} + \cdots$

8. $\dfrac{1}{2 \log 2} - \dfrac{1}{3 \log 3} + \dfrac{1}{4 \log 4} - \dfrac{1}{5 \log 5} + \cdots$

9. $-\log \dfrac{1}{2} + \log \dfrac{1}{3} - \log \dfrac{1}{4} + \log \dfrac{1}{5} - \cdots$ **10.** $\dfrac{1}{2} - \dfrac{2}{5} + \dfrac{3}{10} - \dfrac{4}{17} + \cdots$

In Exercises 11–14, show that the series converge. Then determine the maximum error introduced by taking the fifth partial sum as representing the sum of the series.

11. $1 - \dfrac{1}{2^3} + \dfrac{1}{3^3} - \dfrac{1}{4^3} + \cdots$

12. $1 - \dfrac{1}{2!} + \dfrac{1}{3!} - \dfrac{1}{4!} + \cdots$

13. $\dfrac{1}{1 \cdot 2} - \dfrac{1}{2 \cdot 2^2} + \dfrac{1}{3 \cdot 2^3} - \dfrac{1}{4 \cdot 2^4} + \cdots$ **14.** $1 - \dfrac{1}{4} + \dfrac{1}{9} - \dfrac{1}{16} + \cdots$

In Exercises 15–18, compute the sum of the given series correct to three decimal places.

15. $1 - \dfrac{1}{2^6} + \dfrac{1}{3^6} - \dfrac{1}{4^6} + \cdots$

16. $1 - \dfrac{1}{2!} + \dfrac{1}{4!} - \dfrac{1}{6!} + \cdots$

17. $1 - \dfrac{1}{2 \cdot 10^2} + \dfrac{1}{3 \cdot 10^3} - \dfrac{1}{4 \cdot 10^4} + \cdots$ **18.** $\dfrac{1}{2} - \dfrac{1}{18} + \dfrac{1}{84} - \dfrac{1}{260} + \cdots$

14-9. Absolute Convergence

Let us now consider the series

$$\sum_{n=1}^{\infty} u_n ,$$

where some of the terms are negative but not necessarily in the pattern of the alternating series. Such a series does not satisfy the hypotheses of the convergence tests developed in the preceding pages so we have to obtain new methods for studying them. For that purpose we now prove the following theorem.

Theorem 14-11. If $\sum_{n=1}^{\infty} |u_n|$ converges, then $\sum_{n=1}^{\infty} u_n$ converges.

Since $\sum_{n=1}^{\infty} |u_n|$ is a convergent series of positive terms, we have

$$\lim_{k \to \infty} \sum_{n=1}^{k} |u_n| = S > 0.$$

Now consider the series

$$\sum_{n=1}^{\infty} U_n = \sum_{n=1}^{\infty} (u_n + |u_n|).$$

Note that

$$0 \le U_n \le 2|u_n|$$

since $U_n = 0$ or $2|u_n|$ depending on whether u_n is negative or positive. Then

$$\sum_{n=1}^{k} U_n \le 2 \sum_{n=1}^{k} |u_n| < 2S,$$

and, by Theorem 14-2,

$$\lim_{k \to \infty} \sum_{n=1}^{k} U_k = S' \le 2S.$$

But

$$\sum_{n=1}^{k} u_n = \sum_{n=1}^{k} (U_n - |u_n|) = \sum_{n=1}^{k} U_n - \sum_{n=1}^{k} |u_n|.$$

Combining these results, we obtain

$$\lim_{k \to \infty} \sum_{n=1}^{k} u_n = \lim_{k \to \infty} \sum_{n=1}^{k} U_n - \lim_{k \to \infty} \sum_{n=1}^{k} |u_n|.$$

$$= S' - S.$$

Hence the series $\sum_{n=1}^{\infty} u_n$ converges by definition and the theorem is established.

Thus the convergence of a series of mixed terms (mixed signs) may sometimes be established by studying the series obtained by replacing each term in the original series by its absolute value.

DEFINITION 14-7. If $\sum_{n=1}^{\infty} |u_n|$ converges, the series $\sum_{n=1}^{\infty} u_n$ is said to converge absolutely.

In the light of this definition, Theorem 14-11 may be restated as follows:

THEOREM 14-11(a). If a series converges absolutely, it converges.

DEFINITION 14-8. If a series converges but does not converge absolutely, it is said to converge conditionally.

Example 14-14. Discuss the convergence of the series

$$1 - \frac{1}{2} + \frac{1}{3} - \frac{1}{4} + \cdots + (-1)^{n+1} \frac{1}{n} + \cdots.$$

This is an alternating series and was shown to be convergent in Example 14-13. If, however, we replace each term by its absolute value we get

$$1 + \frac{1}{2} + \frac{1}{3} + \frac{1}{4} + \cdots + \frac{1}{n} + \cdots$$

which is a divergent series (p-series, $p = 1$). Thus the original series *converges conditionally*.

Example 14-15. Discuss the convergence of the series

$$1 - \frac{1}{2\sqrt{2}} - \frac{1}{3\sqrt{3}} + \frac{1}{4\sqrt{4}} - \frac{1}{5\sqrt{5}} - \frac{1}{6\sqrt{6}} + \cdots.$$

We have no means of determining directly whether or not this series converges. It is possible for it to converge since it satisfies the necessary condition, $\lim_{n \to \infty} u_n = 0$, but so may divergent series. The only tool we have to bring to bear on this problem is Theorem 14-11. Replacing each term in the series by its absolute value, we have

$$1 + \frac{1}{2\sqrt{2}} + \frac{1}{3\sqrt{3}} + \frac{1}{4\sqrt{4}} + \frac{1}{5\sqrt{5}} + \frac{1}{6\sqrt{6}} + \cdots,$$

which is the p-series, $p = 3/2$. Hence this series converges, that is, the original series *converges absolutely*, and therefore, converges.

If this series had only converged conditionally, the tests we have developed in this book would have been inadequate to cope with this situation.

With the concept of absolute convergence we are able to state a *new ratio test* for convergence (see Theorem 14-8).

THEOREM 14-12. Given the series of nonzero terms $\sum_{n=1}^{\infty} u_n$, and let

$$\lim_{n \to \infty} \frac{|u_{n+1}|}{|u_n|} = \sigma.$$

If $\sigma < 1$, the series converges absolutely; if $\sigma > 1$, the series diverges; and if $\sigma = 1$, no information regarding convergence is provided.

The proof of this theorem follows from precisely the same arguments as the proof of Theorem 14-8 and will not be repeated here.

Exercises 14-6

Determine whether the following series are conditionally or absolutely convergent.

1. $\frac{1}{2} - \frac{1}{4} + \frac{1}{6} - \frac{1}{8} + \cdots$

2. $1 - \frac{1}{3} + \frac{1}{5} - \frac{1}{7} + \cdots$

3. $\frac{1}{9} - \frac{1}{25} + \frac{1}{49} - \frac{1}{81} + \cdots$

4. $\dfrac{1}{1 \cdot 2} - \dfrac{1}{2 \cdot 3} + \dfrac{1}{3 \cdot 4} - \dfrac{1}{4 \cdot 5} + \cdots$

5. $1 - \dfrac{1}{\sqrt{3}} + \dfrac{1}{\sqrt{5}} - \dfrac{1}{\sqrt{7}} + \cdots$

6. $\dfrac{1}{2!} - \dfrac{2}{3!} + \dfrac{3}{4!} - \dfrac{4}{5!} + \cdots$

7. $\dfrac{1}{\sqrt{3}} - \dfrac{1}{\sqrt{8}} + \dfrac{1}{\sqrt{15}} - \dfrac{1}{\sqrt{24}} + \cdots$

8. $\dfrac{1}{2^2} - \dfrac{2}{2^3} + \dfrac{3}{2^4} - \dfrac{4}{2^5} + \cdots$

9. $3 - \dfrac{3^2}{2!} + \dfrac{3^3}{3!} - \dfrac{3^4}{4!} + \cdots$

10. $\dfrac{2}{1 \cdot 3} - \dfrac{3}{2 \cdot 4} + \dfrac{4}{3 \cdot 5} - \dfrac{5}{4 \cdot 6} + \cdots$

14-10. Power Series

So far we have considered only series of *constant terms*. However, a series

$$\sum_{n=0}^{\infty} u_n(x) = u_0(x) + u_1(x) + u_2(x) + \cdots + u_n(x) + \cdots,$$

where $\{u_n(x)\}$ is any sequence of functions, is much more general. Here we have an infinite series of *variable terms*. The following are examples of such series.

$$1 + \frac{x}{1!} + \frac{x^2}{2!} + \frac{x^3}{3!} + \cdots + \frac{x^n}{n!} + \cdots;$$

$$\frac{\sin x}{1} - \frac{\sin 2x}{2} + \frac{\sin 3x}{3} - \cdots + (-1)^{n+1} \frac{\sin nx}{n} + \cdots;$$

$$e^x + e^{x/2} + e^{x/3} + \cdots + e^{x/n} + \cdots.$$

DEFINITION 14-7. The series of variable terms $\sum_{n=0}^{\infty} u_n(x)$ is said to converge for $x = x_1$ if the series of constant terms $\sum_{n=0}^{\infty} u_n(x_1)$ converges.

Thus the first example above converges for $x = 1$ because the series of constant terms

$$1 + \frac{1}{1!} + \frac{1}{2!} + \frac{1}{3!} + \cdots + \frac{1}{n!} + \cdots$$

can easily be shown to be a convergent series. (Try the ratio test.)

The first example also illustrates a special class of infinite series known as *power series*. Power series have proven to be a very useful mathematical tool so we shall look a little further into series of this type.

DEFINITION 14-8. An infinite series of the form

$$\sum_{n=0}^{\infty} a_n(x - b)^n = a_0 + a_1(x - b) + a_2(x - b)^2 + \cdots + a_n(x - b)^n + \cdots$$

is called a power series.

The first example above, already mentioned twice, is such a series with $a_n = 1/n!$ and $b = 0$.

First we note that a power series always converges for at least one value of x, $x = b$. This follows from the fact that, when $x = b$, all of the terms in the series become zero except the first, which is a_0. Hence all of the partial sums are

$$S_n = a_0, \qquad n = 0, 1, 2, 3, \ldots$$

and

$$S = \lim_{n \to \infty} S_n = a_0.$$

Thus, for $x = b$, the power series converges to the sum a_0.

In order to investigate other values of x for which we may have convergence, we consider the series obtained by taking absolute values of the terms in the power series

$$|a_0| + |a_1| |x - b| + |a_2| |x - b|^2 + \cdots + |a_n| |x - b|^n + \cdots$$

and use the ratio test (Theorem 14-12). We have

$$\lim_{n \to \infty} \left| \frac{u_{n+1}}{u_n} \right| = \lim_{n \to \infty} \frac{|a_{n+1}| |x - b|^{n+1}}{|a_n| |x - b|^n} = |x - b| \lim_{n \to \infty} \frac{|a_{n+1}|}{|a_n|}.$$

Let

$$\lim_{n \to \infty} \frac{|a_{n+1}|}{|a_n|} = \rho.$$

Then, by Theorem 14-12, if $|x - b|\rho < 1$, or equivalently, if

$$|x - b| < \frac{1}{\rho},$$

the power series converges absolutely; and by the same theorem, if

$$|x - b| > \frac{1}{\rho},$$

the power series diverges.

Let us state the foregoing results as a theorem. We have

THEOREM 14-13. If

$$\lim_{n \to \infty} \frac{|a_{n+1}|}{|a_n|} = \rho, \qquad 0 < \rho < \infty,$$

the power series $\sum_{n=0}^{\infty} a_n(x - b)^n$ converges absolutely for

$$|x - b| < \frac{1}{\rho} \dagger$$

and diverges for

$$|x - b| > \frac{1}{\rho}.$$

If $\rho = 0$, it converges absolutely for all x.

Note that nothing is said about $|x - b| = 1/\rho$. The ratio test gives no information in this case. Each particular series must be examined individually for convergence at these two points. Figure 14-1 shows the open intervals on which we know we have convergence and divergence and also the two doubtful points.

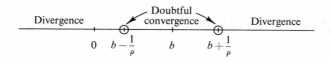

Fig. 14-1

---

† The student should recall that $|x - b| < 1/\rho$ implies $-1/\rho < x - b < 1/\rho$ or the equivalent $b - 1/\rho < x < b + 1/\rho$.

DEFINITION 14-9. The number $1/\rho$ of Theorem 14-13 is called the radius of convergence.

DEFINITION 14-10. The complete interval over which a series converges is called the interval of convergence.

Example 14-16. Determine the radius and complete interval of convergence of the series

$$\sum_{n=0}^{\infty} \frac{x^n}{(n+1)3^n} = 1 + \frac{x}{2\cdot 3} + \frac{x^2}{3\cdot 3^2} + \frac{x^3}{4\cdot 3^3} + \cdots + \frac{x^n}{(n+1)3^n} + \cdots.$$

Applying Theorem 14-13, we have

$$\lim_{n\to\infty} \left| \frac{u_{n+1}}{u_n} \right| = \lim_{n\to\infty} \left| \frac{x^{n+1}/(n+2)3^{n+1}}{x^n/(n+1)3^n} \right| = \lim_{n\to\infty} |x| \left(\frac{n+1}{3(n+2)} \right) = \frac{|x|}{3}.$$

Hence $\rho = \frac{1}{3}$ and the radius of convergence is 3.

Since $b = 0$, the open interval on which we know we have convergence is $-3 < x < 3$ (Fig. 14-2). The only remaining thing necessary is to check the end points of this interval, $x = \pm 3$.

Convergent

−3   0   3

Fig. 14-2

If $x = -3$, the series reduces to

$$1 - \tfrac{1}{2} + \tfrac{1}{3} - \tfrac{1}{4} + \tfrac{1}{5} - \cdots$$

which is easily shown to be convergent by the alternating series test.

On the other hand, if $x = 3$, the series becomes

$$1 + \tfrac{1}{2} + \tfrac{1}{3} + \tfrac{1}{4} + \tfrac{1}{5} + \cdots$$

which is a p-series, $p = 1$,† and therefore diverges.

Example 14-17. Determine the complete interval of convergence of the series

$$\sum_{n=0}^{\infty} (-1)^n \frac{(x-2)^n}{n!} = 1 - \frac{x-2}{1!} + \frac{(x-2)^2}{2!} - \cdots + (-1)^n \frac{(x-2)^n}{n!} + \cdots.$$

We have

$$\lim_{n\to\infty} \left| \frac{u_{n+1}}{u_n} \right| = \lim_{n\to\infty} \left| \frac{(x-2)^{n+1}/(n+1)!}{(x-2)^n/n!} \right| = |x-2| \lim_{n\to\infty} \frac{1}{n+1} = 0.$$

---

† We have also called the p-series, $p = 1$, the harmonic series (Example 14-2).

Thus no restriction on x is necessary in order to make

$$\lim_{n \to \infty} \left| \frac{u_{n+1}}{u_n} \right| < 1,$$

and we conclude that the series converges absolutely for all x.

Exercises 14-7

Determine the interval of convergence of the following power series.

1. $1 - x + \dfrac{x^2}{2} - \dfrac{x^3}{3} + \dfrac{x^4}{4} - \cdots$

2. $1 + \dfrac{x}{1!} + \dfrac{x^2}{2!} + \dfrac{x^3}{3!} + \dfrac{x^4}{4!} + \cdots$

3. $\dfrac{1}{1 \cdot 2} + \dfrac{x}{2 \cdot 3} + \dfrac{x^2}{3 \cdot 4} + \dfrac{x^3}{4 \cdot 5} + \dfrac{x^4}{5 \cdot 6} + \cdots$

4. $x - \dfrac{x^3}{3!} + \dfrac{x^5}{5!} - \dfrac{x^7}{7!} + \dfrac{x^9}{9!} + \cdots$

5. $1 - \dfrac{x^2}{2!} + \dfrac{x^4}{4!} - \dfrac{x^6}{6!} + \dfrac{x^8}{8!} - \cdots$

6. $x + \dfrac{x^2}{3} + \dfrac{x^3}{4} + \dfrac{x^4}{5} + \cdots$

7. $1 - \dfrac{x}{3 \cdot 2} + \dfrac{x^2}{5 \cdot 2^2} - \dfrac{x^3}{7 \cdot 2^3} + \dfrac{x^4}{9 \cdot 2^4} - \cdots$

8. $x + \dfrac{4x^2}{3} + \dfrac{9x^3}{5} + \dfrac{16x^4}{7} + \cdots$

9. $1 - \dfrac{3}{2^3}x + \dfrac{5}{3^3}x^2 - \dfrac{7}{4^3}x^3 + \cdots$

10. $x + \dfrac{x^2}{5} + \dfrac{x^3}{10} + \dfrac{x^4}{17} + \dfrac{x^5}{26} + \cdots$

11. $(x - 1) + 4(x - 1)^2 + 9(x - 1)^3 + 16(x - 1)^4 + \cdots$

12. $\dfrac{(x + 2)}{1 \cdot 2} + \dfrac{(x + 2)^2}{2 \cdot 3} + \dfrac{(x + 2)^3}{3 \cdot 4} + \dfrac{(x + 2)^4}{4 \cdot 5} + \cdots$

13. $1 + (x - 3) + \dfrac{(x - 3)^2}{2!} + \dfrac{(x - 3)^3}{3!} + \dfrac{(x - 4)^4}{4!} + \cdots$

14. $1 + (x - 2) + \dfrac{(x - 2)^2}{2^2} + \dfrac{(x - 2)^3}{3^2} + \dfrac{(x - 2)^4}{4^2} + \cdots$

15. $(x+1) + \dfrac{(x+1)^2}{2} + \dfrac{(x+1)^3}{3} + \dfrac{(x+1)^4}{4} + \cdots$

14-11. Functions Represented by Power Series

Let the open interval of convergence of the power series

$$\sum_{n=0}^{\infty} a_n(x-b)^n = a_0 + a_1(x-b) + a_2(x-b)^2 + \cdots + a_n(x-b)^n + \cdots$$

be $|x-b| < r$. Then for each value of x on this interval the series converges to a unique sum. Hence the series defines a function f on this interval, and we write

$$f(x) = a_0 + a_1(x-b) + a_2(x-b)^2 + \cdots + a_n(x-b)^n + \cdots, \qquad |x-b| < r.$$

The domain of f is the stated interval and its range is the set of sums of the series for these values of x.

The following theorem, stated without proof, forms the basis of operations with functions represented in this manner.

THEOREM 14-14. Let f be defined by

$$f(x) = a_0 + a_1(x-b) + a_2(x-b)^2 + \cdots + a_n(x-b)^n + \cdots$$

and let the open interval of convergence be $|x-b| < r$. Then

(a) f is continuous at each point of $|x-b| < r$;

(b) $f'(x) = \dfrac{d}{dx}(a_0) + \dfrac{d}{dx}[a_1(x-b)] + \dfrac{d}{dx}[a_2(x-b)^2] + \cdots$

$$+ \dfrac{d}{dx}[a_n(x-b)^n] + \cdots$$

$$= a_1 + 2a_2(x-b) + 3a_3(x-b)^2 + \cdots + na_n(x-b)^{n-1} + \cdots$$

at each point of $|x-b| < r$;

(c) $\displaystyle\int_b^x f(t)\,dt = \int_b^x a_0\,dt + \int_b^x a_1(t-b)\,dt + \int_b^x a_2(t-b)^2\,dt$

$$+ \cdots + \int_b^x a_n(t-b)^n\,dt + \cdots$$

$$= a_0(x-b) + \dfrac{a_1}{2}(x-b)^2 + \dfrac{a_2}{3}(x-b)^3$$

$$+ \cdots + \dfrac{a_n}{n+1}(x-b)^{n+1} + \cdots,$$

at each point of $|x-b| < r$.

In essence, this theorem states, in addition to the continuity of f, that the derivative or integral of f may be expressed as a power series, and that this series may be obtained by performing the given operation on the individual terms of the series defining f. However, it is to be emphasized that *these operations are valid only for x in the interval of convergence.*

To provide us with an example, we recall that, for $|r| < 1$, the sum of the infinite geometric series (Sec. 14-4)

$$\sum_{n=0}^{\infty} ar^n = a + ar + ar^2 + \cdots + ar^n + \cdots$$

is

$$S = \frac{a}{1 - r}.$$

Hence we may write

$$\frac{1}{1 - x} = 1 + x + x^2 + \cdots + x^n + \cdots, \qquad |x| < 1.$$

Then, from Theorem 14-14,

$$\frac{d}{dx}\left(\frac{1}{1 - x}\right) = \frac{1}{(1 - x)^2} = 1 + 2x + 3x^2 + \cdots + nx^{n-1} + \cdots, \qquad |x| < 1,$$

and

$$\int_0^x \frac{dt}{1 - t} = -\log|1 - x| = x + \frac{x^2}{2} + \frac{x^3}{3} + \cdots + \frac{x^n}{n} + \cdots, \qquad |x| < 1.$$

Thus, we may use this last series to compute $\log|1 - x|$ for $|x| < 1$. Let $x = -\frac{1}{4}$ for example. We have

$$-\log\left|1 + \tfrac{1}{4}\right| = -\tfrac{1}{4} + \tfrac{1}{32} - \tfrac{1}{192} + \tfrac{1}{1025} - \cdots.$$

If we add the four terms shown, we get

$$\log 1.25 = 0.223,$$

rounding to three decimal places. This result may be verified in a table of natural logarithms. A higher degree of accuracy may be obtained by adding in more terms in the series.

It should be noted, however, that this series may not be used to compute any logarithms of numbers greater than 2 since this would require $|x| > 1$.

In the foregoing we have defined a function by means of a power series. Now let us look at the other side of the coin. Let us consider the problem of obtaining a power series which represents a given function. That is, let us start with a defined function and attempt to find a power series which also defines it, at least on some interval.

Suppose a function f can be defined by a power series, that is,

$$f(x) = a_0 + a_1(x - b) + a_2(x - b)^2 + \cdots + a_n(x - b)^n + \cdots$$

for which the interval of convergence is $|x - b| < r$ $(b - r < x < b + r)$. Then, by (b) of Theorem 14-14,

$$f'(x) = a_1 + 2a_2(x - b) + 3a_3(x - b)^2 + \cdots + na_n(x - b)^{n-1} + \cdots,$$

$$f''(x) = 2a_2 + 3 \cdot 2a_3(x - b) + \cdots + n(n - 1)a_n(x - b)^{n-2} + \cdots,$$

$$f'''(x) = 3 \cdot 2a_3 + 4 \cdot 3 \cdot 2a_4(x - b) + \cdots + n(n - 1)(n - 2)(x - b)^{n-3} + \cdots,$$

. . . .

$$f^{(n)}(x) = n!\, a_n + (n + 1)!\, a_{n+1}(x - b) + \frac{(n + 2)!}{2!}(x - b)^2 + \cdots,$$

. . . . ,

all of which have the same interval of convergence $|x - b| < r$. If we set $x = b$ in each of these series, we obtain

$$f(b) = a_0, \qquad f'(b) = a_1, \qquad f''(b) = 2!\, a_2, \qquad f'''(b) = 3!\, a_3,$$

$$\ldots, f^{(n)}(b) = n!\, a_n, \ldots,$$

or, solving for the a's,

$$a_0 = f(b), \qquad a_1 = \frac{f'(b)}{1!}, \qquad a_2 = \frac{f''(b)}{2!}, \qquad a_3 = \frac{f'''(b)}{3!},$$

$$\ldots, a_n = \frac{f^{(n)}(b)}{n!}, \ldots.$$

Thus we conclude that if a function is represented by a series in powers of $x - b$, the series must be of the form

$$f(x) = f(b) + \frac{f'(b)}{1!}(x - b) + \frac{f''(b)}{2!}(x - b)^2 + \cdots + \frac{f^{(n)}(b)}{n!}(x - b)^n + \cdots.$$

This is called *Taylor's series in powers of* $(x - b)$. We may now state the following theorem.

THEOREM 14-15. If f is represented by a power series in powers of $(x - b)$, this series is Taylor's series.

This is to say that however a power series representation for f is obtained, the result is unique—always Taylor's series.

Example 14-18. Assume $f(x) = 1/x$ has a power series representation in powers of $(x - 1)$. Find it by two different methods.

First we use Taylor's series. For this purpose we compute successive derivatives of $1/x$. We have

$$f(x) = \frac{1}{x}, \qquad f'(x) = -\frac{1}{x^2}, \qquad f''(x) = \frac{2}{x^3}, \qquad f'''(x) = -\frac{2 \cdot 3}{x^4},$$

$$\dots, f^{(n)}(x) = (-1)^n \frac{n!}{x^{n+1}}, \dots,$$

or, setting $x = 1$,

$$f(1) = 1, \qquad f'(1) = -1, \qquad f''(1) = 2!, \qquad f'''(1) = -3!,$$

$$\dots, f^{(n)}(1) = (-1)^n n!, \dots .$$

When we substitute these in the general form of Taylor's series we get

$$f(x) = \frac{1}{x} = 1 - (x-1) + (x-1)^2 - (x-1)^3 + \cdots + (-1)^n (x-1)^n + \cdots.$$

As a second method, we write

$$\frac{1}{x} = \frac{1}{1 + (x-1)} = \frac{1}{1+t},$$

setting $t = x - 1$. Now by continued algebraic division we obtain

$$\frac{1}{1+t} = 1 - t + t^2 - t^3 + \cdots + (-1)^n t^n + \cdots$$

or, replacing t,

$$\frac{1}{x} = 1 - (x-1) + (x-1)^2 - (x-1)^3 + \cdots + (-1)^n (x-1)^n + \cdots,$$

which was to be expected, since the power series representation of a function is unique.

The radius of convergence of this series is readily found to be 1 so it cannot represent $1/x$ for $|x-1| > 1$.

Note that the preceding example *assumed* that $1/x$ had a representation in powers of $(x-1)$. The second method indicates somewhat the means used to answer the question as to whether or not a function may be represented by a power series.

If we stop dividing in this example after n terms in the quotient have been obtained, we have a remainder $(-1)^{n+1} t^{n+1}$ and we may write

$$\frac{1}{1+t} = 1 - t + t^2 - \cdots + (-1)^n t^n + \frac{(-1)^{n+1} t^{n+1}}{1+t}.$$

Consider the *remainder term*

$$R_n(t) = (-1)^{n+1} \frac{t^{n+1}}{1+t},$$

or

$$R_n(x-1) = (-1)^{n+1} \frac{(x-1)^{n+1}}{x}.$$

If $|x - 1| < 1$ (the interval of convergence),

$$\lim_{n \to \infty} R_n(x - 1) = 0.$$

This is the criterion for deciding the "representability" of a function by a power series. *If the remainder after n terms approaches zero as n approaches* ∞, *the answer is affirmative, otherwise, no.*

In more complete books on calculus various methods are developed for obtaining the remainder after n terms of a power series and in a number of forms. These are used for a variety of purposes including this question of "representability." However, in this abbreviated treatment, we shall not enter into a discussion of these matters. We shall simply assume that the functions with which we have to deal have, for a suitably selected b, a power series representation on some interval. Happily this assumption is basically true!

What is the criterion for a "suitably selected b"? As an example of an unsuitable one consider the function f where $f(x) = x^{7/3}$ and let $b = 0$. We have

$$f(x) = x^{7/3}, \qquad f'(x) = \tfrac{7}{3}x^{4/3}, \qquad f''(x) = (\tfrac{7}{3})(\tfrac{4}{3})\mathbf{x}^{1/3},$$

from which

$$f(0) = f'(0) = f''(0) = 0,$$

but each derivative beyond order two is undefined for $x = 0$. Hence there can be no Taylor's series for this function in powers of x. Obviously a "suitably selected b" will be such that $f^{(n)}(b)$ is defined for all n.

If we take $b = 0$, the Taylor's series reduces to

$$f(x) = f(0) + \frac{f'(0)}{1!} x + \frac{f''(0)}{2!} x^2 + \cdots + \frac{f^{(n)}(0)}{n!} x^n + \cdots$$

which is usually called *Maclaurin's series* or a *Maclaurin series*.

Example 14-19. Obtain a Maclaurin series for $f(x) = \sin x$.

Computing successive derivatives, we have

$$f(x) = \sin x, \qquad f'(x) = \cos x, \qquad f''(x) = - \sin x,$$

$$f'''(x) = - \cos x, \qquad \text{etc.}$$

Since $\sin 0 = 0$ and $\cos 0 = 1$,

$$f(0) = f''(0) = f''''(0) = \cdots = f^{(2k)}(0) = 0, \cdots$$

and

$$f'(0) = 1, \qquad f'''(0) = - 1, \cdots, f^{(2n+1)}(0) = (- 1)^n, \cdots.$$

Hence the required Maclaurin series is

$$\sin x = x - \frac{x^3}{3!} + \frac{x^5}{5!} - \cdots + (-1)^n \frac{x^{2n+1}}{(2n+1)!} + \cdots.$$

This is an example of how one proceeds directly to obtain a power series representation of a given function. When successive derivatives are easy to compute and evaluate, it is a good method. However, this is not always the case. Frequently we find it profitable to take a more devious route, relying on the uniqueness of the power series representation of a function to assure us that our result is valid. Consider the following:

Example 14-20. Obtain a Maclaurin series representation of Arctan x.

If we attempt to do this by formula, we soon become discouraged by the complexity of the successive derivatives, although the result could be reached in this manner. A simpler method can be developed by recalling that (Example 14-18)

$$\frac{1}{1+x} = 1 - x + x^2 - x^3 + \cdots + (-1)^n x^n + \cdots$$

for $|x| < 1$. Then, replacing x by x^2, we may write

$$\frac{1}{1+x^2} = 1 - x^2 + x^4 - x^6 + \cdots + (-1)^n x^{2n} + \cdots,$$

valid for $x^2 < 1$. Hence, appealing to Theorem 14-14(c),

$$\int_0^x \frac{dt}{1+t^2} = \text{Arctan } x = \int_0^x dt - \int_0^x t^2\, dt + \int_0^x t^4\, dt - \cdots$$

$$= x - \frac{x^3}{3} + \frac{x^5}{5} - \cdots + (-1)^n \frac{x^{2n+1}}{2n+1} + \cdots,$$

the representation being valid for $x^2 < 1$.

Devices similar to this, applied to known series, will frequently prove to be more effective than the direct approach.

Exercises 14-8

Use the formula to obtain directly the Maclaurin series for the functions in Exercises 1–6 and in each case state the interval in which the representation is valid.

1. e^x **2.** $\cos x$

3. $\log(1 + x)$ **4.** $\log(1 - x)$

5. $(1 + x)^{-1}$ **6.** $(1 - x)^{-1}$

Use the formula to obtain directly Taylor's series for the functions in Exercises 7–12 for the given b, stating in each case the interval of validity.

7. $\sin x$, $b = \pi/4$ 8. $\cos x$, $b = \pi/2$

9. e^x, $b = 1$ 10. $\log x$, $b = 1$

11. $(1 - x)^{-1}$, $b = 2$ 12. x^{-1}, $b = -1$

13. Use the result in Exercise 1 to solve Exercise 9.

14. Use the result in Exercise 5 to compute $\log 0.75$, correct to two decimal places.

15. Use the result in Exercise 11 to compute $\log 2.25$, correct to two decimal places.

16. Use the result in Example 14-19 to compute $\sin \pi/4$, correct to two decimal places.

17. Use the result in Example 14-19 to solve Exercise 2.

18. Use the results of Exercises 3 and 4 to write the Maclaurin series for $\log [(1 + x)/(1 - x)]$.

19. Obtain the Maclaurin series for $(1 - x)^{-1/2}$.

20. Use the result of Exercise 19 to obtain the Maclaurin series for $(1 - x^2)^{-1/2}$ and then use this series coupled with Theorem 14-14 to get the Maclaurin series for Arcsin x. In what interval is it valid?

Appendix A

TABLE OF INTEGRALS

General Forms

1. $\displaystyle\int du = u + c$

2. $\displaystyle\int a\,du = a\int du$

3. $\displaystyle\int (du + dv) = \int du + \int dv$

4. $\displaystyle\int u\,dv = uv - \int v\,du$

Fundamental Forms

5. $\displaystyle\int u^n\,du = \frac{u^{n+1}}{n+1} + c, \quad n \neq -1$

6. $\displaystyle\int \frac{du}{u} = \log|u| + c$

7. $\displaystyle\int e^u\,du = e^u + c$

8. $\displaystyle\int a^u\,du = \frac{a^u}{\log a} + c$

9. $\displaystyle\int \sin u\,du = -\cos u + c$

10. $\displaystyle\int \cos u\,du = \sin u + c$

11. $\displaystyle\int \sec^2 u\,du = \tan u + c$

12. $\displaystyle\int \csc^2 u\,du = -\cot u + c$

13. $\displaystyle\int \sec u \tan u\,du = \sec u + c$

14. $\displaystyle\int \csc u \cot u\,du = -\csc u + c$

15. $\displaystyle\int \frac{du}{\sqrt{a^2 - u^2}} = \text{Arcsin}\,\frac{u}{a} + c, \quad |u| < |a|$

16. $\displaystyle\int \frac{du}{a^2 + u^2} = \frac{1}{a}\,\text{Arctan}\,\frac{u}{a} + c$

17. $\displaystyle\int \frac{du}{u\sqrt{u^2 - a^2}} = \frac{1}{a}\,\text{Arcsec}\,\frac{u}{a} + c, \quad |u| > |a|$

Forms Containing $a + bu$

18. $\displaystyle\int (a + bu)^n \, du = \frac{(a + bu)^{n+1}}{b(n + 1)} + c, \qquad n \ne -1$

19. $\displaystyle\int \frac{du}{a + bu} = \frac{1}{b} \log |a + bu| + c$

20. $\displaystyle\int \frac{u \, du}{a + bu} = \frac{1}{b^2} [a + bu - a \log |a + bu|] + c$

21. $\displaystyle\int \frac{u^2 \, du}{a + bu} = \frac{1}{b^3} [\tfrac{1}{2}(a + bu)^2 - 2a(a + bu) + a^2 \log |a + bu|] + c$

22. $\displaystyle\int \frac{u \, du}{(a + bu)^2} = \frac{1}{b^2} \left[\frac{a}{a + bu} + \log |a + bu| \right] + c$

23. $\displaystyle\int \frac{u^2 \, du}{(a + bu)^2} = \frac{1}{b^3} \left[a + bu - \frac{a^2}{a + bu} - 2a \log |a + bu| \right] + c$

24. $\displaystyle\int \frac{du}{u(a + bu)} = -\frac{1}{a} \log \left| \frac{a + bu}{u} \right| + c$

25. $\displaystyle\int \frac{du}{u^2(a + bu)} = -\frac{1}{au} + \frac{b}{a^2} \log \left| \frac{a + bu}{u} \right| + c$

26. $\displaystyle\int \frac{du}{u(a + bu)^2} = \frac{1}{a(a + bu)} - \frac{1}{a^2} \log \left| \frac{a + bu}{u} \right| + c$

27. $\displaystyle\int u\sqrt{a + bu} \, du = -\frac{2(2a - 3bu)(a + bu)^{3/2}}{15b^2} + c$

28. $\displaystyle\int u^2\sqrt{a + bu} \, du = \frac{2(8a^2 - 12abu + 15b^2u^2)(a + bu)^{3/2}}{105b^3} + c$

29. $\displaystyle\int \frac{u \, du}{\sqrt{a + bu}} = -\frac{2(2a - bu)\sqrt{a + bu}}{3b^2} + c$

30. $\displaystyle\int \frac{u^2 \, du}{\sqrt{a + bu}} = \frac{2(8a^2 - 4abu + 3b^2u^2)\sqrt{a + bu}}{15b^3} + c$

31. $\displaystyle\int \frac{du}{u\sqrt{a + bu}} = \frac{1}{\sqrt{a}} \log \left| \frac{\sqrt{a + bu} - \sqrt{a}}{\sqrt{a + bu} + \sqrt{a}} \right| + c, \qquad a > 0$

32. $\displaystyle\int \frac{du}{u\sqrt{a + bu}} = \frac{2}{\sqrt{-a}} \operatorname{Arctan} \sqrt{\frac{a + bu}{-a}} + c, \qquad a < 0$

33. $\displaystyle\int \frac{du}{u^2\sqrt{a + bu}} = -\frac{\sqrt{a + bu}}{au} - \frac{b}{2a} \int \frac{du}{u\sqrt{a + bu}}$

34. $\displaystyle\int \frac{\sqrt{a+bu}}{u}\,du = 2\sqrt{a+bu} + a\int \frac{du}{u\sqrt{a+bu}}$

35. $\displaystyle\int \frac{\sqrt{a+bu}}{u^2}\,du = -\frac{(a+bu)^{3/2}}{au} + \frac{b}{2a}\int \frac{\sqrt{a+bu}}{u}\,du$

Forms Containing $\sqrt{a^2 - u^2}$

36. $\displaystyle\int \sqrt{a^2 - u^2}\,du = \frac{u}{2}\sqrt{a^2 - u^2} + \frac{a^2}{2}\text{ Arcsin }\frac{u}{a} + c$

37. $\displaystyle\int (a^2 - u^2)^{n/2}\,du = \frac{u(a^2 - u^2)^{n/2}}{n+1}$

$\displaystyle\qquad\qquad + \frac{a^2 n}{n+1}\int (a^2 - u^2)^{n/2 - 1}\,du, \qquad n \neq -1$

38. $\displaystyle\int u^2\sqrt{a^2 - u^2}\,du = -\frac{u}{4}(a^2 - u^2)^{3/2} + \frac{a^2}{8}u\sqrt{a^2 - u^2} + \frac{a^4}{8}\text{ Arcsin }\frac{u}{a} + c$

39. $\displaystyle\int u^m(a^2 - u^2)^{n/2}\,du = -\frac{u^{m-1}(a^2 - u^2)^{n/2 + 1}}{n + m + 1}$

$\displaystyle\qquad\qquad + \frac{a^2(m-1)}{n + m + 1}\int u^{m-2}(a^2 - u^2)^{n/2}\,du$

40. $\displaystyle\int \frac{du}{(a^2 - u^2)^{3/2}} = \frac{u}{a^2\sqrt{a^2 - u^2}} + c$

41. $\displaystyle\int \frac{u^2\,du}{\sqrt{a^2 - u^2}} = -\frac{u}{2}\sqrt{a^2 - u^2} + \frac{a^2}{2}\text{ Arcsin }\frac{u}{a} + c$

42. $\displaystyle\int \frac{u^m\,du}{\sqrt{a^2 - u^2}} = -\frac{u^{m-1}\sqrt{a^2 - u^2}}{m} + \frac{(m-1)a^2}{m}\int \frac{u^{m-2}\,du}{\sqrt{a^2 - u^2}}$

43. $\displaystyle\int \frac{u^2\,du}{(a^2 - u^2)^{3/2}} = \frac{u}{\sqrt{a^2 - u^2}} - \text{Arcsin }\frac{u}{a} + c$

44. $\displaystyle\int \frac{u^m\,du}{(a^2 - u^2)^{n/2}} = -\frac{u^{m-1}}{(m - n + 1)(a^2 - u^2)^{n/2 - 1}}$

$\displaystyle\qquad\qquad + \frac{a^2(m-1)}{m - n + 1}\int \frac{u^{m-2}\,du}{(a^2 - u^2)^{n/2}}$

45. $\displaystyle\int \frac{du}{u\sqrt{a^2 - u^2}} = -\frac{1}{a}\log\left|\frac{a + \sqrt{a^2 - u^2}}{u}\right| + c$

46. $\displaystyle\int \frac{du}{u^2\sqrt{a^2 - u^2}} = -\frac{\sqrt{a^2 - u^2}}{a^2 u} + c$

47. $\displaystyle\int \frac{du}{u^m(a^2 - u^2)^{n/2}} = -\frac{1}{a^2(m-1)u^{m-1}(a^2 - u^2)^{n/2 - 1}}$

$$+ \frac{m+n-3}{a^2(m-1)} \int \frac{du}{u^{m-2}(a^2 - u^2)^{n/2}}, \qquad m \neq 1$$

Forms Containing $\sqrt{u^2 \pm a^2}$

48. $\displaystyle\int \sqrt{u^2 \pm a^2}\, du = \frac{u}{2}\sqrt{u^2 \pm a^2} \pm \frac{a^2}{2}\log|u + \sqrt{u^2 \pm a^2}| + c$

49. $\displaystyle\int u^2\sqrt{u^2 \pm a^2}\, du = \frac{u}{4}(u^2 \pm a^2)^{3/2} \mp \frac{a^2}{8}u\sqrt{u^2 \pm a^2}$

$$- \frac{a^4}{8}\log|u + \sqrt{u^2 \pm a^2}| + c$$

50. $\displaystyle\int u^m(u^2 \pm a^2)^{n/2}\, du = \frac{u^{m-1}(u^2 \pm a^2)^{n/2 + 1}}{n + m + 1}$

$$\mp \frac{a^2(m-1)}{n + m + 1}\int u^{m-2}(u^2 \pm a^2)^{n/2}\, du$$

51. $\displaystyle\int \frac{du}{\sqrt{u^2 \pm a^2}} = \log|u + \sqrt{u^2 \pm a^2}| + c$

52. $\displaystyle\int \frac{u^2\, du}{\sqrt{u^2 \pm a^2}} = \frac{u}{2}\sqrt{u^2 \pm a^2} \mp \frac{a^2}{2}\log|u + \sqrt{u^2 \pm a^2}| + c$

53. $\displaystyle\int \frac{u^m\, du}{\sqrt{u^2 \pm a^2}} = \frac{u^{m-1}\sqrt{u^2 \pm a^2}}{m} \mp \frac{(m-1)a^2}{m}\int \frac{u^{m-2}\, du}{\sqrt{u^2 \pm a^2}}$

54. $\displaystyle\int \frac{du}{(u^2 \pm a^2)^{3/2}} = \pm \frac{u}{a^2\sqrt{u^2 \pm a^2}} + c$

55. $\displaystyle\int \frac{u^2\, du}{(u^2 \pm a^2)^{3/2}} = -\frac{u}{\sqrt{u^2 \pm a^2}} + \log|u + \sqrt{u^2 \pm a^2}| + c$

56. $\displaystyle\int \frac{u^m\, du}{(u^2 \pm a^2)^{n/2}} = \frac{u^{m-1}}{(m - n + 1)(u^2 \pm a^2)^{n/2 - 1}}$

$$\mp \frac{a^2(m-1)}{m - n + 1}\int \frac{u^{m-2}\, du}{(u^2 \pm a^2)^{n/2}}$$

57. $\displaystyle\int \frac{du}{u\sqrt{u^2 + a^2}} = -\frac{1}{a}\log\left|\frac{a + \sqrt{u^2 + a^2}}{u}\right| + c$

58. $\displaystyle\int \frac{du}{u^2\sqrt{u^2 \pm a^2}} = \mp \frac{\sqrt{u^2 \pm a^2}}{a^2 u} + c$

59. $\displaystyle\int \frac{du}{u^m\sqrt{u^2 \pm a^2}} = \mp \frac{\sqrt{u^2 \pm a^2}}{(m-1)a^2 u^{m-1}}$

$$\mp \frac{m-2}{a^2(m-1)}\int \frac{du}{u^{m-2}\sqrt{u^2 \pm a^2}}, \qquad m \neq 1$$

60. $\displaystyle\int \frac{du}{u^m(u^2 \pm a^2)^{n/2}} = \mp \frac{1}{a^2(m-1)u^{m-1}(u^2 \pm a^2)^{n/2-1}}$

$$\mp \frac{m+n-3}{a^2(m-1)}\int \frac{du}{u^{m-2}(u^2 \pm a^2)^{n/2}}, \qquad m \neq 1$$

Trigonometric Forms

61. $\displaystyle\int \tan u\, du = -\log|\cos u| + c = \log|\sec u| + c$

62. $\displaystyle\int \cot u\, du = \log|\sin u| + c = -\log|\csc u| + c$

63. $\displaystyle\int \sec u\, du = \log|\sec u + \tan u| + c$

64. $\displaystyle\int \csc u\, du = -\log|\csc u + \cot u| + c$

65. $\displaystyle\int \sin^2 u\, du = \frac{u}{2} - \frac{1}{4}\sin 2u + c$

66. $\displaystyle\int \cos^2 u\, du = \frac{u}{2} + \frac{1}{4}\sin 2u + c$

67. $\displaystyle\int \sin^3 u\, du = -\cos u + \frac{1}{3}\cos^3 u + c$

68. $\displaystyle\int \cos^3 u\, du = \sin u - \frac{1}{3}\sin^3 u + c$

69. $\displaystyle\int \sin^n u\, du = -\frac{1}{n}\sin^{n-1} u \cos u + \frac{n-1}{n}\int \sin^{n-2} u\, du$

70. $\displaystyle\int \cos^n u\; du = \frac{1}{n}\cos^{n-1} u \sin u + \frac{n-1}{n}\int \cos^{n-2} u\; du$

71. $\displaystyle\int \tan^n u\; du = \frac{\tan^{n-1} u}{n-1} - \int \tan^{n-2} u\; du, \qquad n \neq 1$

72. $\displaystyle\int \cot^n u\; du = -\frac{\cot^{n-1} u}{n-1} - \int \cot^{n-2} u\; du, \qquad n \neq 1$

73. $\displaystyle\int \sec^n u\; du = \frac{1}{n-1}\sec^{n-2} u \tan u + \frac{n-2}{n-1}\int \sec^{n-2} u\; du, \qquad n \neq 1$

74. $\displaystyle\int \csc^n u\; du = -\frac{1}{n-1}\csc^{n-2} u \cot u + \frac{n-2}{n-1}\int \csc^{n-2} u\; du, \qquad n \neq 1$

75. $\displaystyle\int \sin mu \sin nu\; du = -\frac{\sin(m+n)u}{2(m+n)} + \frac{\sin(m-n)u}{2(m-n)} + c, \qquad |m| \neq |n|$

76. $\displaystyle\int \sin mu \cos nu\; du = -\frac{\cos(m+n)u}{2(m+n)} - \frac{\cos(m-n)u}{2(m-n)} + c, \qquad |m| \neq |n|$

77. $\displaystyle\int \cos mu \cos nu\; du = \frac{\sin(m+n)u}{2(m+n)} + \frac{\sin(m-n)u}{2(m-n)} + c, \qquad |m| \neq |n|$

78. $\displaystyle\int \sin^m u \cos^n u\; du = \frac{\sin^{m+1} u \cos^{n-1} u}{m+n}$

$$+ \frac{n-1}{m+n}\int \sin^m u \cos^{n-2} u\; du, \qquad m \neq -n$$

79. $\displaystyle\int \sin^m u \cos^n u\; du = -\frac{\sin^{m-1} u \cos^{n+1} u}{m+n}$

$$+ \frac{m-1}{m+n}\int \sin^{m-2} u \cos^n u\; du, \qquad m \neq -n$$

80. $\displaystyle\int u \sin u\; du = \sin u - u \cos u + c$

81. $\displaystyle\int u \cos u\; du = \cos u + u \sin u + c$

82. $\displaystyle\int u^n \sin u\; du = -u^n \cos u + n \int u^{n-1} \cos u\; du$

83. $\displaystyle\int u^n \cos u\; du = u^n \sin u - n \int u^{n-1} \sin u\; du$

Miscellaneous Forms

84. $\displaystyle\int ue^u \, du = e^u(u - 1) + c$

85. $\displaystyle\int u^n e^u \, du = u^n e^u - n \int u^{n-1} e^u \, du$

86. $\displaystyle\int \log u \, du = u(\log u - 1) + c$

87. $\displaystyle\int u^n \log u \, du = \frac{u^{n+1}}{(n+1)^2} [(n+1)\log u - 1] + c, \qquad n \neq -1$

88. $\displaystyle\int e^{au} \sin bu \, du = \frac{e^{au}(a \sin bu - b \cos bu)}{a^2 + b^2} + c$

89. $\displaystyle\int e^{au} \cos bu \, du = \frac{e^{au}(a \cos bu + b \sin bu)}{a^2 + b^2} + c$

90. $\displaystyle\int \text{Arcsin } u \, du = u \text{ Arcsin } u + \sqrt{1 - u^2} + c$

91. $\displaystyle\int \text{Arctan } u \, du = u \text{ Arctan } u - \frac{1}{2} \log(1 + u^2) + c$

SOME FUNCTION VALUES

Values of e^x and e^{-x}

x	e^x	e^{-x}	x	e^x	e^{-x}	x	e^x	e^{-x}
0.00	1.000	1.000	0.1	1.105	0.905	1	2.72	0.368
0.01	1.010	0.990	0.2	1.221	0.819	2	7.39	0.135
0.02	1.020	0.980	0.3	1.350	0.741	3	20.09	0.0498
0.03	1.030	0.970	0.4	1.492	0.670	4	54.60	0.0183
0.04	1.041	0.961	0.5	1.649	0.607	5	148.4	0.00674
0.05	1.051	0.951	0.6	1.822	0.549	6	403.4	0.00248
0.06	1.062	0.942	0.7	2.014	0.497	7	1097.	0.000912
0.07	1.073	0.932	0.8	2.226	0.449	8	2981.	0.000335
0.08	1.083	0.923	0.9	2.460	0.407	9	8103.	0.000123
0.09	1.094	0.914	1.0	2.718	0.368	10	22026.	0.000045

Trigonometric Functions

Degrees	Radians	Sin	Tan	Cot	Cos		
0°	0.000	0.000	0.000		1.000	1.571	90°
1°	0.017	0.017	0.017	57.29	1.000	1.553	89°
2°	0.035	0.035	0.035	28.64	0.999	1.536	88°
3°	0.052	0.052	0.052	19.081	0.999	1.518	87°
4°	0.070	0.070	0.070	14.301	0.998	1.501	86°
5°	0.087	0.087	0.087	11.430	0.996	1.484	85°
6°	0.105	0.105	0.105	9.514	0.995	1.466	84°
7°	0.122	0.122	0.123	8.144	0.993	1.449	83°
8°	0.140	0.139	0.141	7.115	0.990	1.431	82°
9°	0.157	0.156	0.158	6.314	0.988	1.414	81°
10°	0.175	0.174	0.176	5.671	0.985	1.396	80°
11°	0.192	0.191	0.194	5.145	0.982	1.379	79°
12°	0.209	0.208	0.213	4.705	0.978	1.361	78°
13°	0.227	0.225	0.231	4.331	0.974	1.344	77°
14°	0.244	0.242	0.249	4.011	0.970	1.326	76°
15°	0.262	0.259	0.268	3.732	0.966	1.309	75°
16°	0.279	0.276	0.287	3.487	0.961	1.292	74°
17°	0.297	0.292	0.306	3.271	0.956	1.274	73°
18°	0.314	0.309	0.325	3.078	0.951	1.257	72°
19°	0.332	0.326	0.344	2.904	0.946	1.239	71°
20°	0.349	0.342	0.364	2.747	0.940	1.222	70°
21°	0.367	0.358	0.384	2.605	0.934	1.204	69°
22°	0.384	0.375	0.404	2.475	0.927	1.187	68°
23°	0.401	0.391	0.424	2.356	0.921	1.169	67°
24°	0.419	0.407	0.445	2.246	0.914	1.152	66°
25°	0.436	0.423	0.466	2.144	0.906	1.134	65°
26°	0.454	0.438	0.488	2.050	0.899	1.117	64°
27°	0.471	0.454	0.510	1.963	0.891	1.100	63°
28°	0.489	0.469	0.532	1.881	0.883	1.082	62°
29°	0.506	0.485	0.554	1.804	0.875	1.065	61°
30°	0.524	0.500	0.577	1.732	0.866	1.047	60°
31°	0.541	0.515	0.601	1.664	0.857	1.030	59°
32°	0.559	0.530	0.625	1.600	0.848	1.012	58°
33°	0.576	0.545	0.649	1.540	0.839	0.995	57°
34°	0.593	0.559	0.675	1.483	0.829	0.977	56°
35°	0.611	0.574	0.700	1.428	0.819	0.960	55°
36°	0.628	0.588	0.727	1.376	0.809	0.942	54°
37°	0.646	0.602	0.754	1.327	0.799	0.925	53°
38°	0.663	0.616	0.781	1.280	0.788	0.908	52°
39°	0.681	0.629	0.810	1.235	0.777	0.890	51°
40°	0.698	0.643	0.839	1.192	0.766	0.873	50°
41°	0.716	0.656	0.869	1.150	0.755	0.855	49°
42°	0.733	0.669	0.900	1.111	0.743	0.838	48°
43°	0.750	0.682	0.933	1.072	0.731	0.820	47°
44°	0.768	0.695	0.966	1.036	0.719	0.803	46°
45°	0.785	0.707	1.000	1.000	0.707	0.785	45°
		Cos	.Cot	Tan	Sin	Radians	Degrees

LOGARITHM TABLES

Logarithms to the Base e

N	.0	.1	.2	.3	.4	.5	.6	.7	.8	.9
0	—	−2.303	−1.609	−1.204	−0.916	−0.693	−0.511	−0.357	−0.223	−0.105
1	0.000	0.095	0.182	0.262	0.336	0.405	0.470	0.531	0.588	0.642
2	0.693	0.742	0.788	0.833	0.875	0.916	0.956	0.993	1.030	1.065
3	1.099	1.131	1.163	1.194	1.224	1.253	1.281	1.308	1.335	1.361
4	1.386	1.411	1.435	1.459	1.482	1.504	1.526	1.548	1.569	1.589
5	1.609	1.629	1.649	1.668	1.686	1.705	1.723	1.740	1.758	1.775
6	1.792	1.808	1.825	1.841	1.856	1.872	1.887	1.902	1.917	1.932
7	1.946	1.960	1.974	1.988	2.001	2.015	2.028	2.041	2.054	2.067
8	2.079	2.092	2.104	2.116	2.128	2.140	2.152	2.163	2.175	2.186
9	2.197	2.208	2.219	2.230	2.241	2.251	2.262	2.272	2.282	2.293
10	2.303	2.313	2.322	2.332	2.342	2.351	2.361	2.370	2.380	2.389

Logarithms to the Base 10

N	0	1	2	3	4	5	6	7	8	9
10	0000	0043	0086	0128	0170	0212	0253	0294	0334	0374
11	0414	0453	0492	0531	0569	0607	0645	0682	0719	0755
12	0792	0828	0864	0899	0934	0969	1004	1038	1072	1106
13	1139	1173	1206	1239	1271	1303	1335	1367	1399	1430
14	1461	1492	1523	1553	1584	1614	1644	1673	1703	1732
15	1761	1790	1818	1847	1875	1903	1931	1959	1987	2014
16	2041	2068	2095	2122	2148	2175	2201	2227	2253	2279
17	2304	2330	2355	2380	2405	2430	2455	2480	2504	2529
18	2553	2577	2601	2625	2648	2672	2695	2718	2742	2765
19	2788	2810	2833	2856	2878	2900	2923	2945	2967	2989
20	3010	3032	3054	3075	3096	3118	3139	3160	3181	3201
21	3222	3243	3263	3284	3304	3324	3345	3365	3385	3404
22	3424	3444	3464	3483	3502	3522	3541	3560	3579	3598
23	3617	3636	3655	3674	3692	3711	3729	3747	3766	3784
24	3802	3820	3838	3856	3874	3892	3909	3927	3945	3962
25	3979	3997	4014	4031	4048	4065	4082	4099	4116	4133
26	4150	4166	4183	4200	4216	4232	4249	4265	4281	4298
27	4314	4330	4346	4362	4378	4393	4409	4425	4440	4456
28	4472	4487	4502	4518	4533	4548	4564	4579	4594	4609
29	4624	4639	4654	4669	4683	4698	4713	4728	4742	4757
30	4771	4786	4800	4814	4829	4843	4857	4871	4886	4900
31	4914	4928	4942	4955	4969	4983	4997	5011	5024	5038
32	5051	5065	5079	5092	5105	5119	5132	5145	5159	5172
33	5185	5198	5211	5224	5237	5250	5263	5276	5289	5302
34	5315	5328	5340	5353	5366	5378	5391	5403	5416	5428
35	5441	5453	5465	5478	5490	5502	5514	5527	5539	5551
36	5563	5575	5587	5599	5611	5623	5635	5647	5658	5670
37	5682	5694	5705	5717	5729	5740	5752	5763	5775	5786
38	5798	5809	5821	5832	5843	5855	5866	5877	5888	5899
39	5911	5922	5933	5944	5955	5966	5977	5988	5999	6010
40	6021	6031	6042	6053	6064	6075	6085	6096	6107	6117
41	6128	6138	6149	6160	6170	6180	6191	6201	6212	6222
42	6232	6243	6253	6263	6274	6284	6294	6304	6314	6325
43	6335	6345	6355	6365	6375	6385	6395	6405	6415	6425
44	6435	6444	6454	6464	6474	6484	6493	6503	6513	6522
45	6532	6542	6551	6561	6571	6580	6590	6599	6609	6618
46	6628	6637	6646	6656	6665	6675	6684	6693	6702	6712
47	6721	6730	6739	6749	6758	6767	6776	6785	6794	6803
48	6812	6821	6830	6839	6848	6857	6866	6875	6884	6893
49	6902	6911	6920	6928	6937	6946	6955	6964	6972	6981
50	6990	6998	7007	7016	7024	7033	7042	7050	7059	7067
51	7076	7084	7093	7101	7110	7118	7126	7135	7143	7152
52	7160	7168	7177	7185	7193	7202	7210	7218	7226	7235
53	7243	7251	7259	7267	7275	7284	7292	7300	7308	7316
54	7324	7332	7340	7348	7356	7364	7372	7380	7388	7396
N	0	1	2	3	4	5	6	7	8	9

Logarithms to the Base 10 (cont.)

N	0	1	2	3	4	5	6	7	8	9
55	7404	7412	7419	7427	7435	7443	7451	7459	7466	7474
56	7482	7490	7497	7505	7513	7520	7528	7536	7543	7551
57	7559	7566	7574	7582	7589	7597	7604	7612	7619	7627
58	7634	7642	7649	7657	7664	7672	7679	7686	7694	7701
59	7709	7716	7723	7731	7738	7745	7752	7760	7767	7774
60	7782	7789	7796	7803	7810	7818	7825	7832	7839	7846
61	7853	7860	7868	7875	7882	7889	7896	7903	7910	7917
62	7924	7931	7938	7945	7952	7959	7966	7973	7980	7987
63	7993	8000	8007	8014	8021	8028	8035	8041	8048	8055
64	8062	8069	8075	8082	8089	8096	8102	8109	8116	8122
65	8129	8136	8142	8149	8156	8162	8169	8176	8182	8189
66	8195	8202	8209	8215	8222	8228	8235	8241	8248	8254
67	8261	8267	8274	8280	8287	8293	8299	8306	8312	8319
68	8325	8331	8338	8344	8351	8357	8363	8370	8376	8382
69	8388	8395	8401	8407	8414	8420	8426	8432	8439	8445
70	8451	8457	8463	8470	8476	8482	8488	8494	8500	8506
71	8513	8519	8525	8531	8537	8543	8549	8555	8561	8567
72	8573	8579	8585	8591	8597	8603	8609	8615	8621	8627
73	8633	8639	8645	8651	8657	8663	8669	8675	8681	8686
74	8692	8698	8704	8710	8716	8722	8727	8733	8739	8745
75	8751	8756	8762	8768	8774	8779	8785	8791	8797	8802
76	8808	8814	8820	8825	8831	8837	8842	8848	8854	8859
77	8865	8871	8876	8882	8887	8893	8899	8904	8910	8915
78	8921	8927	8932	8938	8943	8949	8954	8960	8965	8971
79	8976	8982	8987	8993	8998	9004	9009	9015	9020	9025
80	9031	9036	9042	9047	9053	9058	9063	9069	9074	9079
81	9085	9090	9096	9101	9106	9112	9117	9122	9128	9133
82	9138	9143	9149	9154	9159	9165	9170	9175	9180	9186
83	9191	9196	9201	9206	9212	9217	9222	9227	9232	9238
84	9243	9248	9253	9258	9263	9269	9274	9279	9284	9289
85	9294	9299	9304	9309	9315	9320	9325	9330	9335	9340
86	9345	9350	9355	9360	9365	9370	9375	9380	9385	9390
87	9395	9400	9405	9410	9415	9420	9425	9430	9435	9440
88	9445	9450	9455	9460	9465	9469	9474	9479	9484	9489
89	9494	9499	9504	9509	9513	9518	9523	9528	9533	9538
90	9542	9547	9552	9557	9562	9566	9571	9576	9581	9586
91	9590	9595	9600	9605	9609	9614	9619	9624	9628	9633
92	9638	9643	9647	9652	9657	9661	9666	9671	9675	9680
93	9685	9689	9694	9699	9703	9708	9713	9717	9722	9727
94	9731	9736	9741	9745	9750	9754	9759	9763	9768	9773
95	9777	9782	9786	9791	9795	9800	9805	9809	9814	9818
96	9823	9827	9832	9836	9841	9845	9850	9854	9859	9863
97	9868	9872	9877	9881	9886	9890	9894	9899	9903	9908
98	9912	9917	9921	9926	9930	9934	9939	9943	9948	9952
99	9956	9961	9965	9969	9974	9978	9983	9987	9991	9996
N	0	1	2	3	4	5	6	7	8	9

Appendix D

SOME FORMULAS FROM GEOMETRY

Let r denote radius, h altitude, b length of base, s slant height, B area of base, θ central angle in radian measure.

1. Circle: area $= \pi r^2$, circumference $= 2\pi r$.
2. Circular arc: length $= r\theta$.
3. Circular sector: area $= \frac{1}{2}r^2\theta$.
4. Triangle: area $= \frac{1}{2}bh$.
5. Trapezoid: area $= \frac{1}{2}(b_1 + b_2)h$.
6. Right circular cylinder: volume $= \pi r^2 h$, curved (lateral) surface $= 2\pi rh$.
7. Sphere: area $= 4\pi r^2$, volume $= \frac{4}{3}\pi r^3$.
8. Right circular cone: volume $= \frac{1}{3}\pi r^2 h$, curved (lateral) surface $= \pi rs$.
9. Pyramid: volume $= \frac{1}{3}Bh$.

ANSWERS TO ODD-NUMBERED EXERCISES

Exercises 1-1

1. (a) $\sqrt{58}$; (b) $\sqrt{82}$; (c) $2\sqrt{58}$; (d) $2\sqrt{41}$.

3. (a) Ordinate zero; (b) abscissa zero.

5. (3/2, 3).　　**7.** (a) Lengths of sides $\sqrt{58}, \sqrt{145}, \sqrt{145}$; (b) Lengths of sides $\sqrt{34}, \sqrt{34}, \sqrt{68}$.

11. (0, 9/2).　　**13.** (2, 3).　　**15.** (a) 18 sq. un; (b) 15 sq. un; (c) 20 sq. un; (d) 105/2 sq. un.

Exercises 1-2

3. (a) 0; (b) -1; (c) 0; (d) $-\frac{6}{7}$.

7. $-\frac{1}{8}$.　　**13.** 8.　　**17.** 60°.　　**19.** 2, $-\frac{1}{2}$.

Exercises 2-1

7. 25.

Exercises 2-2

1. (a) $3x - y + 7 = 0$;　　(b) $2x + 3y - 21 = 0$;
(c) $2x + y + 18 = 0$;　　(d) $5x - 2y - 24 = 0$.

3. (a) $3x + 5y - 14 = 0$;　　(b) $4x + y - 19 = 0$;
(c) $x + 3y + 19 = 0$;　　(d) $x = 5$.

5. $11x - 4y + 43 = 0$; $x - 9y - 22 = 0$; $2x + y - 6 = 0$.

7. $9x + y + 1 = 0$; $4x + 11y + 6 = 0$; $x - 2y - 1 = 0$.

9. $2x - 3y + 12 = 0$.　　**11.** $2x - 9y + 85 = 0$.

13. $2x + 3y + 5 = 0$.　　**15.** $(-\frac{18}{7}, \frac{48}{7})$.

17. $x + 2y - 1 = 0$.   **19.** $86°38'$.

21. Two solutions:

$$\begin{cases} 7x - 3y + 2 = 0 \\ 3x + 7y + 5 = 0. \end{cases} \quad \begin{cases} 3x + 7y - 24 = 0 \\ 7x - 3y + 31 = 0. \end{cases}$$

Exercises 3-1

1. Sym. to y-axis.
Int. $x = 0$; $y = 0$.
Excl. Val. $y < 0$.

3. Sym. to y-axis.
Int. $x = \pm\sqrt{6}$; $y = 2$.
Excl. Val. $y > 2$.

5. Sym. to x-axis, y-axis, origin.
Int. $x = \pm 4$; $y = \pm 4$.
Excl. Val. $|x| > 4$; $|y| > 4$.

7. Sym. to x-axis, y-axis, origin.
Int. $x = \pm 3$; $y = \pm\frac{12}{5}$.
Excl. Val. $|x| > 3$; $|y| > \frac{12}{5}$.

9. Sym. to x-axis, y-axis, origin.
Int. $x = \pm 3$.
Excl. Val. $|x| < 3$.

11. Sym. to x-axis.
Int. $x = 0$, -10; $y = 0$.
Excl. Val. $x > 0$, $x < -10$; $|y| > 5$.

13. Sym. to origin.

15. Int. $x = 3$; $y = 2$.

17. Sym. to x-axis.
Int. $x = -4$.
Excl. Val. $-4 < x < 0$.

Exercises 3-2

1. (a) $(2, -3)$, $r = 2$; (c) $(-4, -2)$, $r = 3$; (e) $(-\frac{2}{3}, 1)$, $r = \sqrt{5}$; (g) $(0, \frac{1}{2})$,
$r = \sqrt{10}$.

2. (a) $x^2 + y^2 - 4x + 10y - 20 = 0$; (c) $x^2 + y^2 - 2x + 2y - 23 = 0$;
(e) $x^2 + y^2 + 14x - 4y + 49 = 0$; (g) $x^2 + y^2 - 2x - 6y = 0$.
(i) $x^2 + y^2 - 10x = 0$; (k) $x^2 + y^2 + 5x + y - 26 = 0$.

3. (a) $(14, 6)$, $(-3, -11)$; (b) $(0, 5)$, $(-3, -4)$; (c) $(-4, 3)$, $(1, 4)$.

5. Center $(-5, 8)$, $r = 10$.

Exercises 3-3

2. (a) $y^2 = -36x$; (c) $3x^2 + 16y = 0$.

3. (a) $x^2 + 2x + y - 6 = 0$; (c) $11x^2 - 9x - 6y - 8 = 0$.

4. (a) $y^2 - 3y + 3x - 10 = 0$; (c) $11y^2 - 19y - 12x - 18 = 0$.

5. $400x^2 + 1125y = 0$.   **7.** 27.5 ft.

Exercises 4-1

1. (a) $|x| \leqslant 4$; (c) $|x - a| < \varepsilon$.
2. (a) $-3 < x < 3$; (c) $-6 \leqslant x \leqslant 14$.
3. (a) $|S\text{-}600| \leqslant 5$; (b) $595 \leqslant S \leqslant 605$.
5. $0, -2, -12, a - a^2, a^2 - a^4, -x^2 + x(1 - 2h) + h(1 - h)$.
7. (a) $-\infty < x < \infty, f(x) \geqslant 2$;   (b) $|x| \geqslant 2, f(x) \geqslant 0$;
   (c) $|x| \leqslant 2, 0 \leqslant f(x) \leqslant 2$;   (d) $-\infty < x < \infty, -4 \leqslant f(x)$;
   (e) $x \neq -2, f(x) \neq -4$.
9. (a) $-\infty < x < \infty, -1 \leqslant F(x) \leqslant 1$; (b) $-1, -1, \frac{1}{2}, 1, 1$.
11. $F(x) = (1 - x)/3$; yes.

Exercises 4-2

1. (a) 1; (c) 3; (e) 0; (g) 1; (i) 2; (k) 1; (m) ∞; (o) 9.
2. (a) $s^2 - s$; (c) ∞; (e) 1.   3. Yes; 0.

Exercises 4-3

1. -1.   3. 3.   5. 0.   7. 0.   9. -4.   11. $\frac{5}{3}$.   13. $-\frac{1}{3}$.
15. 125.   17. 2.   19. 0.   21. $\sqrt{2}$.   23. ∞.   25. 0.   27. 2.
29. ∞.

Exercises 4-4

1. Cont. for all x.   3. Discont. for $x = -1$.   5. Discont. for $x = \pm 3$.
7. Discont. for $x = \pm 1$.   9. Discont. for $x = 0$.
11. Cont. for $0 \leqslant x \leqslant 1000$.

Exercises 5-1

1. 2.   3. $2x - 2$.   5. $-1/x^2$.   7. $-1/(x - 1)^2$.   9. $1/2\sqrt{x}$.
11. $3\sqrt{x}/2$.   13. $8x + y - 9 = 0$.   15. $x - 3y + 5 = 0$.
17. $3x - y - 2 = 0$.   19. 10.   21. (a) (1) 64; (2) 48; (3) 40; (4) 36;
   (b) 32.
23. (a) $-32t + 64$; (b) 64; (c) -32; (d) -96; (e) 2.
25. $50; - \$50$.   27. $5.60, \$5.63$.

Exercises 5-2

1. $6x - 5$.   **3.** $-5 - 2r - 30r^2$.   **5.** $\dfrac{4}{5} - \dfrac{6x}{17}$.   **7.** $4x + \dfrac{3}{x^2} - \dfrac{22}{x^3}$.

9. $\dfrac{2t^2 - 3}{4t^2}$.   **11.** $6x^5 - 12x^3 + 6x$.   **13.** $\dfrac{-r^2 + 6r - 6}{r^4}$.

15. $\dfrac{1}{2\sqrt{x}}$.   **17.** $\dfrac{3\sqrt{t}}{2} - \dfrac{1}{t\sqrt{t}}$.   **19.** $\dfrac{5x^3 - 3x^2 + x + 1}{2x\sqrt{x}}$.

21. (a) 40,000; (b) 1440.   **23.** $3x - 16y + 24 = 0$.

Exercises 5-3

1. $18x(2 + 3x^2)^2$.   **3.** $60x(3x^2 - 5)^4$.

5. $\dfrac{-x}{\sqrt{4 - x^2}}$.   **7.** $\dfrac{3x^2}{2\sqrt{x^3 + 9}}$.

9. $\dfrac{1}{(x + 1)^2}$.   **11.** $(x^3 + 3)(4x - 3) + (2x^2 - 3x + 1)(3x^2)$.

13. $\dfrac{-2x^2 + 2}{(x^2 + 1)^2}$.   **15.** $\dfrac{2x^2 - 2x - 8}{(4 + x^2)^2}$.

17. $\dfrac{2x^2 - x - 1}{\sqrt{x^2 - 1}}$.   **19.** $\dfrac{5x^2 - 4x - 1}{2\sqrt{x - 1}}$.

21. $(9 - 4x^2)\sqrt{9 - x^2}$.   **23.** $\dfrac{-(4 + 3x)}{\sqrt{(x^2 - 4)^3(x - 2)^2}}$.

25. $\dfrac{\sqrt{x}(4x + 9)}{2\sqrt{(2x + 3)^3}}$.   **27.** $\dfrac{1 + \sqrt{x + 1}}{\sqrt{x + 1}}$.

29. $(0, 0)$.   **31.** Arctan $1.4545 \cong 55°29'$.

33. $(0, 0)$.

Exercises 5-4

1. 4.   **3.** $k(k - 1)(k - 2) \cdots 3 \cdot 2 \cdot 1$.   **5.** $-\dfrac{x}{y}$.   **7.** $\dfrac{-2x - y}{x + 2y}$.

9. $\dfrac{-3x^2 - y^3}{3xy^2}$.   **11.** $\dfrac{-4xy}{4x^2 + 3y}$.   **13.** $\dfrac{3x^2 - 2xy - y^2}{x^2 + 2xy}$.   **15.** $-\dfrac{1}{4y^3}$.

17. $\dfrac{a^{2/3}}{3x^{4/3}y^{1/3}}.$ 19. $\dfrac{2x^3 - 3x}{(1 - x^2)^{3/2}}.$ 21. $78°41'$ (approx).

23. 23 ft/sec; 18 ft/sec/sec.   25. -3 ft/sec; 16ft/sec/sec.
27. -12 ft/sec/sec; 12 ft/sec/sec.   29. $t > 4, 1 < t < 5/2.$

Exercises 6-1

1. (a) Yes; (c) $\eta = \frac{1}{2}$.
3. (a) No; (b) f undefined at $x = -1$.
5. (a) No; (b) $f'(0)$ does not exist.
7. Increasing $x < -3, x > 2$; decreasing $-3 < x < 2$.
9. Increasing $-1 < x < 1$; decreasing $x < -1, x > 1$.

Exercises 6-2

1. (a) $x > -\frac{1}{2}$; (b) $x < -\frac{1}{2}$; (c) all x; (e) none.
3. (a) $x < -3, x > 0$; (b) $-3 < x < 0$; (c) $x > -\frac{3}{2}$; (d) $x < -\frac{3}{2}$; (e) $(-\frac{3}{2}, \frac{9}{4})$.
5. (a) $x < -3, x > 1$; (b) $-3 < x < 1$; (c) $x > -1$; (d) $x < -1$; (e) $(-1, 2)$.
7. (a) $|x| > 1$; (b) $|x| < 1$; (c) $x > 0$; (d) $x < 0$; (e) none.

Exercises 6-3

1. None.   3. $(2, -18)$ rel min; $(-1, 9)$ rel max.   5. $(1, -1)$ rel min; $(-1, 3)$ rel max.   7. 2 rel min; -2 rel max; no pts of infl.
9. 2 rel min for $x = \pm 1$; no pts of infl.   11. $(1, 3)$ max pt; $(2, 0)$ min pt.

Exercises 6-4

1. 7, 7.   3. 10, 5.   5. $6000.   9. 432 cu in.
11. $20\sqrt{2}$ ft long, $15\sqrt{2}$ ft wide.   13. 8 in. long, 4 in. wide.
15. Base 6 in., depth 3/2 in.   17. $(2, 2)$ 19. $r = h$.   21. 11.18 ft.
23. 23.6 miles from D.   25. 3500.   27. 15.   29. 6.   31. 120.

Exercises 6-5

1. 24π sq ft/min.   **3.** 288 sq in./min.   **5.** -2 ft/min.

7. $20\sqrt{3}$ sq in./min.   **9.** $-25/13$ mph.   **11.** 0.5 in/.min.

13. $-10/3$ cu ft/min.   **15.** 16/5 ft/sec.   **17.** -1 sq in./min.

19. 25/13 ft/sec; yes; 5 ft/sec.

Exercises 6-6

1. $\dfrac{-x\,dx}{\sqrt{9-x^2}}$.   **3.** $\dfrac{-x^2-2x-2}{(x^2-2)^2}\,dx$.   **5.** $\dfrac{-3x^2-y}{x+3y^2}\,dx$.   **7.** (a) -1;

(b) 1; (c) -1.5.   **9.** $-.04$.   **11.** $-.0333$.

Exercises 6-7

1. 9.9.   **3.** 3.037.   **5.** .000995.   **7.** 2.44.   **9.** Approx error $-.48$; rel. error .003; percent error .3%.   **11.** Rel. error .0067.

13. $.24\pi$ sq in.; $.36\pi$ sq in.; $.54\pi$ cu in.   **15.** Error of $\frac{4}{3}\pi\,(3rt^2+t^3)$.

17. .2%.   **19.** .75 ft.   **21.** .009 ft (approx).

Exercises 7-1

1. 3/2.   **3.** 8/3.   **5.** 0.

Exercises 7-2

1. 2.   **3.** $\dfrac{\sqrt[3]{2}}{2}$.   **7.** $9, 9\sqrt{2}$.

Exercises 7-3

1. $(1/6)\,(4x^3-9x^2+24x)+C$.   **3.** $(1/3)\,(u+3)^3+C$.

5. $(1/15)\,(12w^5+100w^3+375w)+C$.   **7.** $(1/16)\,(2x^2+5)^4+C$.

9. $2r\sqrt{3r}/3+C$.   **11.** $v^{3/2}\left(\dfrac{8v^2}{7}+\dfrac{24v}{5}+6\right)+C$.

13. $(3/4)\sqrt[3]{(x^2+5)^2}+C$.   **15.** $(2/9)\sqrt{(t^3+10)^3}+C$.

17. $(-1/3)\sqrt{(9-x^2)^3} + C.$ **19.** 8/3.   **21.** 51/64.   **23.** 44/3.   **25.** 1.
27. $2(\sqrt{11} - 1).$ **29.** 31/1800.

Exercises 7-4

1. 6. sq un.   **3.** 32/3 sq un.   **5.** 8/3 sq un.   **7.** 8 sq un.
9. $a^2/6$ sq un.   **11.** $\sqrt{19} - 2$ sq un.

Exercises 7-5

1. 36 sq un.   **3.** $8\sqrt{3}$ sq un.   **5.** 16/3 sq un.   **7.** 97/12 sq un.
9. 4/3 sq un.   **11.** 9/2 sq un.   **13.** 343/12 sq un.   **15.** 148/3 sq un.
17. 148/3 sq un.   **19.** 108 sq un.

Exercises 7-6

1. 24 in. lbs.   **3.** (a) 900 ft lbs, (b) $6\frac{1}{2}$ ft, (c) 9 ft.   **5.** 65,000 ft lbs.
7. 550 ft lbs.   **9.** 300π ft tons.   **11.** 65.3 ft tons.   **13.** 89.9 ft tons.

Exercises 7-7

1. $y = x^2 + x - 4.$ **3.** $x^3 - 2xy + x + 2 = 0.$
5. $3x^2 + 4x + 3y - 2 = 0.$ **7.** 4 ft.   **9.** 125/2 ft.
11. 5 sec; 24 ft/sec.   **13.** 400 ft; 160 ft/sec.   **15.** Both reach the ground
in 3 sec.   **17.** $C(x) = \frac{1}{3}(4x^{3/2} + 42).$ **19.** 51,200.   **21.** 4000;
Infinite population.

Exercises 8-1

1. (a) $-\pi/2$; (b) $11\pi/6$; (c) $-3\pi/4$; (d) $9\pi/4$; (e) $3\pi/2$; (f) $7\pi/6$; (g) $\pi/18$;
(h) .26552; (i) .43915.   **11.** (a) $\frac{1}{2}\sqrt{2 - \sqrt{2}}$; (b) $\frac{1}{2}\sqrt{2 + \sqrt{3}}$; (c) $2 + \sqrt{3}$.
13. $\pm\pi/4$; $\pm 3\pi/4$; $\pm 5\pi/4$; $\pm 7\pi/4$.   **15.** 0; $\pm\pi$; $\pm 2\pi$.   **17.** $\pi/6$;
$5\pi/6$; $-11\pi/6$; $-7\pi/6$.

Exercises 8-2

1. 2.   **3.** 1.   **5.** 0.   **9.** $-3 \sin(3x + 2)$.   **11.** $6 \tan 3x \sec^2 3x$.

13. $\cot x - x \csc^2 x$.   **15.** $-6 \csc^3 2x \cot 2x$.

17. $\cos 2x - 4 \cot 2x \csc^2 2x$.

19. $\dfrac{x}{(1 + x^2)^{1/2}} \cos(1 + x^2)^{1/2}$.   **21.** $\dfrac{\sin x - x \cos x}{\sin^2 x}$.

23. $-2(\csc^2 2x + \sec^2 2x)$.   **25.** $\dfrac{3x - 2}{2\sqrt{1 - x}} \sin(x\sqrt{1 - x})$.

27. $-\dfrac{2x(1 + \tan y)}{x^2 \sec^2 y - 2y}$.   **29.** $\dfrac{\sec y \csc^2 x - y \cos xy}{\cot x \sec y \tan y + x \cos xy}$.

31. No rel. max or min.   **33.** 500π mi/min.

35. $r = a\sqrt{2/3}$, $h = 2a/\sqrt{3}$.   **37.** $\frac{4}{3}a$.   **39.** 125/6 ft.

Exercises 8-3

1. (a) $\pi/4$; (b) $\pi/3$; (c) $\pi/4$; (d) $-\pi/4$.   **3.** $3/\sqrt{1 - 9x^2}$.   **5.** $1/2x\sqrt{x - 1}$.

7. $1/x\sqrt{x^2 - 1}$.   **9.** $\dfrac{x^2 - 1}{x^4 + 3x^2 + 1}$.   **11.** $\dfrac{x^2 + 1}{x\sqrt{3x^2 - x^4 - 1}}$.

13. $2x \text{ Arcsin } 3x + \dfrac{3x^2}{\sqrt{1 - 9x^2}}$.   **15.** $\sqrt{x}/(1 + x)$.   **17.** .05 rad/min.

19. -0.005 rad.   **21.** 0.092 rad/sec.   **23.** 0.097 rad/sec.

Exercises 8-4

1. $\frac{1}{3} \sin 3x + C$.   **3.** $\frac{1}{4} \tan 4x + C$.   **5.** $\frac{1}{2} \sec 2x + C$.

7. $-\frac{1}{4} \cos 4x + C$.   **9.** $-\sin(1 - x) + C$.   **11.** $-\frac{1}{2} \csc(2x - 1) + C$.

13. $\frac{1}{2} \sin x^2 + C$.   **15.** $\frac{1}{3} \tan x^3 + C$.   **17.** $-\frac{1}{4} \cos 2x + C$.

19. $2\sqrt{1 - \cos x} + C$.   **21.** $\dfrac{1}{6 \cos^2 3x} + C$.   **23.** $\text{Arcsin} \dfrac{x}{2} + C$.

25. $\frac{1}{2} \text{Arctan} \dfrac{x}{2} + C$.   **27.** $\frac{1}{6} \text{Arctan} \dfrac{2x}{3} + C$.   **29.** $\text{Arcsec } 2x + C$.

31. 0.   **33.** 3/8.   **35.** 3/8.   **37.** $\frac{1}{8}[2\sqrt{3} - 1]$.   **39.** 1 sq un.

41. $1 + \sqrt{2}$ sq un.   **43.** π sq un.

Exercises 9-1

1. (a) 100; (b) .01; (c) $\sqrt{10}$.   **3.** 6.   **5.** $\dfrac{2\sqrt{2}}{5}$.

7. $2\log(x^2 + 2) - \log(x - 1)$.

9. $\frac{1}{2}[\log(x + 1) + \log(x - 1) - \log(x^2 + 1)]$.   **11.** $-\frac{3}{5}$.   **13.** $\log_5 4$.

15. $\sqrt{2}$.   **17.** $-\log_3 2$.   **19.** $\log_3 y$.   **21.** y.   **23.** $\log(y + \sqrt{y^2 + 1})$.

Exercises 9-2

1. $1/x$.   **3.** $\log 2x + 1$.   **5.** $-3\tan 3x$.   **7.** $\dfrac{-2}{x^2 - 1}$.

9. $\dfrac{2}{3(x + 1)}\log_{10} e$.   **11.** $2\csc 2x$.   **13.** $\dfrac{x - 16}{(x + 1)(x^2 + 16)}$.

15. $\dfrac{15x^3 - 2x^2 + 11x + 30}{6(3x + 2)^{1/2}(x^2 + 1)^{4/3}}$.   **17.** $-4\tan 2x \log_{10} e$.

19. $\dfrac{2(1 - \log x^2)}{x^3}$.   **21.** $\dfrac{(1 - x)y}{(y - 1)x}$.   **23.** $\dfrac{2\sin 2x - 2}{\csc y}$.

31. $y = 3x - (1 + \log 3)$.

Exercises 9-3

1. $-e^{1/x}/x^2$.   **3.** $2\sin 4x \exp[\sin^2 2x]$.   **5.** $\dfrac{1}{1 + x^2}\exp[\text{Arctan } x]$.

7. $2e^{2x}(\sin 2x + \cos 2x)$.   **9.** $e^{x/2} - e^{-x/2}$.   **11.** e^{ex+1}.

13. $\dfrac{e^{2x} - 1}{e^{2x} + 1}\log_a e$.   **15.** $\dfrac{-2xe^y - y^2e^x}{x^2e^y + 2ye^x}$.   **17.** \$62 per year (approx).

19. $6,182,535\,(e^{0.01567t} + e^{-0.01567t})^{-2}$.   **25.** min 0; max e^{-2}.

Exercises 9-4

1. $\frac{1}{3}e^{3x} + C$.   **3.** $-\frac{1}{2}e^{-2x} + C$.   **5.** $-\frac{1}{2}e^{-x^2} + C$.   **7.** $\dfrac{2^x e^x}{1 + \log 2} + C$.

9. $3x + C$.   **11.** $\frac{1}{2}\log(e^{2x} - 1) + C$.   **13.** $\frac{1}{2}\log(x^2 + 1) + C$.

15. $-\log|1 - x| + C$.   **17.** $\log|\sec x| + C$.   **19.** $\log|x^2 + x| + C$.

21. $\sqrt{x^2 - 9} + C.$ **23.** $x + 3 \log|x| - \dfrac{1}{x} + C.$

25. $\tfrac{1}{2}e^{2x} + 2e^x + x + C.$ **27.** $\log|\sec x + \tan x| + C.$ **29.** $-e^{\cos x} + C.$

31. $\log 2.$ **33.** $\tfrac{1}{3}(e - 1).$ **35.** $\dfrac{3}{2 \log 2}.$ **37.** $\dfrac{26}{\log 3} + \dfrac{9}{2}.$

39. $\tfrac{1}{2} + \log \tfrac{9}{16}.$ **41.** $\tfrac{1}{2}(e - 1).$ **43.** $\dfrac{e^4 - 1}{2e^2}.$ **45.** $\gamma = k \log \dfrac{\beta}{\beta_0}.$

47. $AP^b = C.$

Exercises 9-5

1. $x^x(\log x + 1).$ **3.** $x^{\sqrt{x}-1/2}(\tfrac{1}{2}\log x + 1).$ **5.** $\cos x \, e^{\sin x}.$

7. $2x^{2\,\text{Arctan}\,x}\left(\dfrac{\log x}{1 + x^2} + \dfrac{\text{Arctan}\,x}{x}\right).$ **9.** $(\cos x)^x(\log \cos x - x \tan x).$

11. $(\tan x)^{\log x}[(1/x) \log \tan x + \sec x \csc x \log x].$ **13.** $x^x \exp[x^x](1 + \log x).$

15. $\dfrac{x(6x^2 + 15x + 8)}{2\sqrt{x + 1}\sqrt{x + 2}}.$ **17.** $\dfrac{x(3x^5 + 8x^3 - 4x^2 + 9x - 16)}{\sqrt[3]{(x^3 - 1)^2\,(x^2 - 1)^2\,(x^2 + 9)^2}}.$

19. $\dfrac{x^2 e^{2x}(-6x^3 - 6x^2 + 2x + 3)}{\sqrt{(1 - 3x^2)^3}}.$ **21.** $.375\%.$

Exercises 9-6

1. $\dfrac{\log 5}{2}.$ **3.** 17.3 yrs. **5.** 310 yrs.

7. $I = I_0 e^{-.048x}; \; I = 0.32\,I_0.$ **9.** 295 (approx).

11. 1270 billions (approx). **13.** $R(x) = a + \dfrac{2a}{\log 5}(1 - 5^{-x/2a}).$

Exercises 10-1

1. $x \log x - x + C.$ **3.** $\sin x - x \cos x + C.$

5. $x\,\text{Arctan}\,x - \tfrac{1}{2}\log(1 + x^2) + C.$ **7.** $e^{2x}\left(\dfrac{x^2}{2} - \dfrac{x}{2} + \dfrac{1}{4}\right) + C.$

9. $2x \sin x - (x^2 - 2)\cos x + C.$ **11.** $e^x(x^3 - 3x^2 + 6x - 6) + C.$

13. $\dfrac{\sqrt{(x^2+1)^3}}{15}(3x^2-2)+C.$ **15.** $\dfrac{e^x}{5}(\cos 2x + 2\sin 2x)+C.$

17. $\frac{1}{4}(2x+\sin 2x)+C.$

Exercises 10-2

1. $-\frac{1}{2}\cos 2x + \frac{1}{6}\cos^3 2x + C.$ **3.** $\frac{1}{3}\sin 3x - \frac{2}{9}\sin^3 3x + \frac{1}{15}\sin^5 3x + C.$

5. $\dfrac{x}{2}+\frac{1}{8}\sin 4x + C.$ **7.** $-\frac{1}{6}\cot^2 3x - \frac{1}{3}\log|\sin 3x| + C.$

9. $-\frac{1}{3}\cos^3 x + \frac{1}{5}\cos^5 x + C.$

11. $-\frac{1}{8}\cot^4 2x + \frac{1}{4}\cot^2 2x + \frac{1}{2}\log|\sin 2x| + C.$

13. $\dfrac{x}{16}-\dfrac{1}{64}\sin 4x + \dfrac{1}{48}\sin^3 2x + C.$

15. $-\frac{1}{8}[2\cot x\csc^3 x + 3\cot x\csc x + 3\log|\csc x + \cot x|] + C.$

17. $\dfrac{x}{16}-\dfrac{1}{192}\sin 12x - \dfrac{1}{144}\sin^3 6x + C.$ **19.** $\frac{2}{3}(\sin x)^{3/2} - \frac{2}{7}(\sin x)^{7/2} + C.$

21. $-\cos 2x + C.$ **23.** $\frac{2}{3}(1-\cos x)^{3/2} + C.$

Exercises 10-3

1. $-\frac{2}{3}(1-x)^{3/2}+C.$ **3.** $\dfrac{2(x+3)^{1/2}}{5}(x^2-4x+24)+C.$

5. $\frac{3}{4}(x+8)^{4/3}+C.$ **7.** $-\frac{2}{3}\sqrt{1-x}(2+x)+C.$ **9.** $\operatorname{Arcsin}\dfrac{x}{3}+C.$

11. $\frac{1}{8}\operatorname{Arctan}\dfrac{x}{2}+C.$ **13.** $-\dfrac{1}{2(x^2+1)}+C.$

15. $\dfrac{-2\sqrt{4-x}}{15}(128+16x+3x^2)+C.$ **17.** $\dfrac{2}{45}(4+x^3)^{3/2}(3x^3-8)+C.$

19. $\dfrac{1}{\sqrt{x^2+9}}(x^2+18)+C.$ **21.** $\frac{1}{3}\operatorname{Arcsin}\dfrac{3x}{4}+C.$

23. $\sqrt{x^2-4}-2\operatorname{Arctan}\dfrac{\sqrt{x^2-4}}{2}+C.$

25. $\dfrac{x\sqrt{x^2+16}}{2}+8\log|\sqrt{x^2+16}+x|+C.$ **27.** $x-4\operatorname{Arctan}\dfrac{x}{4}+C.$

29. $\frac{1}{4}$ Arctan $\frac{x}{4} + C$.   **31.** $\frac{-2(5-x)^{3/2}}{15}(10+3x)+C$.

33. $\log|\sqrt{x^2+4x+13}+x+2|+C$.   **35.** Arcsin $\frac{x-2}{2}+C$.

37. $\frac{x}{9\sqrt{x^2+9}}+C$.   **39.** $9\pi/2$.   **41.** $\frac{1}{4}$.   **43.** $\frac{8}{3}(2-\sqrt{2})$.

45. $\frac{2(2p-1)}{\sqrt{p(1-p)}}+C$.

Exercises 10-4

1. $\frac{1}{2}\log\left|\frac{x-1}{x+1}\right|+C$.   **3.** $\log[(x+3)^2(x-2)^2]+C$.

5. $\frac{x^2}{2}-x+\log\frac{(x-2)^2}{|x+2|}+C$.   **7.** $\log\frac{x^2}{x^2+4}-\frac{1}{2}$ Arctan $\frac{x}{2}+C$.

9. $\log\frac{|x-5|(x+1)^2}{x^4}+C$.   **11.** $\log\frac{|x|}{\sqrt{|x+1|}\sqrt[4]{x^2+1}}-\frac{1}{2}$ Arctan $x+C$.

13. $\log\left|\frac{x-1}{x+2}\right|-\frac{6}{x^2+x-2}+C$.

15. $\log\frac{\sqrt{x^2+4}}{|x-1|}+\frac{3x-10}{4(x^2+4)}+\frac{7}{8}$ Arctan $\frac{x}{2}+C$.   **17.** a/b.

19. $x=\frac{nx_0}{(n-x_0)\exp[-knt]+x_0}$.

Exercises 10-5

1. $\log|x+\sqrt{x^2-5}|+C$.   **3.** $\frac{1}{2}\log|\sqrt{9+4x^2}+2x|+C$.

5. $\frac{1}{2}(x^2+1)$ Arctan $x-\frac{x}{2}+C$.   **7.** $\frac{2}{5}(x-9)^{1/2}(x^2+12x+216)+C$.

9. $\frac{1}{8}(2x^2-2x\sin 2x-\cos 2x)+C$.

11. $-\frac{1}{120}\cos^6 5x(4-3\cos^2 5x)+C$.

13. $x(\log^2 2x-2\log 2x+2)+C$.

15. $-\frac{1}{15}\csc x(3\csc^4 x-10\csc^2 x+15)+C$.

17. $\frac{1}{4}(2x\tan 2x+\log|\cos 2x|)+C$.

19. $\frac{1}{4}(\tan^2 2x+2\log|\cos 2x|)+C$.

21. $\frac{3}{28}(x + 1)^{4/3}(4x - 3) + C.$ **23.** $\frac{2}{27}(3x^3 + 1)^{3/2} + C.$

25. $-\frac{1}{4}e^{-2x}(2x^2 + 2x + 1) + C.$ **27.** $\sin(\log x) + C.$

29. $-\frac{4}{81}(5 - 3x)^{5/4}(3x + 4) + C.$ **31.** $\frac{1}{5}e^{2x}(2 \sin x - \cos x) + C.$

33. $\text{Arctan } e^x + C.$ **35.** $\frac{1}{6} \log \left| \frac{x^3(x - 2)}{(x + 1)^4} \right| + C.$

37. $\frac{1}{8}(\sin 2x - 2x \cos 2x) + C.$ **39.** $\frac{1}{12} \log \left| \frac{2x - 3}{2x + 3} \right| + C.$

41. $\text{Arctan } x + \frac{1}{2} \log(1 + x^2) + C.$ **43.** $\text{Arctan}(\cos x) + C.$

45. $\frac{1}{13}e^{3x}(3 \cos 2x + 2 \sin 2x) + C.$

47. $\frac{1}{2}[x \cos(\log x) + x \sin(\log x)] + C.$

49. $\frac{1}{3} \log |\sqrt{16 + 4x^2} + 3x| + C.$

Exercises 10-6

1. $20; \frac{1}{4}[1 + 2x - \log|1 + 2x|] + C.$ **3.** $75; -\frac{\sin 5x}{10} + \frac{\sin x}{2} + C.$

5. $31; \frac{1}{\sqrt{3}} \log \left| \frac{\sqrt{3 - x} - \sqrt{3}}{\sqrt{3 - x} + \sqrt{3}} \right| + C.$

7. $45; -\frac{1}{\sqrt{3}} \log \left| \frac{\sqrt{3} + \sqrt{3 - x^2}}{x} \right| + C.$

9. $90; x \text{ Arcsin } \frac{2x}{3} + \frac{1}{2}\sqrt{9 - 4x^2} + C.$

11. $83, 80; (2x^2 - 16) \sin \frac{x}{2} + 8x \cos \frac{x}{2} + C.$

13. $71; \frac{1}{9}[\tan^3 3x - 3 \tan 3x + 9x] + C.$

15. $21; \frac{1}{54}[(2 + 3x)^2 - 8(2 + 3x) + 8 \log|2 + 3x|] + C.$

17. $52; \frac{1}{54}[3x\sqrt{9x^2 + 4} - 4 \log|3x + \sqrt{9x^2 + 4}|] + C.$

19. $82, 81; (16 - 2x^2) \cos \frac{x}{2} + 8x \sin \frac{x}{2} + C.$

21. $22; \frac{1}{1 + e^x} + \log(1 + e^x) + C.$ **23.** $40; \frac{\log|x|}{\sqrt{1 - \log^2|x|}} + C.$

25. $31; \frac{1}{2}\log\left|\dfrac{\sqrt{1 + \tan 2x} - 1}{\sqrt{1 + \tan 2x} + 1}\right| + C.$ **27.** $5; \frac{1}{3}(\text{Arcsin } 2x)^{3/2} + C.$

29. $44; \dfrac{2 - x^2}{\sqrt{1 - x^2}} + C.$

Exercises 11-1

1. (a) 16 cu un; (b) 8 cu un; (c) 4 cu un.

3. $\dfrac{1024}{3}$ cu in.   **5.** 576 cu in.

Exercises 11-2

1. (a) 16π cu un; (b) $64\pi/5$ cu un.   **3.** $96\pi/5$ cu un.

5. (a) $9\pi/2$ cu un; (b) $48\pi/7$ cu un; (c) $96\pi/35$ cu un.

7. $(\pi/2)(1 - e^{-2})$ cu un.   **9.** $\pi[3(\log 3)^2 - 6\log 3 + 4]$ cu un.

11. $16\pi/3$ cu. un.   **13.** $(\pi/2)(8 - \pi)$ cu un.   **15.** π^2 cu un.

17. $6\pi^2$ cu un.   **19.** $(\pi/3)[472 - 450 \text{ Arcsin}(4/5)]$ cu un.

Exercises 11-3

1. 8π cu un.   **3.** 15π cu un.   **5.** $128\pi/3$ cu un.   **7.** 216π cu un.

9. $101\pi/10$ cu un.   **11.** $(2\pi/e)(e - 1)$ cu un.   **13.** π cu un.

15. $2\pi^2$ cu un.   **17.** $40\pi/3$ cu un.   **19.** $4\pi\sqrt{3}$ cu un.

Exercises 11-4

1. $\sqrt{2} + \log(\sqrt{2} + 1)$ lin un.   **3.** 335/27 lin un.

5. $(a/e)(e^2 - 1)$ lin un.   **7.** 33/16 lin un.   **9.** $2 + 12\log\frac{5}{3}$ lin un.

Exercises 11-5

1. $4\sqrt{2}\pi$ sq un.   **3.** $2\pi\sqrt{5}$ sq un.   **5.** $2\pi\sqrt{5}$ sq un.

7. $(8\pi/3)(2\sqrt{2} - 1)$ sq un.   **9.** $\pi\left[2\sqrt{5} - \sqrt{2} + \log\dfrac{\sqrt{5} + 2}{\sqrt{2} + 1}\right]$ sq un.

11. $\dfrac{\pi a^2}{2e^2}\,(e^4 + 4e^2 - 1)$ sq un. **13.** $\dfrac{\pi}{6}\,[3\sqrt{10} + \log(\sqrt{10} + 3)]$.

15. $2\pi\left[e^2\sqrt{e^4 + 1} - \sqrt{2} + \log\dfrac{\sqrt{e^4 + 1} + e^2}{\sqrt{2} + 1}\right]$.

Exercises 11-6

1. 24 tons.   **3.** 62 ft.   **5.** 3281.25 lbs.   **7.** $\frac{9}{16}[5\pi + 1]$ tons.
9. 266.67 lbs.   **11.** 200π lbs.

Exercises 12-1

1. $x + 2y = 2$.   **3.** $y^2 = 16x$.   **5.** $\dfrac{(x - 1)^2}{9} + \dfrac{(y - 2)^2}{16} = 1$.

7. $y = 1 + 3x + 3x^2 + x^3$.   **9.** $x^2 y = 1$.   **11.** $x = t,\ y = \frac{1}{3}(t - 7)$.
13. $x = t,\ y = \frac{1}{2}(3t^2 + 1)$.   **15.** $x = 2\cos t,\ y = 3\sin t$.
17. $x = x_1 + (x_2 - x_1)t,\ y = y_1 + (y_2 - y_1)t$.

Exercises 12-2

1. $t, \frac{1}{2}$.   **3.** $\dfrac{-t}{1 - 3t^2},\ \dfrac{3t^2 + 1}{(3t^2 - 1)^3}$.   **5.** $-\dfrac{(v - 1)^2}{(v + 1)^2},\ 4\dfrac{(v - 1)^3}{(v + 1)^3}$.

7. $2\cot s,\ -2\csc^3 s$.   **9.** $\dfrac{1}{t\log t},\ -\dfrac{1 + \log t}{t^2 \log^3 t}$.

11. Min $(-\frac{9}{8}, -\frac{1}{4})$; pts of inf $(-2, 0), (-1, 0)$; concave upward $-1 < t < 0$;
concave downward $t < -1, t > 0$.   **13.** Min $(5, -16)$; no pts of in-
flection; concave upward $u > 0$.

Exercises 12-3

1. (a) $(-2, -\pi/2), (2, 5\pi/2)$; (c) $(4, -11\pi/6), (-4, 7\pi/6)$; (e) $(2, 3\pi/4)$,
$(-2, -\pi/4)$.   **3.** (a) $(2\sqrt{2}, \pi/4), (2\sqrt{2}, -7\pi/4), (-2\sqrt{2}, -3\pi/4)$,
$(-2\sqrt{2}, 5\pi/4)$; (c) $(2\sqrt{2}, -\pi/4), (2\sqrt{2}, 7\pi/4), (-2\sqrt{2}, 3\pi/4)$,
$(-2\sqrt{2}, -5\pi/4)$; (e) $(2, -\pi/3), (2, 5\pi/3), (-2, 2\pi/3), (-2, -4\pi/3)$.

5. (a) $x^2 + y^2 = a^2$; (b) $x^2 + y^2 - 4x = 0$; (c) $2xy = 9$; (d) $x = 2$.
7. All lie on the curve.

Exercises 12-5

1. $(5/2, \pi/6)$, $(5/2, 5\pi/6)$.   **3.** $(6, \pi/6)$, $(6, 5\pi/6)$.   **5.** $(2 - \sqrt{2}, \pi/4)$,
$(2 + \sqrt{2}, 5\pi/4)$, $(0, 0)$.   **7.** $(1, \pi/2)$.   **9.** $(\sqrt{3}/2, \pi/3)$, $(\sqrt{3}/2, 2\pi/3)$,
the pole.   **11.** $(\sqrt{3}/2, 5\pi/6)$, $(\sqrt{3}/2, -5\pi/6)$.   **13.** $(1, \pm\pi/3)$,
$(-1, \pm\pi/3)$.   **15.** $(\pm1, \pi/4)$, $(\pm1, 3\pi/4)$.   **17.** $(2, 0)$, the pole.

Exercises 12-6

1. 6π sq un.   **3.** $\pi/2$ sq un.   **5.** $\pi/3$ sq un.
7. $\frac{1}{12}(10\pi + 3\sqrt{3})$ sq un.   **9.** $\frac{1}{3}(3\sqrt{3} - \pi)$ sq un.   **11.** 2 sq un.

Exercises 12-7

1. 16 lin un.   **3.** $3\pi\sqrt{4\pi^2 + 1} + \frac{3}{2}\log(2\pi + \sqrt{4\pi^2 + 1})$ lin un.

5. $6a$ lin un.   **7.** $2[2 - \sqrt{2} + \log\dfrac{\sqrt{3}}{3}(1 + \sqrt{2})]$ lin un.

9. $2\pi a^2 (2 - \sqrt{2})$ sq un.   **11.** $\pi/3$ sq un.   **13.** $12\,a^2\pi/5$ sq un.
15. $5\pi^2 a^3$ cu un.

Exercises 12-8

1. $-2\sqrt{3}/3$; $-\sqrt{3}/5$.   **3.** $\sqrt{3}$; undef.   **5.** $\frac{1}{2}$; $\frac{1}{2}$.   **9.** $\pi/3$.

Exercises 13-1

1. $f_x = 2x + y, f_y = x + 2y, f_{xx} = 2, f_{xy} = 1, f_{yy} = 2$.
3. $h_x = \frac{1}{2}(x - y)^{-1/2}, h_y = -\frac{1}{2}(x - y)^{-1/2}, h_{xx} = -\frac{1}{4}(x - y)^{-3/2}$,
$h_{xy} = \frac{1}{4}(x - y)^{-3/2}, h_{yy} = -\frac{1}{4}(x - y)^{-3/2}$.   **5.** $G_r = 1/r, G_s = -1/s$,
$G_{rr} = -1/r^2, G_{rs} = 0, \quad G_{ss} = 1/s^2$.   **7.** $f_u = -\sin(u - v)$,
$f_v = \sin(u - v), f_{uu} = -\cos(u - v), f_{uv} = \cos(u - v), f_{vv} = -\cos(u - v)$.

9. $F_x = \dfrac{-x}{F(x, y, z)}, F_y = \dfrac{-y}{F(x, y, z)}, F_z = \dfrac{-z}{F(x, y, z)}$.

11. $U_r = \dfrac{st}{1 + (rst)^2}$, $U_s = \dfrac{rt}{1 + (rst)^2}$, $U_t = \dfrac{rs}{1 + (rst)^2}$.

13. $2(30 - E)$, $2(40 - I)$.   **15.** (a) 2; (b) -6.

17. $g_{xx} = (yz)^2 e^{xyz}$, $g_{xy} = ze^{xyz}(xyz + 1)$, $g_{xz} = ye^{xyz}(xyz + 1)$, etc.

19. $F_{yy} = -\dfrac{1}{(x - y + 2z)^2}$, $F_{yx} = \dfrac{1}{(x - y + 2z)^2}$, $F_{yz} = \dfrac{2}{(x - y + 2z)^2}$, etc.

21. $z_{xy} = -\dfrac{xy}{(x^2 + y^2)^{3/2}}$.   **23.** $z_{xy} = e^x/y$.   **25.** $z_{xyy} = -\cos y$.

Exercises 13-2

1. $\Delta f = -.0033003$, $df = -.0033333$.   **3.** $\Delta g = -.0001$ approx, $dg = 0$.

5. $\dfrac{1}{2\sqrt{xy + y^2}}[y\, dx + (x + 2y)\, dy]$.   **7.** $\dfrac{1}{s^2 + r^2}\left[-r\, dr + \dfrac{r^2}{s}\, ds\right]$.

9. $(2x - 2yz)\, dx + (2y - 2xz)\, dy - 2xy\, dz$.   **11.** π cu in. incr.

13. 240.92 cu in.; 240.918594 cu in.   **15.** .39 sq in.   **17.** 1.5%

19. 0.15 cu ft.   **21.** 0.3%

Exercises 13-3

1. $-\dfrac{1}{t^2}[4x + y - 3t^4(x + 2y)]$.   **3.** $\dfrac{1}{\sqrt{r^2 + s^2}}(2s - r)$.

5. $\dfrac{e^x}{r}(\cos y - \pi r \sin y)$.   **7.** $\dfrac{2e^t}{r^2 - w^2}(r + rt - w)$.

9. $\dfrac{1}{w^2 + u^2 v^2}(vw \cos t - uw \sin t - 2uvt)$.

11. $\dfrac{1}{r} - \dfrac{s}{t^2} + 4t\left(\dfrac{1}{s} - \dfrac{t}{r^2}\right) - \dfrac{1}{2t^2}\left(\dfrac{1}{t} - \dfrac{r}{s^2}\right)$.   **13.** $\dfrac{\sin 2\theta - r^2}{z^2} - \dfrac{2\cos 2\theta}{z(1 + z^2)} - 4r$.

15. $\dfrac{\partial f}{\partial r} = \dfrac{1}{s\sqrt{x^2 - y^2}}(x - s^2 y)$; $\dfrac{\partial f}{\partial s} = \dfrac{-r}{s^2\sqrt{x^2 - y^2}}(x + ys^2)$.

17. $\dfrac{\partial h}{\partial x} = -\dfrac{1}{t^3}(t \sin w + 2 \cos w)$; $\dfrac{\partial h}{\partial z} = \dfrac{1}{t^3}(t \sin w - 2 \cos w)$.

19. $\dfrac{\partial F}{\partial r} = \dfrac{1}{\sqrt{x^2 + y^2 - z^2}} (x \cos t + y \sin t);$

$\dfrac{\partial F}{\partial t} = \dfrac{1}{\sqrt{x^2 + y^2 - z^2}} (-xr \sin t + yr \cos t - 2tz).$

Exercises 13-4

1. $\dfrac{\partial^2 f}{\partial r^2} = 2 \sin 2\theta + 2 \sin^2 \theta.$

$\dfrac{\partial^2 f}{\partial r \partial \theta} = 2r \cos 2\theta + r \sin 2\theta + 2y(\cos \theta - \sin \theta) + 2x \cos \theta.$

$\dfrac{\partial^2 f}{\partial \theta^2} = 2r^2 \cos^2 \theta - 2r^2 \sin 2\theta - 2r\,(x + y) \sin \theta - 2yr \cos \theta.$

2. $\dfrac{\partial^2 f}{\partial r^2} = 2e^{-x} \sin y; \quad \dfrac{\partial^2 f}{\partial r \partial t} = -2e^{-x} \cos y; \quad \dfrac{\partial^2 f}{\partial t^2} = -2e^{-x} \sin y.$

Exercises 13-5

1. $\dfrac{x + y}{2y - x}.$ **3.** $-\sqrt{y/x}.$ **5.** $\dfrac{y + x\sqrt{x^2 + y^2}}{x - y\sqrt{x^2 + y^2}}.$ **7.** $\dfrac{y(1 - y^2)}{x(1 + y^2)}.$

9. $\dfrac{\partial z}{\partial x} = \dfrac{x}{z}; \dfrac{\partial z}{\partial y} = \dfrac{y}{z}.$ **11.** $\dfrac{\partial z}{\partial x} = \dfrac{x}{z} + \dfrac{(x^2 - z^2)^2}{2zy^2}; \dfrac{\partial z}{\partial y} = \dfrac{z^2 - x^2}{yz}.$

13. $\dfrac{\partial z}{\partial x} = -\dfrac{y\sqrt{y^2 - z^2}}{x^2 + y^2}; \dfrac{\partial z}{\partial y} = \dfrac{xy\sqrt{y^2 - z^2} + z(x^2 + y^2)}{y(x^2 + y^2)}.$

15. $\dfrac{\partial z}{\partial x} = -\dfrac{\cos(x + y) + \cos(x + z)}{\cos(y + z) + \cos(x + z)}; \dfrac{\partial z}{\partial y} = -\dfrac{\cos(x + y) + \cos(y + z)}{\cos(y + z) + \cos(x + z)}.$

Exercises 14-1

1. $1 + \dfrac{1}{4} + \dfrac{1}{9} + \dfrac{1}{16} + \dfrac{1}{25} + \cdots.$

3. $-\dfrac{2}{2\cdot3}+\dfrac{4}{3\cdot4}-\dfrac{6}{4\cdot5}+\dfrac{8}{5\cdot6}-\dfrac{10}{6\cdot7}+\cdots.$

5. $\dfrac{1}{2\sqrt{1}}-\dfrac{2}{3\sqrt{2}}+\dfrac{3}{4\sqrt{3}}-\dfrac{4}{5\sqrt{4}}+\dfrac{5}{6\sqrt{5}}-\cdots.$ **7.** $1/n^3$.   **9.** $\dfrac{n+2}{n(n+1)}.$

11. $(-1)^{n-1}\dfrac{2n-1}{(n+1)(n+2)}.$ **13.** Yes.   **15.** No; the series diverges.

Exercises 14-2

1. Conv; sum $=3/2$.   **3.** Div.   **5.** Conv.   **7.** Div.
9. Conv; Sum $=6$.   **11.** 10,000, 9000, 8100, 7290, 6561; 100,000.

Exercises 14-3

1. Conv.   **3.** Div.   **5.** Div.   **7.** Conv.   **9.** Conv. **11.** Conv.

Exercises 14-4

1. Conv.   **3.** Conv.   **5.** Div.   **7.** Conv.   **9.** Conv.
11. Conv.   **13.** Div.   **15.** Conv.

Exercises 14-5

1. Conv.   **3.** Conv.   **5.** Conv.   **7.** Conv.   **9.** Div.
11. .0046.   **13.** .0026.   **15.** .986.   **17.** .995.

Exercises 14-6

1. Conv. cond.   **3.** Conv. abs.   **5.** Conv. cond.   **7.** Conv. cond.
9. Conv. abs.

Exercises 14-7

1. $-1<x\le1$.   **3.** $-1\le x\le1$.   **5.** $-\infty<x<\infty$.
7. $-2\le x<2$.   **9.** $-1\le x\le1$.   **11.** $0<x<2$.
13. $-\infty<x<\infty$.   **15.** $-2\le x<0$.

Exercises 14-8

1. $1 + \dfrac{x}{1!} + \dfrac{x^2}{2!} + \dfrac{x^3}{3!} + \cdots + \dfrac{x^n}{n!} + \cdots, \ |x| < \infty.$

3. $x - \dfrac{x^2}{2} + \dfrac{x^3}{3} - \dfrac{x^4}{4} + \cdots + (-1)^{n+1} \dfrac{x^n}{n} + \cdots, \ -1 < x \le 1.$

5. $1 - x + x^2 - x^3 + \cdots + (-1)^n x^n + \cdots, \ -1 < x \le 1.$

7. $\dfrac{\sqrt{2}}{2}\Bigg(1 + \dfrac{(x - \pi/4)}{1!} - \dfrac{(x - \pi/4)^2}{2!} - \dfrac{(x - \pi/4)^3}{3!} + \dfrac{(x - \pi/4)^4}{4!}$

$$+ \cdots\Bigg), \ |x| < \infty$$

9. $e\Bigg(1 + \dfrac{(x - 1)}{1!} + \dfrac{(x - 1)^2}{2!} + \dfrac{(x - 1)^3}{3!} + \cdots\Bigg), \ |x| < \infty.$

11. $-1 + (x - 2) - (x - 2)^2 + (x - 2)^3 - \cdots, \ 1 < x \le 3.$ **15.** $0.81.$

19. $1 + \dfrac{1}{2} x + \dfrac{1 \cdot 3}{2 \cdot 4} x^2 + \dfrac{1 \cdot 3 \cdot 5}{2 \cdot 4 \cdot 6} x^3 + \cdots, \ -1 \le x < 1.$

INDEX